Free Atoms, Clusters, and Nanoscale Particles

Kenneth J. Klabunde

Department of Chemistry

Kansas State University

Manhattan, Kansas

ACADEMIC PRESS

A Division of Harcourt Brace & Company

San Diego   New York   Boston   London   Sydney   Tokyo   Toronto

Academic Press, Inc.
525 B Street, Suite 1900, San Diego, California 92101-4495

United Kingdom Edition published by
Academic Press Limited
24–28 Oval Road, London NW1 7DX

Library of Congress Cataloging-in-Publication Data

Klabunde, Kenneth J.
    Free atoms, clusters, and nanoscale particles / Kenneth J.
  Klabunde.
      p.   cm.
    Includes index.
    ISBN 0-12-410760-5
    1. Chemical reaction, Conditions and laws of. 2. Atoms. 3. Metal
  crystals. I. Title.
  QD501.K75453   1994
  546'.269--dc20   93-38294
                                                                CIP

PRINTED IN THE UNITED STATES OF AMERICA
94 95 96 97 98 99   BB   9 8 7 6 5 4 3 2 1

To John

Contents

CHAPTER 3

Alkali and Alkaline Earth Elements (Groups 1 and 2)

CHAPTER 4

Early Transition-Metal Elements (Groups 3–7)

CHAPTER 5

Late Transition Metals (Groups 8–10)

CHAPTER 6

Copper and Zinc Group Elements (Groups 11 and 12)

CHAPTER 7

Boron Group (Group 13)

CHAPTER 8

Carbon Group (Group 14)

CHAPTER 9

Phosphorus and Sulfur Groups (Heavier Elements of Groups 15 and 16)

CHAPTER 10

Lanthanides and Actinides

Preface

During the 1960s and 1970s the field of metal atom chemistry, sometimes called metal vapor synthesis (MVS), was initially developed. I reviewed the field in 1980 in *Chemistry of Free Atoms and Particles* (Academic Press). Progress in this field coupled with new developments in lasers, laser spectroscopy, and mass spectrometry laid the foundation for the blossoming of a major new area of science in which free atoms as well as clusters of atoms and molecules are involved. The periodic table of the elements has literally become a playground for chemists and physicists pursuing new clusters and bi-element and tri-element free molecules and clusters.

The descriptor "free" is an important one and differentiates this field of free atoms, clusters, and nanoscale particles from the chemical field of molecular clusters (ligand-stabilized metallic clusters) and solid-state chemistry/physics. "Free" atoms and clusters are generated by high-temperature methods and are detected in the gas phase or in cold, condensed matrices. Particles are defined as free, high-temperature molecules or molecular clusters, for example, molecular P_2O_5 or $(MoO_3)_n$ clusters.

As we consider clusters of varying numbers of atoms, bimetallic combinations, and bi-element and tri-element "particles," the number of possible combinations becomes extremely large. Furthermore, we should not separate the field by techniques employed: metal atom/vapor synthesis, matrix isolation spectral studies, gas phase spectral studies, or use of atoms/clusters in synthesis of new molecules, materials, or films. If we are to truly codify this new field of science all the vastly different techniques and interests must be included. This book attempts to do that in an organized way, and thereby should be of interest to chemists, physicists, material scientists, and engineers. It serves as a review of the literature since 1980. Because the coverage is broad, the literature review could not be totally exhaustive, but I hope the attempt to organize all aspects of this mushrooming field will bring order out of chaos and help stimulate the thinking of a wide cross section of scientists.

The coverage of this volume is as follows: First, the most important experimental techniques developed since 1980 are described. Then metal atoms, clusters, and particles are covered in accordance with the periodic table. Species such as hydrogen, oxygen, halogen atoms, organic free radicals, and carbenes are not covered. Low-temperature clusters of stable molecules or noble gases are not discussed. The main theme deals with "high-temperature species" since a high-energy process is generally needed to produce them, although subsequent cluster growth may occur at low temperature.

Literature before 1980 is rarely discussed since the earlier book dealt with this and is frequently referred to.

Grateful acknowledgment must be given to the author's students who carried out some of the work described herein. They also provided help in organizing and reading manuscripts, proofreading drafts, and carrying out their research work quite independently as much of the author's time was expended on the manuscript. These students are George Glavee, Kim Norkjaer (with special thanks for proofreading), Eckhardt Schmidt, Kathie Easom, Balaji Jagirdar, Jane Stark, Olga Koper, Sam Asirvatham, Yong-Xi Li, Paul Hooker, Michael Henricksen, Abbas Khaleel, Ray Bouchier, Gi Ho Jeong, Cathy Mohs, and Dajie Zhang.

Sincere acknowledgment is also given to Jan Vaughn, Stacey Cornelius, and Kelly Meyeres who did the skillful typing and revisions. The author also gratefully acknowledges the National Science Foundation for research support for the past 20 plus years, and his family for their love and patience.

Kenneth J. Klabunde

Free Atoms, Clusters, and Nanoscale Particles

Introduction

I. Free Atoms, Clusters, and Nanoscale Particles

This book deals with the generation, physical properties, and chemistry of free atoms, clusters, and nanoscale particles. Included are high-temperature species such as free metal atoms (e.g., Ti or Pd atoms), clusters and bimetallic clusters (e.g., Cr_3, C_{60}, $(FeLi)_n$), and heterobinuclear/trinuclear particles (e.g., CS, CaOH). Larger particles and films formed from such species are also briefly covered.

In the past 13 years we have witnessed an unprecedented and spectacular growth of a new field of science. Metal atoms and clusters, clusters of nonmetals and semiconductors, and a host of nanoscale particles are now available to us. Literally thousands of new species have received study and thousands more await our probing. This new field of free atoms, clusters, and nanoscale particles touches on the disciplines of chemistry, physics, electronics, astronomy, mathematics, and engineering.

Figure 1-1 illustrates the size of atoms, clusters, and particles under consideration. With regard to scientific investigation, the realm of 1×10^2 to 1×10^6 atoms/cluster has been a barren landscape until now. But as creative approaches to their synthesis have made their study possible it has been found that such clusters/particles exhibit a host of new properties, depending on particle size, including magnetic,[1] optical,[2] and physical[3] properties as well as surface reactivity.[4] The material in this book generally covers the lower end of this scale, from atoms to particles/clusters to about the 1000 atom limit.

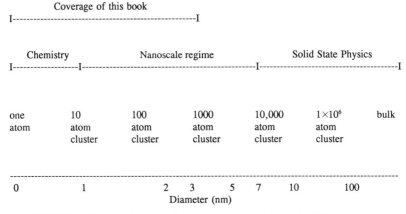

FIGURE I-I   Size relationships of chemistry, nanoparticles, and solid state physics.

II. Extremes in Temperatures and Energies

In order to produce the species of interest it is usually necessary to vaporize elements, alloys, or ionic solids, and a great deal of energy is required. Heats of formation of the metallic elements from the atoms range from 60 kJ/mol for mercury to 900 kJ/mol for tungsten.[5] Similarly, vaporization of solids such as $MgCl_2$, MoO_3, TiO, $ZnCl_2$, or many other high-melting solids requires similar inputs of energy. So the initial step in producing the reactive atoms or coordination-deficient molecules requires high-temperature methods. Resistive heating of crucibles, electron beams, arcs, lasers, or discharges provide such temperatures (500–2000° C).[5] In more recent years especially the laser vaporization method, which when coupled with advances in spectroscopic methods has allowed dramatic progress, has been greatly improved (see Chapter 2).

III. Organization of the Book

Literature is reviewed based on the periodic table of the elements. Thus, each group of elements is separated into a discussion of free atoms, followed by free clusters, followed by bimetallic clusters, followed by bielement molecules and clusters, followed by a brief section on films formed from these reactive species. These sections are each broken down further into subsections on "Occurrence and Techniques," "Physical Properties and Theoretical Studies," and "Chemistry." This organizational pattern is used throughout this volume. The "Chemistry" subsections are further divided as described below when appropriate. Thus, if specific headings are not applicable for a certain group of species, then these headings are not included.

IV. Chemistry Headings

A. Abstraction Processes

A fragment of a reactant molecule is removed by the reactive species in question, for example:

$$Mg + CO_2 \longrightarrow MgO + CO.$$

B. Electron Transfer Processes

$$Ca + C_5H_6 \longrightarrow Ca^+ C_5H_5^- + H.$$

C. Oxidative Addition

$$Rh + CH_4 \longrightarrow CH_3RhH.$$

D. Orbital Mixing (σ- or π-Complex Formation)

References

1. S. Gangopadhyay, G. C. Hadjipanayis, B. Dale, C. M. Sorensen, K. J. Klabunde, V. Papaefthymiou, and A. Kostikas, *Phys. Rev. B.*, **45**, 9778 (1992).
2. M. Steigerwald, A. P. Alivisatos, J. M. Gibson, T. D. Harris, R. Kortan, A. J. Mueller, A. M. Thayer, T. M. Duncan, D. C. Douglass, and L. E. Brus, *J. Am. Chem. Soc.*, **110**, 3046 (1988).

3. R. P. Andres, R. S. Averback, W. L. Brown, L. E. Brus, W. A. Goddard, A. Kaldor, S. G. Louis, M. Moskovits, P. S. Peercy, S. J. Riley, R. W. Siegel, F. Sapaepan, and Y. Wang, *J. Mater. Res.*, **4,** 704 (1989).
4. H. Itoh, S. Utamapanya, J. V. Stark, K. J. Klabunde, and J. R. Schup, *Chem. Mater.* **5,** 71 (1993).
5. K. J. Klabunde, ''Chemistry of Free Atoms and Particles,'' Academic Press, New York, 1980.

2

New Laboratory Techniques and Methods

I. Introduction

In this chapter are collected schematic diagrams of special apparatus developed for studies of the species of interest. Brief explanations for their operation are given.

II. "Optical Molasses" Apparatus (Atomic Beam Cooling)

Figure 2-1 describes an instrument capable of trapping the vapors of relatively volatile metals in the gas phase and holding them essentially motionless (thereby at micro-Kelvin temperatures) in a vacuum.[1] This is done by the use of focused laser beams arranged in such a way that metal atoms that drift into a laser beam are "pushed back" by the photon energies. With several such lasers, the atom/atoms can be trapped for long periods of time, allowing many spectroscopic measurements to be carried out on motionless, gas-phase atoms.

Basically "light pressure" forces atoms to absorb photons and their momentum from a light beam. In a short time (~10 nsec) the atoms spontaneously remit a photon (fluorescence). Since the direction of the emitted photon is random, overall the absorbed photon can cause directional change. With the advent of tunable dye lasers, this hitherto theoretical approach to laser cooling and storage of atoms has become a reality.

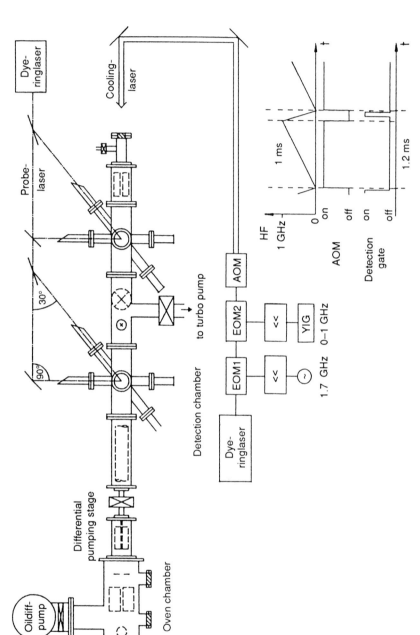

FIGURE 2-1 Optical molasses (atomic beam cooling) apparatus for isolation of gas-phase metal atoms (after Ertmer[1]).

TABLE 2-1
Cooling Limits of Some Atoms by Laser Cooling[a]

Element	λ (nm)	t (ns)	T_r (μK)
Na	589	16	0.8
Rb	780	27	0.1
Cs	852	32	0.07
Ca	423	4.6	0.9
Ca	657	4×10^5	0.4
Mg	285	2	3.3
Mg	457	2.3×10^6	1.3

[a] λ is the cooling (laser) wavelength, t denotes the natural lifetime of
the upper level of the cooling transition, and T_r refers to temperature
motion because of photon recoil.

The basic scheme consists of an atomic beam and a counterpropagating laser beam. The atoms absorb momentum and are decelerated. Thus, a Na atom can lose an average of 3 cm/sec speed per absorption, when the laser is tuned to the sodium D_2 line. Typically, a Na atom traveling at 600 m/sec needs about 20,000 absorptions to be stopped. The time scale for complete deceleration is about 1 msec over a distance of less than 1 m.

In one scheme of operation, a Na atomic beam was cooled below 10 mK with a density of 10^6 atoms/cm^3. Table 2-1 shows the cooling limits of some elements (atoms) studied when such a polarized cooling laser beam is carefully mode-matched to the weakly diverging atomic beam.

The interest in producing such cold gas-phase atoms is clearly for the ability to obtain exact spectroscopic information. Optical or microwave frequency standards are of obvious interest. Other applications may be in the areas of collision physics, surface physics, photon statistics, quantum effects, and isotope separation schemes.

III. Pulsed Cluster Beam (PCB)

First developed by Smalley and co-workers,[2-4] this technique utilizes a laser pulse to evaporate any desired element. The vapor plume is ejected into a flow tube where a pulse of cold He is simultaneously injected. This supersonic beam of atoms/inert gas finds itself in a relatively high pressure of the inert gas. The atoms begin to aggregate and are cooled to about 1–20 K as they form.

The cluster growth can be moderated by the He pressure, flow rate, and laser pulse power. The flowing clusters and inert gas enter a skimmer where a small amount is differentially pumped so that a portion is led into a chamber where the clusters are ionized by a second laser and then mass analyzed.

In more elaborate setups, certain ionized clusters can be magnetically separated and held in a vacuum cell for subsequent further study.

Figure 2-2 shows a schematic cross section of an improved, miniaturized version of the CB apparatus, especially constructed for generation of cluster ions that can be trapped and studied by FT–ICR (Fourier transform–ion cyclotron resonance).[4] The laser target rod is rotated and translated under computer control so that fresh surface is always available for vaporization. The vaporization laser (second harmonic of a Nd-YAG, 10–30 mJ/pulse, 5 nsec pulse length focused on a 0.07-cm-diam spot) is fired on the leading edge of the rising carrier gas pulse. This allows the vapor plume to expand unimpeded for a short while before it is entrained in the rising density of the carrier gas pulse.

Operating with a He backing pressure of 10 atm, the pulsed value is capable of putting out 0.05 Torr liter in a 125-μsec pulse. In a 3-liter

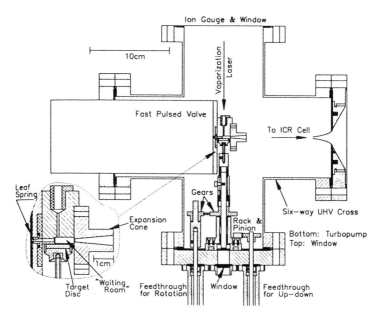

FIGURE 2-2   Pulsed cluster beam apparatus for production of gas-phase clusters (supersonic cluster beam source) (after Smalley and co-workers[2–4]).

chamber such a fast pulse temporarily raises the pressure to 2×10^{-2} Torr.

In this design the "waiting room" is the zone in the nozzle where clusters are formed and thermalized. The main flow of the carrier gas then passes through a 2.0-cm-long conical expansion zone. The gas can then undergo a free supersonic expansion with the central 0.2-cm-diameter section of the jet being skimmed about 8.4 cm downstream. After passing through the skimmer the clusters can be ionized by a second laser and trapped or directly analyzed by MS.

IV. Continuous Flow Cluster Beam (CFCB)

A variation on this theme due to Riley and co-workers[5,6] utilizes a continuous flow of He or Ar. This apparatus allows more control of pressure and temperature, and thus more meaningful kinetic analyses. The main disadvantage is the need for large pumping capacity in order to move the large volume of He gas rapidly enough.

A cross section of the central part of the apparatus is shown in Fig. 2-3. An aluminum block with three inserting channels allows a pulsed laser beam to hit the sample rod ejecting vapor into the continuous main flow. (The target rod is continually rotated and translated automatically so that a fresh surface is available as vaporization continues). The ejected vapor is rapidly cooled and nucleation and cluster growth occurs quickly. (An additional flow of gas over the laser window is needed to prevent metal film formation on the window).

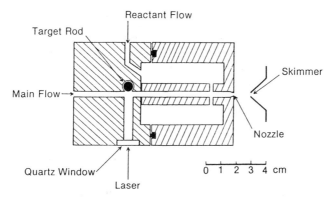

FIGURE 2-3   Continuous flow cluster beam (CB) apparatus (after Riley and co-workers[6]). Cross section of the cluster source/flow tube reactor.

Reagent gas can be inlet into the metal cluster/carrier gas mixture through four inlets equally spaced down the circumference of the main channel. Any reaction that is taking place downstream of these inlets is quenched where the mixture expands into vacuum through a 0.1-cm-diameter nozzle at the end of the main channels.

A portion of the expanded sample is collected by a skimmer and this passes through several stages of differential pumping to arrive at a time-of-flight mass spectrometer (40 cm downstream from nozzle). Ionization of a portion of the sample is achieved by a second pulsed laser.

The vaporization laser used is a XeCl excimer typically operating with a pulse energy of 50 mJ and a repetition rate of 20 Hz. Carrier gas flows used were 500 std cm^3/min (sccm) He in the main channel, yielding a pressure of about 30 Torr in the reaction zone. The reagents are usually added as 1–10% diluted in He or as pure gases. Flows ranged from 5 to 310 sccm. Thus, partial pressures of reagent gases were varied from 3 μm to 12 Torr. Flow velocities were typically 5×10^3 cm/sec. Total pressure of cluster species were small, about 0.01 to 0.1 μm.

The ionization laser was a collimated ArF excimer laser and fluence was kept low (<1 mJ/pulse).

V. Ionized Cluster Beam (ICB)

Another major innovation, due to Takagi and co-workers[7,8] is the development of the ionized cluster beam source. Mainly used as a method for producing high quality films, this technique utilizes the vaporization of elements and expansion of the vapor through a nozzle into a region where clustering takes place.

The ICB method is entirely different from conventional vacuum or sputtering techniques since the kinetic energy of the ion clusters can be controlled by the acceleration voltage. In this way, increased implantation energies can be achieved. ICB techniques are characteristic due to enhanced migration of adatoms on the substrate surface.

Figure 2-4 shows an example of an ion source with a direct-heating type of crucible.[8] The clusters formed after nozzle vapor ejection are multiply or singly charged. The charged clusters are accelerated and deposited. Film deposition rates can vary from 10 Å to several micrometers per minute. Upon impacting the substrate, the kinetic energy of the accelerated ion clusters may be converted to thermal energy, sputtering energy, implantation energy, or adatom migration energy. These processes can lead to high-quality, strongly adhering films that are denser and

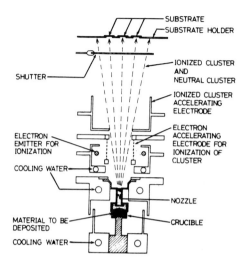

FIGURE 2-4   Ionized cluster beam apparatus, schematic of the vaporized-cluster ion source (after Takagi *et al.*[8]).

smoother. In addition, good quality crystalline films of metals and semiconductors can be prepared at pressures of 10^{-6} to 10^{-7} Torr and a substrate temperature below 750°C. Some typical parameter for silver clusters/films are: cluster size, 10^3–10^4 atoms with crucible temperatures of 1200–1600°C; nozzle diameter, 0.3 to 1.5 mm with vacuum of 1×10^{-5} Torr; acceleration voltage, 1000–5000 V. Reports on the use of this method for many elements and surface substrates are discussed toward the end of the chapters that follow.

VI. Gas Evaporation Method (GEM)

The cluster beam sources and ionized cluster beam sources are derivatives of an older method that we refer to as the gas evaporation method. The GEM simply refers to vaporization of elements in a pressure of static inert gas. Cluster growth occurs in the gaseous medium, and the clusters/particles eventually deposit on the inner walls of the evaporation chamber. This approach has been used for many years and on very large scale.[9–11]

Generally a vacuum chamber is evacuated to less than 1×10^{-3} Torr (1 Torr = 133.322 Pa) and then He gas inletted in the range of 10–50 Torr.

In this static environment the metal is vaporized from a crucible by resistive heating.

Cluster growth occurs just above the vaporization source, and ultimate cluster size depends on pressure and gas temperature. Clusters are collected usually on a grid above the vaporization source, or on the walls of the chamber.

VII. Laser Plume Method (LPM)

This method is dependent on laser evaporation (pulsed or continuous) of the element in question, usually in a pressure of static inert gas. The atoms form clusters, and the clusters/particles collect on the inner walls of the evaporation chamber. The apparatus is similar to the GEM system, but a window must be provided for the laser beam, and a gas stream is usually needed to keep the window from being covered with a metal film.

VIII. Preparative-Scale Matrix Isolation Apparatus for Use with Small Molecules (N_2, O_2, H_2, CO, CH_4, NO, C_2H_4)[12]

Earlier reviews of metal atom-vapor chemistry[13,14] have described apparatus for both microscale matrix isolation studies as well as macroscale synthetic work. Static as well as rotary reactors were described using resistive heating electron beam and laser vaporization methods.

One of the disadvantages of all of these systems is that very volatile ligands could not be used on preparative scale since they could not be condensed at 77 K, the lowest achievable temperature for macroscale experiments.

Ozin and co-workers have reported an apparatus design so that lower temperatures can be achieved (down to 15 K).[12] This development allows the synthesis of metal complexes of CO, for example, as well as studies of N_2, O_2, H_2, CH_4, NO, and C_2H_4. Shown in Fig. 2-5, the heart of this preparative-scale equipment is a high-capacity refrigeration cryopump produced by Air Products. A reverse polarity, electrostatically focused 3.5-kW electron gun is the vaporization source, and metal deposition rates were monitored using a quartz crystal microbalance. About 10–100 mg of metal (e.g. V, Cr, Mn, Fe) with 10–100 g of CO over a 1- to 6-hr period could be codeposited.

Another key feature is a radiation shield to cut down on heat transfer to the matrix and a liquid N_2-cooled electron gun cryoshield. The various

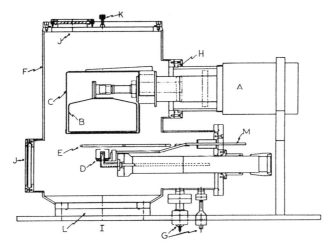

FIGURE 2-5   Preparative-scale metal atom-vapor reactor for use with volatile ligands such as CO, N_2, NO. (A) High-refrigeration-capacity cryopump (Air Products); (B) reaction cryoshield on first stage (4 W at 10 K); (C) radiation shield on 77 K second stage (70 W at 77 K); (D) reverse-polarity, electrostatically focused, quartz-crystal, mass-monitored (resolution 20 ng), 3.5-kW electron gun; (E) mass-flow-controlled ligand inlet (Vacuum General); (F) stainless-steel vacuum chamber; (G) cold-cathode, thermocouple vacuum gauges (Varian); (H) rotation seal for cryopump; (I) high-capacity diffusion pump (Edwards, 2300 liter s^{-1}); (J) observation windows; (K) arrangement of temperature-controlled Schlenk cannulas; (L) pneumatic slide valve (Airco); (M) electron gun liquid N_2 cryoshield. [Reprinted with permission from Godber, Huber, and Ozin, *Inorg. Chem.* **25**, 1675 (1986). Copyright 1986 American Chemical Society.]

components are described in Fig. 2-5. During codeposition of CO and metal vapor a vacuum below 1×10^{-5} Torr could be achieved. After deposition the cryoshield can be rotated 180° to aid in removal of unreacted ligand. The product can then be extracted with added solvent and brought out via a temperature-controlled cannula. Product yields of 60% based on metal vaporized could be obtained (eq. $Ru(CO)_5$, $Mn_2(CO)_{10}$, $V(CO)_6$, $Cr(CO)_6$, and $Fe_3(CO)_{12}$).

IX. Solvated Metal Atom Dispersion (SMAD) Apparatus for Producing SMAD Catalysts

Another recent innovation has been the construction of an apparatus for producing bimetallic solvated metal atom dispersed catalysts. The SMAD method[13,15] involves the codeposition of metal vapors with weakly coordinating solvents at 77 K, producing upon meltdown a solution of solvated metal atoms. The thermal stability of these solvates varies depending on the metal–solvent pair. Ideally, the solvates are stable in the 150–300 K region and as liquids can be allowed to permeate high-surface-area catalyst supports such as SiO_2, Al_2O_3, or MgO. Upon further warming the solvate decomposes, depositing very small metal particles on the support surface. In this way highly dispersed, ready-to-use, very active catalysts are obtained.

The most recent innovation has been to use this SMAD methodology to produce half-SMAD[16] and full-SMAD bimetallic catalysts.[17] Half-SMAD refer to deposition of a second metal onto a preformed catalyst (e.g. Sn onto Pt/Al_2O_3). In this way a surface coating can be formed. Full-SMAD refers to simultaneous deposition of two metals with the solvent of choice (usually toluene or THF), matrix formation, warming, and formation of bimetallic clusters on the catalyst support.

Figure 2-6 illustrates an apparatus that can be used for half-SMAD or full-SMAD catalyst preparations. With this example, both Pt and Sn were

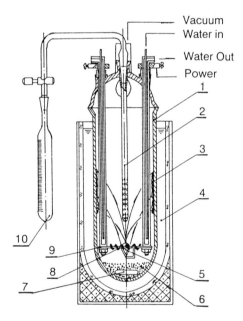

FIGURE 2-6 A solvated metal atom dispersion (SMAD) reactor for preparing highly dispersed metallic catalysts. Use for half-SMAD or full-SMAD is possible. (1) reactor chamber; (2) solvent vapor inlet; (3) matrix; (4) liquid N_2; (5) metal vaporization crucible (e.g. for Sn); (6) catalyst support (eq. Al_2O_3 powder); (7) magnetic stirring bar; (8) second metal (e.g. Pt); (9) tungsten rod as a second metal vaporization source; (10) solvent (after Li and Klabunde[17]).

vaporized simultaneously and deposited at 77 K with solvent vapors. After deposition the matrix is allowed to warm and the liquid solvate flows down onto a stirred slurry of catalyst support. Further warming allows thermal decomposition of the solvate and impregnation of the support with Pt–Sn clusters. The solvent is removed and the catalyst is thereby ready for study.

With experience various metal:metal:solvent ratios can be employed and catalytic properties changed.

References

1. W. Ertner, *Phys. Scr.* **36,** 306 (1987).
2. T. G. Dietz, M. A. Ducan, D. E. Powers, and R. E. Smalley, *J. Chem. Phys.* **74,** 6511 (1981).
3. M. D. Morse, M. E. Geusic, J. R. Heath, and R. E. Smalley, *J. Chem. Phys.* **83,** 2293 (1985).
4. S. Maruyama, L. R. Anderson, and R. E. Smalley, *Rev. Sci. Instrum.* **61,** 3686 (1990).
5. E. K. Parks, K. Liu, S. C. Richtsmeier, L. G. Pobo, and S. J. Riley, *J. Chem. Phys.* **82,** 5470 (1985).
6. S. C. Richtsmeier, E. K. Parks, K. Liu, L. G. Pobo, and S. J. Riley, *J. Chem. Phys.* **82,** 3659 (1985).
7. T. Takagi, I. Y. M. Kunori, and S. Kobiyama, *in* "Proceedings, 2nd International Conference on Ion Sources, Vienna, 1972," p. 790.
8. T. Takagi, I. Yamada, and A. Sasaki, *Thin Solid Films,* **45,** 569 (1977).
9. C. Hayashi, *Phys. Today* **December** 44 (1987); *J. Vac. Sci. Technol.* **A5,** 1375 (1987).
10. G. A. Niklasson, *J. Appl. Phys,* **62,** 258 (1987); and references therein.
11. (a) K. Kimura, and S. Bandow, *J. Chem. Soc., Jpn.* **56,** 3578 (1983); (b) S. Yatsuya, S. Kasukabe, and R. Uyeda, *Jpn. J. Appl. Phys.* **12,** 1675 (1973).
12. J. Godber, H. X. Huber, and G. A. Ozin, *Inorg. Chem.* **25,** 2909 (1986).
13. K. J. Klabunde, "Chemistry of Free Atoms and Particles," Academic Press, New York, 1980.
14. M. Moskovits, G. A. Ozin, Eds. "Cryochemistry," Wiley–Interscience, New York, 1976.
15. K. J. Klabunde, Y. X. Li, and B. J. Tan, *Chem. Mater.* **3,** 30 (1991).
16. Y. X. Li, Y. F. Zhang, and K. J. Klabunde, *Langmuir* **4,** 385 (1988).
17. Y. X. Li and K. J. Klabunde, *J. Catal.* **126,** 173 (1990).

CHAPTER

3

Alkali and Alkaline Earth Elements (Groups 1 and 2)

I. Free Atoms (Li, Na, K, Rb, Cs, Be, Mg, Ca, Sr, Ba)

A. Occurrence and Techniques

The natural occurrence of the atoms/vapors of Li, Na, K, Mg, Ca, Sr in the upper atmosphere, flames, and in stars was discussed earlier.[1] Techniques for their study were also described: diffusion flame method and life-period method, gas-phase flow system, rotating cryostat, and stationary matrix isolation methods.[1] The ease with which these elements can be vaporized has encouraged extensive gas phase and low temperature matrix studies.[1]

A very intriguing development in recent years has been the gas-phase trapping of alkali metal atoms in so-called "optical molasses." This exciting development allows the storage and "cooling" of free atoms[2] and is characterized by evaporation of a low-boiling metal, such as Na, and forming an atom beam of low velocity. Some of the atoms are trapped in a cell that is equipped with lasers arranged so that as the metal atoms drift into the laser beam, the photons absorbed stop the atoms' movement (see Chapter 2, Fig. 2-1). This counterpropagating laser light is capable of cooling these trapped atoms to milli-Kelvin temperatures. The cold beam of atoms can be further manipulated by these laser beams. Highly precise spectroscopic measurements are now possible. It also appears that the technique can be applied to particles as well as atoms.[2b] Many future applications await further development.

B. Physical Properties and Theoretical Studies

Recent years have brought considerable advances in the theory of inner electron-binding energies for the alkali and alkaline earth metal atoms and ions.[3] In particular, core level shifts and relaxation energies for cations and anions of many metallic and semimetallic elements were calculated using nonrelativistic, numerical Hartree–Fock procedures. Ionization potentials were also calculated and comparison with experimental values tabulated. Similarly, kinetic energies of Auger transitions were calculated and compared with experiment.[4] A Dirac–Fock procedure embodying relativistic theory was used. Since the Auger energies of the free atoms of alkali-alkaline-earth metals are reliable and well studied, a critical comparison between calculated and experimental data was possible.

Further progress on determining theoretical values of inner electron-binding energies was possible by taking differences in total energies; Na, Na^+, K, K^+, Cs, Cs^+ (others included carbon, oxygen, fluorine, and neon).[5] Quantitative calculations of Auger lineshapes of free atoms (atom ionization) have been studied.[6] It was found that lineshapes deviated from Lorentzian for all impact energies. This was believed to be due to long-range continuum interactions between the Auger electron and unobserved collision fragments. The magnitude of these effects was significant and not realized before.

Experimental results on K-shell-binding energies of Be atoms were obtained by applying Auger spectroscopy to fast Be atoms in collision with CH_4 and He.[7] These values, which are really the lowest ionization energies of the atoms studied, were higher than values experimentally determined from solid samples and found good agreement with free atom theoretical data. In closely related work, free atoms have been ionized in a static electric field. Data on the field ionization of Li atoms suggested that the first ionization process is due to tunnelling.[8] The second ionization, important for all atoms except H, is due to level mixing and autoionization.

Rigorous studies of gas-phase photoelectron spectra of Na, K, Rb, Cs, and Mg have been reported[9] and compared with solid-state electron-binding energies, the differences being substantial, for example $K(2p) = 303.2$ eV in the gas phase, but only 300.5 in solid potassium metal.

Synchrotron and laser radiation for photoelectron spectroscopy of free atoms of some of these elements have also been reported.[10] Polarization of these radiation sources and alignment of the atoms under study have a strong influence on the absorption processes and on the angular distribution of the photoelectrons.

A useful review of electron spectroscopy of free atoms discusses autoionization, electron–electron coincidence, shake-up processes, and angle-resolved electron spectroscopy of aligned atoms.[11]

Other physical phenomena reported include the changes in electronic charge distribution during condensation of free atoms into a crystal, calculated for Li, Na, and K.[12] Changes in valence electron distributions were related to orbital overlays at lattice points, and distribution changes were expressed in terms of atomic orbital compression.

Interaction of these atoms with rare gas atoms allows the formation of weak bonding interactions, for example, K and Mg atom van der Waals interaction with argon and xenon.[13] Breckenridge and co-workers[14] have examined the Mg–Ar van der Waals molecule and found an internuclear separation of 3.63 Å and dissociation energy around 300 cm.$^{-1}$ The $E(^3\Sigma^+)$ state is more deeply bound with a much smaller separation of 2.38 Å.

Laser ablation of Li metal and trapping of atoms/clusters in Ar, Kr, and Xe has been described by Fajardo and co-workers.[15] The UV-visible absorption spectra of Li/Ar and Li/Kr matrices prepared in this way were dominated by a "blue-shifted triplet" not previously observed. These new absorption features were attributed to Li atoms trapped in unusual sites in Ar and Kr. This difference must be due to deeper penetration of Li atoms into previously deposited close-packed Ar and Kr structures, resulting in their trapping in unusually tight sites. Thus, laser ablation produces more kinetically hot atoms than simple thermal evaporation, and this is manifested in unusual trapping mechanisms. Fajardo has also reported Li atom trapping in Ne, H_2, and D_2.[16] Upon codeposition, often a "doublet" visible absorption band was observed. Upon annealing, a "triplet" was produced with absorptions at 626, 640, and 649 nm in Ne. However, the doublet persisted in H_2 and D_2. These studies suggested that Li atoms could be trapped and maintained a 5 K or below in H_2. The work also further supported the idea that laser ablation is a better way to trap atoms in cryogenic matrices, due to the higher kinetic energy of laser ablated atoms vs thermal (oven)-generated atoms. These studies lend some credence to the proposal/goal to prepare and handle high-energy metal atom-doped cryogenic propellants.

Laser excitation of Rb and Cs atoms trapped in frozen argon has also been of interest[17] and has allowed identification of absorption bands associated with transitions from the ground state to the lowest d state and to the second excited p state (see Fig. 3-1).

Closely related work has been reported on Li-doped argon matrices studied by both optical absorption and ESR.[18] Broad optical absorptions for the Li atom was in reality a superposition of two absorptions ("blue and red" triplets). These two absorptions were correlated with two ESR

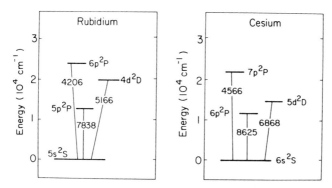

FIGURE 3-1   Grotian diagrams showing the relevant energy levels of Rb and Cs. Wavelengths are in angstroms (after Balling and Wright[17]).

lines. Spin relaxation times were obtainable, and temperature dependence seemed to depend on local structural change that has an extremely low activation enthalpy. Similar observations on K atoms in frozen Ar have been reported.[19] The influence of X rays and light irradiation caused both the blue and red triplets to change into a new absorption, which disappeared on annealing at 15 K, and the original peaks reappeared. These results were explained in terms of exciton production, self-trapping, and deexcitation.[19] A master equation was offered as a rationale for time and temperature dependencies. Detailed understanding of optical and ESR absorption spectra of hydrogen and alkali metal atoms in frozen matrices can allow studies of basic phenomena of the matrix–guest interaction.[20] Using potassium as an example, the so-called red triplet belongs to the absorption of rather weakly bound K atoms, and it disappears irreversibly upon warmup at 11–13 K. The blue triplet is thermally stable up to 25 K, and it is shifted about 80 nm to shorter wavelengths compared with the gas-phase atomic lines. It is believed that the K atoms giving rise to the blue triplet are located in the center of a fourfold vacancy in the Ar crystal, which is a very stable trapping site.

By being able to simultaneously observe optical and esr spectral changes upon light irradiation and/or annealing, Schrimpt and co-workers[20a] determined that the Ar matrices formed upon slow deposition and lowest temperature were amorphous. During preparation of the matrix, atoms froze just where they impinged on the cold substrate without any rearrangement. By annealling above 12 K this amorphous state changed to a microcrystalline one. Thus, the spectra of alkali atoms can serve as a sensitive probe of matrix environment, and this has been further demonstrated by Pellow and Vala.[20b] These studies are helping to sort out the seemingly mysterious matrix effects often observed.

A useful summary of resonance radiation and excited atoms has been presented by Mitchell and Zemansky.[21]

C. Chemistry

1. Abstraction Processes

The chemistry of these atoms has been a rich field in the past,[1] and recent work has continued to be fruitful. On synthetic scale these metals have been used as reagents for reduction of other metal halides. A good example is the synthesis of ultrafine niobium particles by the reaction of $NbCl_5$ with gas-phase Na atoms.[22] Argon carried the gaseous $NbCl_5$ into an 800°C reaction chamber where it was mixed with Na vapor at a Na : $NbCl_5$ ratio of 1.2 : 1. The products of Nb, Na, and NaCl were mixed with ethanol to remove Na and then ethylene glycol to remove NaCl. Ethanol washing and air drying yielded surface oxide-stabilized Nb particles of 50–100 nm.

In a similar way, but under low-temperature liquid conditions, Hawker and Timms[23] have allowed K atoms to react with transition-metal halides in the presences of arenes. In this way a new preparation of $(arene)_2M$ was developed:

$$MoCl_5 + 5K_{vapor} + 2\,\bigcirc\!\!\bigcirc \longrightarrow \left(\bigcirc\!\!\bigcirc\right)_2 Mo + 5KCl$$

The ease with which potassium is evaporated compared with evaporation of the transition metals makes this an attractive method. The fact that the same reaction cannot be carried out with bulk K metal suggests that the presence of excess K metal in localized areas causes the destruction of the $(arene)_2M$ product. Thus, reduction using one atom at a time is a necessary condition.

In tribute to Richard Bernstein,[24] several alkaline-earth metal atom reactions with halides have come under investigation, particularly in crossed-beam gas-phase experiments.[25-27] The steric effects of the approach and reaction of excited $Ca(^1D_2)$ with $CH_3F \rightarrow CaF(A) + CH_3$ have been examined, using a state-selected and oriented CH_3F molecular beam.[25] Interestingly, the steric effects increased with increase in collision energy, and this was rationalized in terms of a necessary reorientation of the initially oriented CH_3F axis due to long-range interaction (atom molecule distances of 5–10 Å).

In addition, the dynamics of the Ba atom + Br_2 chemiionization reaction have been explored.[26] The formation of $BaBr^+$ in a crossed-beam experiment was monitored, and it appeared that head on, colinear collisions and proximal crossing of the potential energy surface are necessary for its formation (as opposed to neutral products).

In a report by Bernstein and co-workers[27] the reaction of Sr atoms with RX to yield SrX and R was explored. The effect of the nature of R ($R = C_2H_5$, $n\text{-}C_3H_7$, $t\text{-}C_4H_9$, $X = Br$, I) was studied. In particular, the reaction cross section dependence on collision energy was examined, and in several cases clear maxima were observed (e.g. at 0.5 eV for $t\text{-}C_4H_9I$). Energy partitioning in the reactions of Ba atoms with normal and branched alkyl iodides and dibromoalkanes has been examined earlier by Chakravorty and Bernstein.[28]

Chemiluminescent reactions of Mg atoms interacting with the O_2F radical have also been observed.[29] The initial $Mg(^1S_0) + O_2F$ reaction produced a long-lived vibrationally excited MgF molecule, which then reacted further to produce electronically excited Mg^*, MgF_2^*, or $MgF^*(A^2\Pi)$.

It is clear that abstraction reactions of alkaline-earth metal atoms have developed into a rich field of gas-phase inorganic chemistry.[30] (This follows much earlier work on the alkali metal atom-halide abstraction work of Steacie and many others).[1] In recent years Be, Mg, Ca, Sr, and Ba have come under intense study. Of course these metals usually yield divalent ions upon oxidation and so gas-phase abstraction reactions are more unusual. Whether paramagnetic M–X or diamagnetic MX_2 are the products depends very much on the conditions of the experiment; gas-phase and frozen argon matrices usually yield M–X, whereas liquid-phase reactions and some matrix reactions yield MX_2 (or RMX).

Summarizing the most recent work in this area should begin by recalling the early theoretical and matrix isolation work on MgOH and CaOH.[31,32] The preparation of MgOH involved the reaction of Mg atoms with H_2O_2 under matrix conditions. It should also be noted that monovalent free radical derivatives of alkaline-earth metals such as CaOH are frequently encountered in high-temperature gaseous environments.[33–35] However, in recent years gas-phase abstraction reactions have considerably broadened our knowledge of M–X species. This is due to new techniques such as flowing afterglows,[36] ion beams,[37] Fourier transform ion cyclotron resonance,[38] and high-pressure mass spectrometry[39] for studying gas-phase ionic reaction products. Moreover, progress on the study of neutrals has been possible due to the use of a beam of neutral atoms excited by laser irradiation so that M^* becomes reactive. For example:[30] $Ca^*(^3P_1) + HNCO \rightarrow CaNCO + H$. In this case the $^3P_1$ state is reactive whereas the ground state $Ca(^1S_0)$ is nearly unreactive.

Many monovalent gas-phase molecules have been prepared in similar ways, including mono-hydroxides, -alkoxides, -carboxylates, -formami-dates, -isocyanates, -azides, -isocyanides, -cyanides, -hydrosulfides, -al-kylamides, -thiolates, -methyls, -acetylides, -cyclopentadienides, -pyrro-lates, and -borohydrides.[30] An emphasis in this work has been to obtain high-resolution electronic and vibrational/rotational spectra of these gas-phase species. One way of viewing the molecules is to consider them as ionic, i.e., Ca^+OH^-. CaOH is linear instead of bent like H_2O.[40–42] The linear geometry maximizes the distance between Ca^+ and $H.^{\delta+}$ However, as the ligand becomes less electronegative (such as SH instead of OH), the degree of covalency in the metal–ligand bond increases, and thus, for example CaSH is bent.

A neutral Ca atom has a $[Ar]4s^2$ configuration. When Ca^+OH^- forms, one electron must be transferred. The remaining $4s$ electron on Ca^+ is nonbonding. The first and second excited states of Ca^+ results from pro-motion of this $4s$ electron to $3d$ and $4p$ states.[30,43,44] Overall, such qualita-tive descriptions of electronic structure are useful, but have been strengthened by quantum chemical calculations of Ortiz.[43,44]

One of the most interesting gas-phase molecules produced is the "half-sandwich" cyclopentadienyl system $Ca(C_5H_5)$:[45]

$$Ca^* + C_5H_6 \rightarrow Ca^+(C_5H_5)^- + H.$$

Electronic spectra suggest that CaCp and SrCp have pentagonal symme-try (C_{5v}). The isoelectronic pyrrolate $C_4H_4N^-$ anion can also yield a half-sandwich structure.

It would appear that there is extensive gas-phase chemistry of mono-valent derivatives of Ca and Sr when excited state atoms are employed.

Some further interesting work with gas-phase Ca and Sr has been reported by Dagdigian.[46] It was found that a gas-phase catalysis process was apparently operating, perhaps as shown below:

$$Ca + HN_3 \rightarrow CaNH + N_2$$

$$\underline{CaNH + HN_3 \rightarrow Ca + 2N_2 + H_2}$$

overall $2HN_3 \rightarrow 3N_2 + H_2 \; \Delta H^\circ_{rxn} = -632 \text{ kJ}$

This process was also used to rationalize why in some instances excited Ca atoms were observed in emission spectra.[47,48]

Hydrogen reactions with Be and Mg atoms also lead to abstraction processes. Complexes of Be and Mg with H_2 have yielded information about the $Mg(^1P) + H_2 \rightarrow MgH(^2\Sigma) + H(^2S)$ process.[49] Likewise, a variety of oxidants have been examined, for example the $Mg + CO_2 \rightarrow MgO + CO$ gas-phase abstraction reaction.[50] The Mg ($3s3p$, $^1P_1$) atom reacted with CO_2 to produce ground-state MgO ($X^1\Sigma^+$), and there was less vibrational and rotational energy in the molecule than expected statistically; this seemed to rule out the existence of a long-lived $MgCO_2$ complex. A "late" potential barrier seems likely in the overall mechanism.

Laser excitation of SF_6 prior to interaction with K vapor has allowed the examination of energy dependence.[51] The reaction rate increased with vibrational energy but was not dependent on translational energy.

2. Electron Transfer Processes

It is perhaps not appropriate to separate abstraction vs. electron transfer processes with the alkali- and alkaline-earth metal atoms. Many of these reactions, especially with alkyl halides, occur by an initial electron transfer followed by atom transfer. In fact, careful studies of this "harpoon mechanism" have been carried out in recent years; for example short distance harpooning reactions such as the $Mg(^1S) + Cl_2$ system have been reported.[52] The MgCl formed was rotationally excited, whereas generally harpooning mechanisms yield vibrationally excited products. It is clear that the energy disposal differed markedly from the analogous $Ca(^1S) + F_2/Cl_2$ reactions. The key difference appears to lie in the electron jump distance, which is likely to be much shorter with Ca than Mg.

Likewise, electron transfer to gas-phase-oriented molecules have been reported; the $K + CF_3I$ and CH_3I systems, in particular.[53] Collisional ionization was found to be favored for both molecules when the fast K atom was incident at the iodide end of the molecule, even though the polarity is different for CF_3I vs CH_3I. In this case the "harpoon" mechanism is *independent* of orientation. For impact at the I end, the I^- is ejected backward toward the incoming K^+.

Electron transfer in matrices has also been reported, for example $Li + O_2 \rightarrow Li^+O_2^-$.[54] In earlier years many similar species have been reported.[1] However, for the $Li^+O_2^-$ species in this case, using ^6Li and ^7Li isotopes, good ESR data were obtainable, and the LiO_2 molecule was well characterized. Complete electron transfer for Li to O_2 occurred to yield $Li^+O_2^-$ of C_{2v} symmetry.

The interaction of Li with CO also results in electron transfer.[55] The $Li_2^+(CO)_2^-$ species may exist as $Li^+ \ ^-O\text{—}C(Li)\text{=}C\text{=}O$. However, a wide variety of $Li(CO)_n$ and $Li_n(CO)$ products were observed. In analogous studies of CO_2, it was found that electron transfer by metal atoms in a frozen matrix is a very facile process. Kafafi and co-workers have reported several examples such as $M^+CO_2^-$, $M_2^{2+}CO_2^{2-}$, and $M^+C_2O_4^-$ in argon and nitrogen matrices.[56,57]

Finally, the use of Na atoms to induce anionic polymerization of styrene should be noted.[58] An easy batch polymerization process that gave high molecular weights was developed, and kinetic analysis of initiation steps was possible.

3. Oxidative Addition

Ab initio self-consistent field calculations on Li atom insertion into a C–H bond of CH_4 have been reported.[59] The geometry of the more stable $^2A_1$ state suggested that the resultant molecule is best described as a methyl radical interacting with LiH, and with the C–Li bond considered as a single electron bond.

In contrast to gas-phase processes, where abstraction and electron transfer processes dominate, low-temperature liquid/solid matrices favor oxidation addition processes with the alkaline-earth metal atoms. Ault[60] has reported IR spectra of unsolvated Grignard reagents formed by the reaction of Mg atoms and alkyl halides. The infrared spectrum showed product bands that were attributed to a C_{3v} methyl group for CH_3MgX. Similar reactions were observed for Ca and Sr atoms, but Zn atoms did not react under these matrix conditions. Interestingly, no evidence for a strong polar covalent C–Mg bond was found, suggesting that the unsolvated Grignard reagent has a structure different from the solvated analog. This work was followed by studies of Mg and Ca atoms as well as Mg_2, Mg_3, Mg_4, Ca_2, Ca_3, and Ca_4 reacting with alkyl halides in matrices.[61,62] Although Mg and Ca atoms were found to react, it was determined that the small clusters of these elements were even more reactive under such conditions. More detail is given under the "Free Clusters" section of this chapter.

An interesting reaction of Ca atoms with ethers was discovered by Billups et al.[63,64] Carbon-carbon bond formation was attributed to initial complexation and insertion into the C–O bond. However, C–C and C–H bonds were also susceptible to oxidative addition processes, and this was obvious from the evolution of C_4–C_{10} hydrocarbons upon hydrolysis. Barium and Sr atoms also caused activation of dimethyl ether. These reactions are quite unexpected and further work is necessary to clarify mechanisms. It is possible that carbides are intermediates. Calcium atoms are extremely reactive with organohalides as well, and the preparation of RCaX by codeposition of Ca vapor with solvents has been reported.[65]

Earlier work with the Ca–C_2F_6 system showed that extensive reaction took place upon codesposition at $-196°C$. Warming and hydrolysis led to a mixture of products including hydrogen, methane, ethyne, and other unsaturated hydrocarbons.[66] And this work was preceded by study of the Ca^- alkenyl fluorides, where for example $FCaCF=CFCF_3$ was formed.[67]

Finally, the photophysics of atomic Mg in solid CH_4 and CD_4 should be mentioned. Optical absorption, luminescence, and photoinsertion of Mg^* into the C–H bond has been reported.[68,69]

4. Simple Orbital Mixing (Complexation)

Theoretical analyses of the interaction of Be and Mg atoms with C_2H_2 and C_2H_4 have been reported by two laboratories.[70,71] Although such interactions were at first believed to be repulsive, distortion of the hydrocarbon allows for deep minima on the potential energy surfaces. Formation of $HBeC\equiv CH$ and $HMgC\equiv CH$ also appears likely. Hisatsune[72] has reported a comparison of Mg and Ni atoms for low-temperature complexation to butenes. Unstable π-complexes were formed when Ni atoms and butenes were codeposited at 77 K. The IR spectra resembled those of Zeise's salts $C_4H_8PtCl_3$. The cis-butene complex underwent some isomerization at fairly low temperature. In contrast, Mg atom–butene complexes did not form at 77 K, but upon warming, extensive isomerization did take place, apparently on the surface of magnesium particles.

Experiments have been carried out where Li atoms have been codeposited with benzene at frozen argon temperatures.[73] IR spectra of both Li–benzene and Li(benzene)$_2$ have been reported and structural assignments made. IR bands suggest that the $Li–C_6H_6$ adduct is axial and that the ligand is distorted from sixfold symmetry (to give C_{2v} or C_{3v}). In the case of $Li(C_6H_6)_2$ a sandwich-like complex with perhaps D_{2h} symmetry is most likely. No adducts with Na atoms were detected.

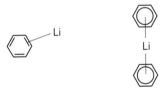

The complexation of Li atoms in matrix-isolated CH_3CN and CH_3NC has also come under study.[74] Photolysis of such matrices caused decomposition of the $Li–CNCH_3$ complex to yield LiNC (which is known to be more stable than LiCN under low-temperature conditions). The interaction of alkali metal atoms with NH_3 in solid argon is of considerable interest since liquid ammonia–Na solutions are well known as yielding solutions of solvated electrons. The IR spectra of 1 : 1 adducts of $M-NH_3$ in frozen matrices demonstrated an interaction where a very small ammonia $\rightarrow$ metal charge transfer took place for Li and Na (weak Lewis acids), and a possible acid–base role reversal is likely.[75] The strongest overall interaction was found for Li.

II. Free Clusters

A. Occurrence and Techniques

Niedermayer[1,76] has discussed theoretical and physical requirements for metal atom cluster growth on surfaces. This area of investigation of alkali metal atoms was an active research topic in past years,[1] as was the study of alkaline-earth metal dimers in frozen matrices, mainly due to Andrews and co-workers.[77] However, the most recent work has dealt with the formation of gas-phase clusters by atom aggregation under pressures of inert gases, usually He or Ar. Basically four new techniques have been developed and have been used for many elements and solid materials; these are illustrated in Chapter 2. The first is referred to as the cluster beam technique. This method, first developed by Smalley and co-workers[78–80] utilizes a pulsed laser to generate a plume of atoms that are ejected into a rather high pressure of He gas by use of a pulsed valve. The He pulse is released at the same instant as the laser pulse. In pulsing/expanding the helium, it is cooled to about 1–20 K, so the atoms are thus in a high-pressure 0.5–6 atm of cold helium. Atoms then can cluster together with the energy of condensation drained off by the cold He. For further discussion, see Chapter 2. A variation on this theme due to Riley and co-workers,[81,82] utilizes a continuous flow of He. This apparatus allows more control of pressure, more meaningful kinetic analyses, and temperature control. The main disadvantage is the need for large pumping capacity to move the large volume of He gas rapidly enough (see Chapter 2).

A third major innovation, due to Takagi and co-workers, is the development of the ionized cluster beam (ICB) technique also discussed in Chapter 2.[83–87] Mainly used as a method for producing high-quality films, this technique utilized the evaporation of elements and expansion of the vapor through a small nozzle into a high-vacuum region where clustering takes place. The clusters are then ionized and, being charged can be accelerated by electrical fields. The ionized cluster beam can then be deposited on a substrate surface at desired speeds.

The cluster beam (CB) and the ionized cluster beam techniques are derivatives of two more general and older methods that we refer to as the gas evaporation method (GEM) and laser plume method (LPM). The GEM simply refers to evaporating elements in a static pressure of inert gas and collecting the fine powders formed on the inside walls of the evaporation chamber. This approach has been used for many years and on very large scale.[88,89] Kimura and co-workers have added a new development, however, that allows the collection of the gas-phase clusters in liquid media.[90] The LPM is very similar to the GEM except a laser is used for vaporization.

B. Properties and Theoretical Studies

1. Small Clusters

The CB method has been employed in several studies of alkali- and alkaline-earth metals. Kappes and co-workers[91] examined continuous cluster beams resulting from supersonic expansions of alkali metal vapor. Judging from the clusters produced (as detected by mass spectrometry after cluster ionization), it was proposed that shell closing and geometric structural effects combine to make it favorable for certain sized clusters to possess enhanced thermodynamic stability. Assumptions inherent in such discussions deal with questions about whether the ionization/detection procedure causes cluster fragmentation. Indeed, a series of experiments gave further insight into unimolecular dissociation processes for alkali cluster ions and into neutral cluster growth.

Comparisons of metal cluster properties in cluster beams vs cold matrices have been reported.[92] Optical and magnetic effects were studied, and the quantum size effect was discussed. It was shown that magnetic resonance is a good method for investigating the structure of small, odd-numbered clusters. In a similar vein, supersonic cluster beams of $K_{(n)}$ ($n=1-100$) showed magnetic deflection patterns with three components; atoms, undeflectable even-numbered clusters, and deflectable odd-numbered clusters. In this way magnetic moments of small, odd-numbered clusters were measured and yielded values of about 1 mμ_B.

Perhaps the "holy grail" for gas-phase or matrix cluster work is to produce mass-selected clusters, or in other words, pure beams of clusters of the same molar mass. Since the clusters are generated as a complex mixture, this becomes a very desirable goal. In principle, since the clusters are ionized at some stage, magnetic/electric field selection should be possible. Then, if a pure beam of neutral clusters is desired, the cluster ions would have to be neutralized. Methods for carrying such experiments have been described by Arnold and co-workers.[93] Charge exchange of mass-selected cluster ions and velocity selection of neutral clusters were involved. Thus, sodium and tin clusters were produced by electron exchange of mass-selected clusters with potassium vapor.

The laser generation and laser probing of Sr_2 has been reported in some detail.[94] Laser-excited photoluminescence of Sr_2 exhibited discrete molecular line spectra, although dominated by bound-free transitions of excimer systems. Some of the spectral lines were attributed to fluorescence from dimers formed by laser-induced photoassociation of free Sr atoms. Cross sections for dissociation by Ar were reported as dependent on the laser excitation wavelength. Breckenridge and co-workers[95] have applied perturbation theory for van der Waals complexes of ground-state Mg_2 and Ar_2 bound by electron correlation effects. A comparison of their theoretical results with experiment showed reasonably good agreement. In Ar_2, of course the binding originates in dispersion forces only, while for Mg_2 dispersion forces are augmented by short-range exchange-deformation effects. The dissociation energy of Mg_2 has been further examined by Partridge and co-workers, using theoretical approaches.[96] Their best estimate of dissociation energy (De) = 464 cm^{-1} at the valence level.

Lithium clusters have been studied from a thermochemical point of view, in particular a stable Li_4 cluster.[97] Using mass spectrometry, gaseous equilibria were analyzed:

$$4Li(s,l) \rightleftarrows Li_4(g)$$

$$Li_4(g) + Li_2(g) \rightleftarrows Li_3(g)$$

$$Li_4(g) + Li(g) \rightleftarrows Li_3(g) + Li_2(g)$$

An atomization energy $D_0(Li_4)$ was estimated as 77.8 ± 3.0 kcal/mol, IP(Li_4) = 4.69 ± 0.3 eV, and compared with $D_0(Li_4^+)$ = 94.0 ± 5.0 kcal/mol. The heat of formation $\Delta H°(Li_4)$ was determined to be 77.0 ± 3.0 kcal/mol.

Theoretical treatments of Li_4 and Li_6 regarding their structure and stability have been reported.[98] Interestingly, planar and less compact sections of the *fcc* lattice, with interatomic distances near the experimental ones in bulk *fcc* Li crystal, are the more stable structures for Li_4 and Li_6 clusters. These are also closed-shell electronic structures with singlet states lying substantially below the triplets. ($Li_4(T_d)$ and $Li_6(O_h)$ are triplets and are less favored). As clusters get larger, the energy difference gets smaller. These results are not in agreement with the calculations of

Richtsmeler and co-workers,[99] where more compact structures were proposed to be lowest in energy.

Closely related is some work dealing with cluster ions such as $(Mg)_n^+$ and $(Mg)_n^{2+}$ when $n \leq 7$.[100] Structure and stability was discussed, and singly charged clusters of Mg atoms are characterized by a great number of possible structures in a narrow energy range and could be viewed as Mg^+ and Mg^{2+} cores surrounded by neutral atoms.

Ab initio calculations dealing with photoelectron spectral predictions for $(Na)_n^-$ and $(Na)_n$ clusters have also been reported.[101] A configuration–interaction study was emphasized, and their calculations account for observed experimental patterns in excitation energies and permit assignment of cluster geometries. Linear geometries are energetically favorable for small anionic clusters because they minimize Coulomb repulsion for the extra charge at the ends of the chain. For Na_5 and Na_5^- the most stable computed structure is the trapezoidal planar form; a linear structure appears less stable.[102]

Experimental photoelectron spectra of alkali metal cluster anions $(Na)_n^-$, $n = 2–7$; $(K)_n^-$, $n = 2–7$; $(Rb)_n^-$, $n = 2–3$; $(Cs)_n^-$, $n = 2–3$ has been reported by Bowen and co-workers.[103] The number of peaks observed for even-numbered clusters was larger than for odd, and this can be explained if the ground states are doublets for even and singlets for odd. Thus, photodetachment transitions to singlets and triplets are allowed from a doublet ground state, while only transitions to doublets are allowed from a singlet ground state. Electron affinities and electron detachment energies were determined, the latter ranging from 0.5 to 1.1ev (which were higher for K_n^- clusters). Ionization energies for $(Na)_n$ and $(K)_n$ clusters were analyzed in terms of the current theoretical MND model.[104] Evidence of many particle effects at core thresholds suggests that the current MND model is incorrect.

A technique that has been particularly helpful for studying small alkali metal clusters has been ESR. Weltner and Van Zee have reviewed some of the work published in this area and have suggested that a number of important properties can be determined:[105] (1) multiplicity or total spin(s); (2) geometry; (3) spin populations; (4) zero-field splitting; (5) g-tensor components; (6) spin rotation constants; and (7) electric quadruple coupling constants.

Lindsay and coworkers showed that by trapping Na vapor in frozen argon allowed formation of some Na_3. According to ESR, this molecule does not have an equilateral triangle geometry.[106] It possessed a spin of $\frac{3}{2}$ and the spin density was almost all distributed over just two of the atoms. Nearly all of the spin apparently resides in $3s$ orbital of the center atom, the rest (about 10%) due to p character.[107] Molecular orbital considerations suggested that the observed spin density corresponded to an odd electron being in an anti-bonding orbital between the terminal atoms with a node at the central atom. The geometry is probably linear or possibly slightly bent.[108] The dynamic spectra are the result of rapid exchange between three equivalent obtuse-angle geometries of the static trimer, and surprisingly both the static and dynamic spectra can be observed simultaneously. This is apparently the result of trapping at rotationally sensitive sites in the frozen matrix material. Gas-phase spectroscopic analysis of Na_3 has also been reported.[109]

The Li_3 molecule at 4 K in argon and 77 K in adamantane exhibited an ESR pattern characteristic of a molecule with three equivalent Li nuclei.[110] Temperature had little effect on the spectra. The large s-p hybridization (~30% p) of the unpaired spin density concurs with the theory that Li_3 should be fluxional, even at such low temperatures.[111-113]

ESR spectra have also been reported for Li_7, Na_7, and K_7. The authors proposed that pentagonal bipyramid structure best explained their findings of two equivalent and five equivalent nuclei, with the two apical atoms possessing much of the unpaired spin density.[114-116] (See Figs. 3-2 and 3-3).

The ESR spectra of Li_5, Na_5, and K_5 were not detected in these studies, perhaps due to complex anisotropy and complex hyperfine structure.[105] Theoretical approaches to predicting the most stable geometries of such alkali metal clusters M_n ($n = 2$–9) and cluster cations M_n^+ ($n = $ 2–9) based on the Hückel model and graph theory have been reported by Wang et al.[117] Such structures are shown in Fig. 3-2.

With the exception of M_5^+ and M_6^+, the Hückel model yielded minimum energy structures the same as those predicted by local-spin-density and configuration interaction calculations. A close analogy was inferred between cluster stability and topology. Indeed, in most cases the stable neutral and cation structures are the same as those predicted by more rigorous ab initio techniques[118-122] and agree with experimental results reported previously.[105-109,114,115]

The principle conclusion of these theoretical studies is that smaller alkali metal clusters tend to adopt planar geometries and that only for larger clusters do three-dimensional structures become important. Extensions of this work by Lindsay et al.[123,124] to still larger clusters ($n \leq 14$) also dealt with orbital energies, shell structures, ionization energies, and electron affinities. Cations and anions were also considered. One very useful aspect of this work is a comparison of Hückel theory with the jellium model and classical drop model, along with comparisons to known experimental data (if known).

Numerous additional theoretical and experimental studies have been reported on the properties of these small clusters; ionization ener-

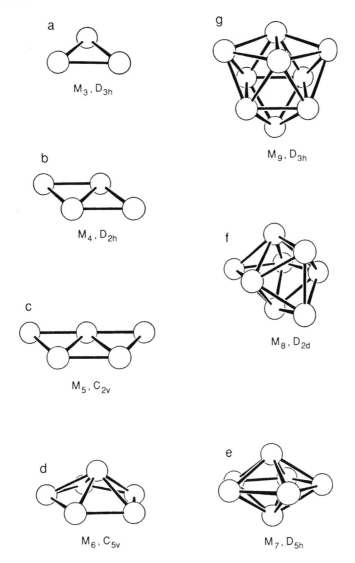

FIGURE 3-2   Most stable structures for metal clusters up to nine atoms (after Lindsay and co-workers[117]).

gies,[125–130] abundances in cluster beams,[131,132] dissociation energies,[133] cluster geometry,[134] spectroscopy,[135] and polarizability.[136]

A very interesting approach to studying magnetic effects of small alkali metal clusters has been reported by de Heer *et al.*[137] Their work is based on the well-known Stern–Gerlach deflection experiments. In alkali

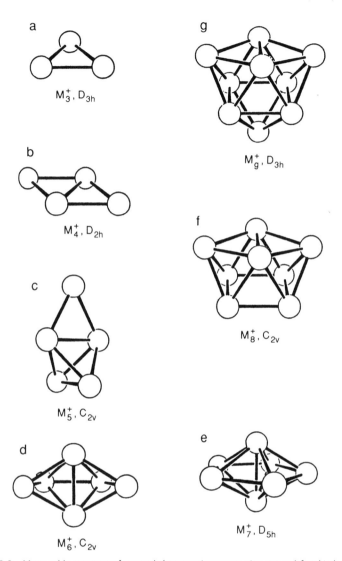

FIGURE 3-3   Most stable structures for metal cluster cations up to nine atoms (after Lindsay and co-workers[117]).

clusters, every atom contributes one valence electron to the conduction band. It would be expected that odd numbered clusters (M_1, M_3, M_5...) would be paramagnetic and could have electronic spin $\frac{1}{2}$ for odd and 0 for even. In a strong magnetic field it would be expected that deflections should consist of two peaks with magnetic moments of $+\mu_B$ and $-\mu_B$.

A cluster beam source was used to produce alkali metal clusters of M_{1-6}, and the collimated cluster beam was passed between the poles of a Stern-Gerlach magnet, which deflected the paramagnetic particles. However, the magnetic moments found were not at all as expected for spin $\frac{1}{2}$ particles (for the odd-numbered clusters). In fact Na_3 showed three peaks, the two expected and a broad central peak, which reflected a distribution of reduced magnetic moments. Similarly, Li_3 showed three peaks with a weaker central peak. However, for K_3 the $\pm\mu_B$ peaks were entirely absent and only a peak for reduced magnetic moments was observed; the magnetic field had no effect on large clusters of sodium. These results were rationalized in terms of coupling of the spin of the rotating cluster with the molecular framework of the cluster. The magnitude of the total spin (about 1 μ_B for alkali clusters) and the coupling strength of the spin with the cluster caused the reduced deflections. A similar model was used to interpret Stern-Gerlach magnetic deflection results for iron clusters.

2. Larger Clusters and Film Forming Processes

Edmonds et al.[138] have reported ESR spectra for much larger clusters of Na, K, and Rb metal, particles of about 5000 atoms. Magnetic susceptibility measurements coupled with conduction ESR of the particles frozen in HMPA solvent allowed the following conclusions: (1) a single magnetic resonance absorption is produced by such particles, unless the microwave energy is unable to fully penetrate the particle whereupon asymmetry is observed; (2) electron-spin relaxation occurs via the effects of electron–phonon interactions and the mixing of spin-up and spin-down wave functions by spin-orbit coupling; and (3) paramagnetic susceptibility is independent of temperature, as for bulk metal.

As described earlier in this chapter, the development of the ionized cluster beam method has opened a new era in film deposition techniques. It has been used to form tailored thin-film materials and machines capable of large area film deposition have been developed.[139] Study of the fundamental processes of ICB continue.[140] Low-energy bombardment enhanced adatom migration. In addition, effects of ICB bombardment energies on epitaxial growth of CaF_2 films have been analyzed. The properties of the cluster state (size, speed, charge) cause the unique film properties observed. Further studies of Mg clusters in an ICB experiment employed another approach. Magnesium was vaporized and forced to undergo supersonic jet expansion. There was an unexpectedly low concentration of large clusters (approaching 100 atoms/cluster).[141] It was determined that large cluster formation becomes appreciable only if stagnation pressure is greater than 10 Torr.

To test the value of the ICB technique, compared with normal physical deposition of metal vapors (atoms), a high vapor pressure of Na was used.[142] Mass distribution of the cluster beam was analyzed by means of an electrostatic energy analyzer. Past results were interpreted by assuming the presence of large clusters of Na, but in this work this interpretation was disputed. Instrumental effects such as internal reflections and desorption phenomena in the ionization source in front of the analyzer could have led to the earlier misinterpretations.

In related work in Cs partially ionized cluster beams, clusters containing several thousand atoms/cluster have been described. The ionized cluster portion yielded a film deposition rate near 4 nm/second at a distance of 15 cm from the nozzle, while the total mass flux was believed to be at least 10 times greater.[143]

It is obvious that the fundamentals of the ICB technique are still not clear. The presence of large ionized clusters is disputed. However, it is clear that, regardless of the size or distribution of ionized clusters, the mix of atoms, neutral clusters, and ionized clusters does lead to high-quality films. Thus, the importance of the contribution of this approach cannot be disputed. But more work on understanding the mechanistic and physical details is certainly warranted.

Abe et al.[144] prepared particles of Li, Na, K, Ca, Cu, and Ag by gas aggregation (GEM). These were about 100 Å in size and were subsequently trapped in frozen argon. Optical absorption spectra were collected on the trapped colloidal particles. Surface plasmon absorption bands in the range of 200–700 nm were observed and compared with calculated values from bulk optical data.

Coalescence growth mechanisms in GEM (see Chapter 2) have been studied.[145] Convection currents of the inert gas cause distinct growth zones for metal nucleation and aggregation. Clear-cut crystal habits from hexagonal plates to multiply twinned particles were observed in the outer reaches of a cluster growth zone. Based on these observations a model of the growth patterns was presented. Size distributions of the particles were calculated by counting the ratio of the number of collisions by using the effective cross section of collisions and the existence probability of the volume of the particle. This model aids understanding of growth rate derived from crystal habit.

In related work, Mg particles have been grown by GEM, and their static magnetic susceptibilities measured as a function of particle diameter. Temperature-independent paramagnetism enhancement was observed, and this was inversely proportional to the cube of the particle size. This was taken as evidence for zero-dimensional magnetism in a small particle. In addition, this enhanced paramagnetism seemed to exhibit a phase transition about 150 K.[146] These quantum size effects were further elucidated in the marked decrease in magnetic susceptibility at low temperature.

Magnesium particles of about 2 nm were prepared in He, and larger particles in Ar (about 100 nm). By varying the pressure and changing from He to Ar, intermediate size particles could also be grown.[147] Further insight into the gas aggregation process was obtained by growing particles of 10 nm size and noting that gravity-driven convection increased the coalescence rate, leading to larger sizes and broader size distributions.[148] The possibilities of reducing coalescence in a microgravity environment are discussed by Webb.[148] Indeed, the GEM method continues to be a very useful but perhaps poorly understood process.[149–150]

Kimura and Bandow have done extensive studies of fine particles prepared by the GEM, followed by trapping in cold matrices and cold liquids.[90,151] Magnesium particles trapped in THF and hexane were examined by conduction ESR (CESR) and static magnetic susceptibility at 2–150 K. In the ESR, a peak at 6 G was correlated with the conduction electrons in the metal particles. Magnetic susceptibility determinations showed the presence of two components, one exhibiting Curie behavior with increased susceptibility at low temperature and the other with decreased susceptibility at low temperature. This latter behavior is characteristic of the quantum effect expected of very small metal particles. In further analysis, it was found that the broader line in the ESR could be attributed to the quantum size effect, which appeared only in small particles, whereas the narrower line was due to a surface impurity.[152]

Granquist and Buhrman have also carried out studies of many metals, including Mg, Zn, Sn, Al, Cr, Fe, Co, Ni, Cu, and Ga,[153] using the GEM. Diameter as a function of evaporation rate, pressure, and inert gas was determined. Particles 20 nm in diameter appeared almost spherical in the electron microscope, although larger particles exhibited pronounced crystal habits. Based on these results, a statistical growth model was proposed. This model was based on liquid like coalescence of particles. Applications of this model to colloids, discontinuous films, and supported catalysts were discussed.

Further ESR measurements have been carried out on Li, Ag, and Au metallic particles (average size 1.7 nm). No ESR evidence of quantum size effects was detected even though temperatures down to 4 K were examined.[154] The ESR linewidth for the Li particles broadened to 14 G for 2.0-nm particles. This broadening is due to surface spin scattering (from 0.4 G for bulk Li to 14 G for 2.0-nm particles). No ESR signals were detected at all for Ag and Au due to extreme line broadening.

Finally, a related method for the preparation of potassium particles by resonant laser excitation of K_2 molecules has been reported. In a heated pipe containing K vapor and a noble gas, excitation of K_2 molecules caused rapid heating of the gas due to the high absorption rate of the laser light. A supersaturated region was created where the vapor condensed to form fine particles.[155]

C. Chemistry

The reactivity of neutral metal clusters is a fascinating topic and one that promises to shed much light on fundamental concepts of chemical reactivity. An example dealing with Mg atoms, Mg_2, and Mg_3 has been reported.[61,62] The interaction of these species with CH_3Br in a low-temperature matrix showed that Mg_2 and Mg_3 were consumed under the same conditions under which Mg atoms were inert. It was proposed that the Mg clusters reacted to $CH_3(Mg)_xBr$, and this is the first example where σ-bond breaking took place on a cluster but not an atom under the same experimental conditions. With theoretical support of ab initio calculations,[156,157] it was proposed that such cluster reactions were more thermodynamically favored. Lower ionization energies of the clusters may also be important since an initial electron transfer may be necessary in the bond-breaking process:

$$Mg_2 + CH_3Br \rightarrow [Mg_2^+CH_3Br^-] \rightarrow CH_3Mg_2Br.$$

Additional investigations demonstrated that Ca clusters behaved similarly.[62] In fact studies on Ca, Ca_2, Ca_x, Mg, Mg_2, Mg_3, Mg_4, and Mg_x in the presence and absence of CH_3F, CH_3Cl, CH_3Br, and CH_3I established trends: clusters were more reactive than atoms; larger clusters were more reactive than smaller ones; and the reactivity trends were $CH_3I > CH_3F > CH_3Br > CH_3Cl$. Bond strengths and cluster ionization energies were used to rationalize these trends.

Clustering of Mg atoms in cold THF and other ethers has led to very active particles of Mg that showed excellent reactivity with hydrogen.[158a] The hydriding behavior of these small particles is so efficient that this material has been proposed for hydrogen storage applications. It is believed that the high reactivity of the $(Mg)_n$-THF system toward H_2 is due to unusual and very rapid surface processes of chemisorption, migration, and chemisorption–adsorption transitions. The chemical bonding of Li_5 to H atoms has also aided our understanding of these systems.[158b]

Chemistry in a cluster beam is a topic Martin has explored for Cs clusters.[159] Gas-phase reactions with sulfur vapor and Cl_2 gas have been reported and showed a wide variety of reaction products often of unexpected stoichiometry. Product ratios could be varied by varying the partial pressures of the reactants. Analysis of the resulting cluster mass spectra allowed certain trends to be deduced for the stability of metal-rich clusters and helped identify new building blocks for the synthesis of solids. For example, in the Cs/Cl system it was found that clusters predominated for Cs_nCl^+ where n is an even number, such as Cs_2Cl^+, Cs_4Cl^+, Cs_6Cl^+, Cs_8Cl^+. However, if two Cl atoms were incorporated, then if n was an odd number, certain clusters were dominant, such as $Cs_3Cl_2^+$, $Cs_5Cl_2^+$, $Cs_7Cl_2^+$. A general rule was deduced: Highly stable clusters (neutral or charged) must contain an even number of electrons. This rule was found to hold for several systems, including subhalides and suboxides, and seems to be general for ionic systems where the ions have rare-gas electronic configurations (alkali metal cations, halide anions, oxide dianions, and sulfur dianions).

III. Bimetallic and Other Binuclear Clusters/Particles

A. Occurrence and Techniques

Laser ablation and electron beam evaporation are the most frequently used techniques for producing vapors of binuclear species. For example, Honda evaporated dielectric and semiconductor materials by electron beam and generated vapors from ZrO_2, TiO_2, SiO_2, CeO_2, Al_2O_3, Te_2O_5, Nd_2O_3, and MgO.[160] Optical thin films were produced. The temperature of the face of the evaporation source was kept uniform by deflecting the electron beam with two coils so as to sweep the entire surface of the target.

B. Physical Properties and Theoretical Studies

1. Bimetallics

A systematic study was made of the mass spectra of particles formed in a supersonic nozzle expansion of Zn, Cd, Hg, and Na vapors together.[161] The particles under study contained one heavy atom (Zn, Cd, or Hg) and up to 18 Na atom ligands. Neutral particles were ionized by laser photoionization at several wavelengths. Ion abundances were corrected to relative differential ionization cross sections. Ionization energies of

Na_nM, with n less than 5 but greater than 10 and M = Zn, Cd, or Hg, were measured. When n was <6, no mixed clusters were found. Cluster abundance occurred for all M at $n = 8$. Another maximum occurred where $n = 16$ or $n = 18$. These selectivity patterns were rationalized by recent ab initio and jellium calculations, which considered a closed shell at $n = 8$ with the occupations $1s^2 \, 2s^2 \, 2p^6$. Simulations of cluster growth with empirical potentials showed that the heavy metal migrates to the center of a cube. For larger central atoms, a square antiprismal arrangement appeared to be most stable.

Ab initio calculations on low-lying states of AlLi have been carried out by Rosenkrantz.[162] State-averaged calculations on the lowest three $^{1,3}\Sigma^+ + {}^{1,3}\Pi$ states were performed; these are the states that dissociate to $Al^2P + Li^2S$ atoms. Three bound $^1\Sigma^+$ states were obtained, one being the ground state of AlLi. However, all three $^3\Sigma^+$ states were repulsive while all three $^3\pi$ states were bound.

Experimental work dealing with the attempted detection of gaseous AlLi has been reported by Carrick and Brazier.[163] Laser ablation of commercial Li–Al alloy (Al : Li 4 : 1) yielded gaseous species that were examined by mass spectrometry. Upon electron impact ionization of the neutral gaseous species, no $AlLi^+$ species were detected. Further experiments with sputtered Al mixed with Li vapor and laser-induced fluorescence yielded a new band at 655 nm that may be due to the AlLi molecule. Further work is needed to confirm this. The driving force for this work is the design of cryogenic propellants with the highest specific impulses. Stabilizing metal atoms or clusters in solid H_2 is a preliminary goal of such experiments. The specific elements of interest include Li, B, Mg, and Al. Intermetallic diatomics of these metals are also of interest. According to theory, one of the few combinations that should yield a singlet ground state (thus presumably more stable) is Al–Li. Thus, the particular interest in the AlLi molecule.

Ionization energies of Al_nNa_m bimetallic clusters have been reported.[164a] Using a tunable ultraviolet laser combined with time-of-flight MS, it was determined that the ionization energies decreased monotonically with the number of Na atoms. In contrast, ionization energies of $Al_{13}Na$ and $Al_{23}Na$ are higher than (or equal to) those of Al_{13} and Al_{23}. These results were rationalized by use of an electronic shell model. Upon addition of one Na atom to the Al_{13} or Al_{23} clusters, the total number of valence electrons strictly satisfies the shell closing of $2p$ and $3s$ shells. Similarly, the electronic and geometrical structure of the B_3Li molecule have been investigated.[164b]

High-resolution laser spectroscopy of the NaRb dimer has also been reported,[165a] employing Doppler-free laser polarization spectroscopy. A comparison of molecular constants for Na_2, NaK, K_2, NaRb, and Rb_2 is presented and literature reviewed. Generally the internuclear distances are in very good agreement with the values calculated by the additivity of atomic radii. In addition, ab initio theoretical studies of the excited states of $LiNa_3$ and Li_2Na_2 have been reported.[165b]

ESR spectra of mixed Group 1 and 2 bimetallic dimers in argon matrices have been reported.[166] Thus, NaMg, NaCa, NaSr, KMg, and KCa radicals were trapped, and the ESR spectra indicated that $^2\Sigma$ is the ground state for these species in which the unpaired electron is substantially localized over the (Mg, Ca, Sr) atom. These results imply chemical bond formation between the metallic atoms.

An earlier study of HCr, LiCr, and NaCr diatomics, using ESR, was reported by Welter and co-workers.[167a] Both metals were simultaneously evaporated from tantalum cells and codeposited with argon at 4–6 K. It was found that LiCr and NaCr possessed similar electronic properties and differ from CrH. Quite simply, the differences can be attributed to ionicity where $Li^{\delta+}Cr^{\delta-}$ and $Na^{\delta+}Cr^{\delta-}$ are polarized differently from $H^{\delta-}Cr^{\delta+}$ (and which is more covalent). Thus, the low ionization energies (IE) of alkali metals leading to this polarization would suggest that most of the unpaired spin resides on the $Cr^{\delta-}$ ion. It should be pointed out that no promotional energy is required for the Cr atom (d^5s, ^7S) to react with such atoms as H or Li.

Ionization potentials of Co_nNa_m clusters are also of interest.[167b] In this study $n = 3–48$ and $m = 1–3$. When $n \le 17$ the IE decrease upon Na doping was large, whereas when $n \ge 18$ the decrease was smaller and constant. This was attributed to a significant geometric change in the cluster at smaller sizes when Na was added. However, when n was large, the geometric effect of added Na was much less.

2. Nonmetallic Systems

Matrix studies of CaF_2 isolated in several host matrices have shown that the F–Ca–F angle is sensitive to the host.[168] This important finding is significant for all such matrix studies for isolated metal halide molecules. For CaF_2 the bond angle was found to vary from 142° to 156° as matrices changed from neon to argon to krypton to nitrogen to carbon monoxide. DeKock et al. have given an excellent overview of this area.[169] A brief summary from this work is shown in Table 3-1, which shows the geometries of the simple MX_2 species.

Two studies of free molecules of MgF_2 have been reported.[170] X-ray emission and X-ray fluorescence spectra of free MgF_2 compared with solid MgF_2 were reported. Analysis of these results indicated that the

TABLE 3-1
Structure of Group 2 Dihalides, MX_2

	F	Cl	Br	I
Be	Linear	Linear	Linear	Linear
Mg	Linear	Linear	Linear	Linear
Ca	140°	Linear	Linear	Linear
Sr	108°	142°	Linear	Linear
Br	100°	127°	—	148°

chemical bond in free MgF_2 is closer to a metallic-covalent bond than an ionic bond (as found in the solid MgF_2 sample). Electronic transitions in gaseous CaS have also been reported recently.[171]

Morphologies of NaCl crystals grown by the GEM technique have been described. It was found that the cubic shape was most prominent at lower pressures of inert gas.[172a] Indeed, a review of properties of alkali halide nanocrystals that nicely summarizes the unusual nature of such small ionic networks has recently appeared.[172b] Other theoretical and experimental studies of Na_nCl_m clusters have also appeared recently.[172c,d]

Cluster beam techniques have been employed for the deposition of films of BaF_2. Different aspects of the deposition process included deposition temperature, cluster source geometry, and acceleration voltage. Studies of the films by X-ray diffraction indicated that above 200–300°C deposition temperatures, dense oriented films of good crystallinity were produced. At lower temperatures the films were of different orientations and were inhomogeneous.[173] This transition temperature could be influenced by cluster source geometry and acceleration voltage.

In addition, specific heats of fine particles of MgO have been measured.[174] Then data were analyzed in terms of Debye temperature and lattice vibrations. The MgO particles were prepared by CO_2 laser heating of bulk MgO, and the particles produced were 10 nm average diameter. The specific heats were regarded as containing two parts, one being due to ordinary lattice vibrations and the other due to resonance.

Thin films of metal oxides of relevance to high-temperature superconductors have been prepared by ionized cluster beam deposition.[175] Thin films of Pb–O, Bi–O, Zn–Fe–O, and Ba–Bi–O were studied. Orientation, crystallinity, and transmittance of the films could be controlled by changing electron current for ionization and acceleration voltage. The amount of charge and the kinetic energy of the ionized clusters could be used to explain these results.

C. Chemistry

1. Bimetallic and Immiscible Bimetallic Species in Matrices

An interesting development regarding the low-temperature growth of bimetallic clusters of immiscible metals has been reported by Glavee et al.[176] Combination of Fe and Li, normally immiscible at elevated temperatures, was achieved by deposition of the vapors of the metals with a hydrocarbon solvent at 77 K. Upon warming, this matrix of Fe and Li atoms in frozen pentane softened and melted. During this time Fe–Li clusters grew yielding crystallites of Fe (3–4 nm) imbedded in a nearly amorphous Li host. Some hydrocarbon was also trapped in these host particles, which were about 20 nm in size. Upon heat treatment and/or controlled oxidation, Fe crystallite size grew and could be controlled in the range of 3–25 nm. Further phase segregation occurred so that the Fe clusters were coated with Li, and upon oxidation a $Li_2O/LiOH/FeO_3$ coating formed and protected the Fe core. These particles were shown to be stable to the atmosphere for months, encapsulated in the robust Li oxide layer. Upon extended exposure to air, Li_2CO_3 formed and this proved to be especially effective in protecting the Fe core. These results demonstrated for the first time that normally immiscible metals can be forced to form metastable clusters using low temperature, kinetic growth control methods. Extension of this approach to Fe–Mg combinations has been reported by Zhang.[177] Excellent control of Fe crystallite size and protection by a Mg/MgO coating has been achieved.

2. Nonmetallic Species and Matrices

The interactions of chemicals with vapors of NaCl has led to the trapping of normally immiscible materials, or "salting" of neutral organic molecules in a NaCl host matrix. This procedure serves to matrix isolate the organics even as high as 450°C.[178] The presence of the guest in the host matrix was detected by IR and UV absorption, fluorescence, and magic angle spinning-NMR spectra of the solids. The salted guest molecules (for example benzene, naphthalene, dichlorophenol, decamethylcyclohexasilane, 1-adamantol, and iron pentacarbonyl) were found to be aggregated in some cases, but single molecule isolation was possible for $Fe(CO)_5$ by depositing $Fe(CO)_5$, n-pentane, and KBr at a 1 : 10 : 2000 ratio. These novel matrices hold promise for unusual photochemical processes. For example, photolysis of aggregated $Fe(CO)_5$ led to $Fe_2(CO)_9$ formation. However, isolated $Fe(CO)_5$ molecules led to CO loss, oxidation, and reduction processes.[179] Defects in the alkali halide matrix seem to play a role in these reactions. Another example is the conversion of "salted" quadricyclane to norbornadiene which may be influenced by alkali halide color centers that can behave as catalysts.[180] Thus, under conditions of rapid CsI or KBr deposition, or UV or X-ray irradiation, "missing electron" centers can form that seem to catalyze the quadricyclane → norbornadiene conversion. Under low-temperature conditions, the normal UV-photochemical pathway was followed.

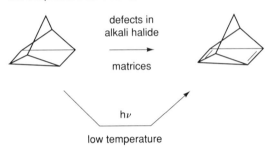

In somewhat related work, Mile and co-workers[181] have cocondensed vapors of NaCl and hydrocarbons at 77 K. Hydrocarbon radicals were formed due to C–H cleavage caused by electrostatic fields generated. For example, cyclohexane and some alkenes reacted on the forming NaCl surface to yield and trap the cyclohexyl- and alkenyl-radicals. Thus, scission of C–H bonds with bond strengths of >400 kJ/mol took place. The very large electrostatic fields (ca. 10^7 V/cm) on the forming NaCl surfaces seem to be responsible for this reactivity.

Polymerization catalysts have been prepared by using $MgCl_2$ vapors.[182] For example, $MgCl_2$ vapor condensed with alkanes yields highly activated, high-surface-area materials that when treated with $TiCl_4$ yield $MgCl_2$ supported, highly active Ziegler–Natta catalysts useful for alkene polymerization processes.

Finally, an ab initio theoretical study of the interaction of gaseous MgO and CaO with CO showed that the energetic differences between

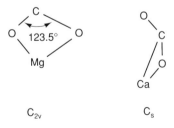

linear and cyclic structures can be considerable.[183] For the MgO case, linear carbonyl (OC–MgO) and isocarbonyl (CO–MgO) minima are higher in energy than a four-membered ring geometry with C_{2v} symmetry. Analogous results were found for the CaO system; however, a second preferred ring structure was found with a Ca–O–C three-membered ring. The cyclic structures resemble coordination of a CO_2^{2-} anion with a M^{2+} cation, and the anion has considerable flexibility in bonding.

References

1. K. J. Klabunde, "Chemistry of Free Atoms and Particles," Academic Press, New York, 1980.
2. (a) W. Ertmer, *Phys. Scr.* **36**, 306 (1987); (b) S. Chu, *Science* **253**, 861 (1991).
3. (a) J. Q. Broughton and P. S. Bagus, *J. Electron Spectrosc. Relat. Phenom.* **20**, 127 (1980); (b) M. T. Djerad, F. Gounand, A. Kumar, and M. Cheret, *J. Chem. Phys.* **97**, 8334 (1992).
4. H. Aksela, S. Aksela, and H. Patana, *Phys. Rev. A.* **30**, 858 (1984).
5. D. R. Beck and C. A. Nicolaides, *J. Electron Spectrosc. Relat. Phenom.* **8**, 249 (1976).
6. W. Sander and M. Voelkel, *Phys. Rev. Lett.* **62**, 885 (1989).
7. P. Bisgaard, R. Bruch, P. Dahl, B. Fastrup, and M. Roedbro, *Phys. Scr.* **17**, 49 (1978).
8. M. G. Littman, M. M. Kash, and D. Kleppner, *Phys. Rev. Lett.* **41**, 103 (1978).
9. M. S. Banna, B. Wallbank, D. C. Frost, C. A. McDowell, and J. S. H. Q. Perera, *J. Chem. Phys.* **68**, 5459 (1978).
10. M. Meyer, M. Pahler, T. Prescher, E. Von Raven, M. Richter, B. Sonntag, S. Baier, W. Fiedler, B. R. Muller, *et al.*, *Phys. Scr. T* **T31** 28 (1990).
11. B. F. Sonntag, *Phys. Scr. T* **T34** 93 (1991).
12. N. A. Shilkova, V. P. Shirokovskii, and G. V. Ganin, *Fiz. Met. Metalloved.* **43**, 685 (1977). [In Russian]
13. K. Haug and H. Metiu, *J. Chem. Phys.* **95**, 5670 (1991).
14. R. R. Bennett, J. G. McCaffrey, and W. H. Breckenridge, *J. Chem Phys.* **92**, 2740 (1990).
15. M. E. Fajardo, P. G. Carrick, and J. W. Kenney, III, *J. Chem. Phys.* **94**, 5812 (1991).
16. (a) M. E. Fajardo, *in* "Proceedings, High Density Matter HEDM Conference, Albuqerque, NM Feb. 24–27, 1991" (M. E. Cordonnier, Ed.), p. 61, special report, Air Force Systems Command, Edward Air Force Base; (b) M. E. Fajardo, *J. Chem. Phys.* **98**, 110 (1993). (c) M. E. Fajardo, *J. Chem. Phys.* **98**, 119 (1993).
17. L. C. Balling and J. J. Wright, *J. Chem. Phys.* **78**, 592 (1983).
18. A. Schrimpf, R. Rosendahl, T. Bornemann, H.-J. Stöckmann, F. Faller, and L. Manceron, *J. Chem Phys.* **96**, 7992 (1992).
19. A. Steinmetz, A. Schrimpf, H.-J. Stöckmann, E. Görlach, R. Dersch, G. Sulzor, and H. Ackermann, *Z. Phys. D: At. Mol. Clusters* **4**, 373 (1987).
20. (a) A. Schrimpf, G. Sulzer, H.-J. Stöckmann, and H. Ackermann, *Z. Phys. B: "Condens. Matter* **67**, 531 (1987); (b) R. Pellow and M. Vala, *J. Chem. Phys.* **90**, 5612 (1989).
21. A. C. G. Mitchell and M. W. Zemansky, "Resonance Radiation and Excited Atoms," Cambridge Univ. Press, London, 1971. [First published in 1934]

22. Toyo Soda Mfg. Co., Ltd., Jpn. KoKa; Tokkyo Koho, JP 60121207; CA 103(24): 199791g, 1985.
23. P. N. Hawker and P. L. Timms, *J. Chem. Soc. Dalton Trans.* 1123 (1983).
24. Memorial issue of *J. Phys. Chem.* **95** (1991).
25. M. H. M. Janssen, D. H. Parker, and S. Stolte, *J. Phys. Chem.* **95,** 8142 (1991).
26. A. G. Suits, H. Hou, H. F. Davis, and Y. T. Lee, *J. Phys. Chem.* **95,** 8207 (1991).
27. R. S. Mackay, Q. X. Xu, F. J. Aoiz, and R. B. Bernstein, *J. Phys. Chem.* **95,** 8226 (1991).
28. K. Chakravorty and R. B. Bernstein, *J. Phys. Chem.* **88,** 3465 (1984).
29. R. D. Coombe and R. K. Horne, *J. Phys. Chem.* **84,** 2085 (1980).
30. (a) P. F. Bernath, *Science,* **254,** 665 (1991); (b) C. N. Jarman and P. F. Bernath, *J. Chem. Phys.* **97,** 1711 (1992); (c) T. C. Steimie, D. A. Fletcher, K. Y. Jung, and C. T. Scurlock, *J. Chem. Phys.* **97,** 2909 (1992); (d) T. C. Steimie, D. A. Fletcher, K. Y. Jung, and C. T. Scurlock, *J. Chem. Phys.* **96,** 2556 (1992).
31. J. M. Brom, Jr. and W. Weltner, Jr., *J. Chem. Phys.* **5,** 5323 (1973).
32. C. W. Bauschlicher, Jr., S. R. Langhoff, T. C. Steimie, and J. E. Shirley, *J. Chem. Phys.* **93,** 4179 (1990).
33. C. G. James and T. M. Sugden, *Nature (London),* **175,** 333 (1955).
34. P. A. Bonczyk, *Combust. Flame* **59,** 143 (1988).
35. T. Tsuji, *Astron. Astrophys.* **23,** 411 (1973).
36. D. Smith and N. Adams, *in* "Gas Phase Ion Chemistry" (M. T. Bowers, Ed.), Vols. 1–3, Academic Press, New York, 1979.
37. P. B. Armentrout, *Science* **251,** 175 (1991).
38. M. V. Buchanan, Ed., "Fourier Transform Mass Spectrometry," Vol. 359, ACS Symposium Series, American Chemical Society, Washington, DC, 1987.
39. P. Kebarle, *Annu. Rev. Phys. Chem.* **28,** 445 (1977).
40. P. F. Bernath, B. Pinchemel, and R. W. Field, *J. Chem. Phys.* **74,** 5508 (1981).
41. S. F. Rice, H. Martin, and R. W. Field, *J. Chem Phys.* **82,** 5073 (1985).
42. T. Törring, W. E. Ernst, and J. Kändler, *J. Chem. Phys.* **90,** 4927 (1989).
43. J. V. Ortiz, *J. Chem. Phys.* **92,** 6728 (1990).
44. J. V. Ortiz, *Chem. Phys. Lett.* **169,** 116 (1990).
45. I. C. O'Brian and P. F. Bernath, *J. Am. Chem. Soc.* **108,** 5017 (1986).
46. P. J. Dagdigian, *in* "Proceedings, High Density Matter (HEDM) Conference, Albuqerque, NM, Feb. 24–27, 1991" (M. E. Cordonnier, Ed.), p. 137, special report Air Force Systems Command, Edward Air Force Base.
47. J. Chen, E. Quinones, and P. J. Dagdigian, *J. Chem. Phys.* **93,** 4033 (1990).
48. J. W. Cox and P. J. Dagdigian, *J. Phys. Chem.* **86,** 3738 (1982).
49. R. P. Bleckersderfer, K. D. Jordan, N. Adams, and W. H. Breckenridge, *J. Phys. Chem.* **86,** 1930 (1982).
50. (a) W. H. Breckenridge and H. Umemoto, *J. Phys. Chem.* **87,** 476 (1983); (b) W. H. Breckenridge and H. Umemoto, *J. Phys. Chem.* **87,** 1804 (1983); (c) M. Helmer, J. M. C. Plane, and M. R. Allen, *J. Chem. Soc., Faraday Trans.,* 763 (1993); (d) A. G. Suits, H. Hou, H. F. Davis, and Y. T. Lee, *J. Chem. Phys.* **96,** 2777 (1992).
51. M. Eyal, F. R. Grabiner, U. Agam, and L. A. Gamess, *J. Phys. Chem.* **87,** 3400 (1983).
52. B. Bourgaignon, M. A. Gargoura, J. Rostas, and G. Taieb, *J. Phys. Chem.* **91,** 2080 (1987).
53. P. W. Harland, H. S. Carmen, Jr., L. F. Phillips, and P. R. Brooks, *J. Chem Phys.* **93,** 1089 (1990).
54. D. M. Lindsay and D. A. Garland, *J. Phys. Chem.* **91,** 6158 (1987).

55. O. Ayed, A. Loutellier, L. Manceron, and J. P. Perchard, *J. Am. Chem. Soc.* **108**, 8138 (1986).
56. Z. Kafafi, R. H. Hauge, W. E. Billups, and J. L. Margrave, *Inorg. Chem.* **23**, 177 (1984).
57. (a) Z. Kafafi, R. H. Hauge, W. E. Billups, and J. L. Margrave, *J. Am. Chem. Soc.* **105**, 3886 (1983); (b) R. H. Hauge, J. L. Margrave, J. W. Kauffman, N. A. Rao, M. M. Konarski, J. P. Bell, and W. E. Billups, *J. Chem. Soc., Chem. Commun.* 1258 (1981).
58. S. A. Heffner, M. P. Andrews, and M. E. Galvin, *Polym. Commun.* **29**, 335 (1988).
59. J. G. McCaffrey, R. A. Poirier, G. A. Ozin, and I. G. Csizmadia, *J. Phys. Chem.* **88**, 2898 (1984).
60. B. S. Ault, *J. Am. Chem. Soc.* **102**, 3480 (1980).
61. Y. Imizu and K. J. Klabunde, *Inorg. Chem.* **23**, 3602 (1984).
62. K. J. Klabunde and A. Whetten, *J. Am. Chem. Soc.* **108**, 6529 (1986).
63. W. E. Billups, M. M. Konarski, R. H. Hauge, and J. L. Margrave, *J. Am. Chem. Soc.* **102**, 3649 (1980).
64. W. E. Billups, M. M. Konoarski, R. H. Hauge, and J. L. Margrave, *J. Organomet. Chem.* **194**, C22 (1980).
65. K. Mochida and Y. Yamanishi, *J. Organomet. Chem.* **332**, 247 (1987).
66. W. E. Billups, J. P. Bell, J. L. Margrave, and R. H. Hauge, *J. Fluorine Chem.* **26**, 165 (1984).
67. K. J. Klabunde, J. Y. F. Low, and M. S. Key, *J. Fluorine Chem.* **2**, 207 (1972).
68. J. G. McCaffrey and G. A. Ozin, *J. Chem. Phys.* **89**, 1839, 1844 (1988).
69. J. G. McCaffrey, J. M. Parnis, and G. A. Ozin, *J. Chem. Phys.* **89**, 1858 (1988).
70. V. Balaji and K. D. Jordon, *J. Phys. Chem.* **92**, 3101 (1988).
71. J. R. Flores and A. Largo, *J. Phys. Chem.* **95**, 9278 (1991).
72. I. C. Hisatsune, *J. Catal.* **75**, 425 (1982).
73. L. Manceron and L. Andrews, *J. Am. Chem. Soc.* **110**, 3840 (1988).
74. Z. Kafafi, R. H. Hauge, and J. L. Margrave, *Polyhedron* **2**, 167 (1983).
75. S. Süzer and L. Andrews, *J. Am. Chem. Soc.* **109**, 300 (1987).
76. R. Niedermayer, *Angen. Chem. Int. Ed. Engl.* **14**, 212 (1975).
77. For example see J. C. Miller and L. Andrews, *J. Chem. Phys.* **69**, 3034 (1978).
78. T. G. Dietz, M. A. Ducan, D. E. Powers, and R. E. Smalley, *J. Chem. Phys.* **74**, 6511 (1981).
79. M. D. Morse, M. E. Geusic, J. R. Heath, and R. E. Smalley, *J. Chem. Phys.* **83**, 2293 (1985).
80. S. Maruyama, L. R. Anderson, and R. E. Smalley, *Rev. Sci. Instrum.* **61**, 3686 (1990).
81. E. K. Parks, S. Liu, S. C. Richtsmeier, L. G. Pobo, and S. J. Riley, *J. Chem. Phys.* **82**, 5470 (1985).
82. S. C. Richtsmeier, E. K. Parks, K. Liu, L. G. Pobo, and S. J. Riley, *J. Chem. Phys.* **82**, 3659 (1985).
83. T. Takagi, I. Y. M. Kunori, and S. Kobiyama, *in* "Proceedings, 2nd International Conference on Ion Sources: Vienna, 1972," p. 790.
84. T. Takagi, I. Yamada, and A. Sasaki, *Thin Solid Films* **45**, 569 (1977).
85. K. H. Guenther, *in* "Proceedings, 31st Annual SPIE, August 16–21, 1987," p. 30.
86. K. H. Mueller, *in* "Proceedings, 31st Annual SPIE, August 16–21, 1987," Paper 821-01, p. 61; P. Meakin, *in* "Proceedings, 31st Annual SPIE, August 16–21, 1987, Paper 821-02, p. 61.
87. W. T. Elam, D. Van Vechten, S. A. Wolf, J. Sprague, D. U. Gubser, G. I. Bartz, and P. Meakin, *Phys. Rev. Lett.* **54**, 701 (1985).

88. C. Hayashi, *Phys. Today* **December** 44 (1987); *J. Vac. Sci. Technol.* **A5,** 1375 (1987).
89. G. A. Niklasson, *J. Appl. Phys.* **62,** 258 (1987); and references therein.
90. (a) K. Kimura and S. Bandow, *J. Chem. Soc., Jpn.* **56,** 3578 (1983); (b) S. Yatsuya, S. Kasukabe, and R. Uyeda, *Jpn. J. Appl. Phys.* **12,** 1675 (1973).
91. M. M. Kappes, P. Radi, M. Schaer, C. Yeretzian, and E. Schumacher, *Z. Phys. D: At. Mol. Clusters* **3,** 115 (1986).
92. W. D. Knight, *Surf. Sci.* **106,** 172 (1981).
93. M. Arnold, J. Kowalski, G. Z. Putlitz, T. Stehlin, and F. Traeger, *Surf. Sci.* **156,** 149 (1985).
94. T. Bergeman and P. F. Liao, *J. Chem Phys.* **72,** 886 (1980).
95. G. Chalasinski, D. J. Funk, J. Simons, and W. H. Breckenridge, *J. Chem Phys.* **87,** 3569 (1987).
96. H. Partridge, C. W. Bauschlicher, Jr., L. G. M. Pettersson, A. D. McLean, B. Liu, M. Yoshimine, and A. Komornicki, *J. Chem. Phys.* **92,** 5377 (1990).
97. C. H. Wu, *J. Phys. Chem.* **87,** 1534 (1983).
98. D. Plavsic, J. Kontecky, G. Pocchioni, and V. Bonacic-Kontecky, *J. Phys. Chem.* **87,** 1096 (1983).
99. S. C. Richtsmeler, D. A. Dixon, and J. M. Gole, *J. Phys. Chem.* **86,** 3942 (1982).
100. G. Durland, *J. Chem. Phys.* **91,** 6225 (1989).
101. V. Bonacic-Kontecky, P. Fantucci, and J. Kontecky, *J. Chem. Phys.* **91,** 3794 (1989).
102. V. Bonacic-Kontecky, P. Fantucci, and J. Kontecky, *Phys. Rev. B.* **37,** 4369 (1988).
103. K. M. McHugh, J. G. Eaton, G. H. Lee, H. W. Sarkas, L. H. Kidder, J. T. Snodgrass, M. R. Manna, and K. H. Bowen, *J. Chem. Phys.* **91,** 3792 (1989).
104. P. A. Bruhwiler and S. E. Schnatterly, *Phys. Rev. B: Condens. Matter* **41,** 8013 (1990).
105. W. Weltner, Jr. and R. J. Van Zec, *in* "Physics and Chemistry of Small Clusters" (P. Jena, B. K. Rao, and S. N. Khanna, Eds.), p. 353, New York, 1987.
106. D. M. Lindsay, D. R. Hershbach, and A. L. Kwiram, *Mol. Phys.* **32,** 1199 (1976); **39,** 529 (1980).
107. D. M. Lindsay and G. A. Thompson, *J. Chem. Phys.* **77,** 1114 (1982).
108. (a) G. A. Thompson and D. M. Lindsay, *J. Chem. Phys.* **74,** 959 (1981); (b) G. A. Thompson, F. Tischler, D. Garland, and D. M. Lindsay, *Surf. Sci.* **106,** 408 (1981).
109. H. A. Eckel, J. M. Gress, J. Biele, and W. Demtröder, *J. Chem. Phys.* **98,** 135 (1993).
110. J. A. Howard, R. Sutcliffe, and B. Mile, *Chem. Phys. Lett.* **112,** 84 (1984).
111. T. C. Thompson, G. Izmirlian, Jr., S. J. Lemon, D. G. Truhlar, and C. A. Mead, *J. Chem. Phys.* **82,** 5597 (1985).
112. J. Kendrick and J. H. Hillier, *Mol. Phys.* **33,** 635 (1977).
113. E. Schumacher, W. H. Gerber, H. P. Harri, M. Hoffman, and E. Scholl, *Am. Chem. Soc. Symp. Ser.* **179,** 83 (1982).
114. D. A. Garland and D. M. Lindsay, *J. Chem. Phys.* **78,** 2813 (1983).
115. G. A. Thompson, F. Tischler, and D. M. Lindsay, *J. Chem. Phys.* **78,** 5946 (1983).
116. P. Fantucci, J. Kontecky, and G. Pacchioni, *J. Chem. Phys.* **80,** 325 (1984).
117. Y. Wang, T. F. George, D. M. Lindsay, and A. C. Beri, *J. Chem. Phys.* **86,** 3493 (1987).
118. H. O. Beckman, J. Kontecky, and V. Bonacic-Kontecky, *J. Chem. Phys.* **73,** 5182 (1980).
119. D. Plavsic, J. Kontecky, G. Pacchioni, and V. Bonacic-Kontecky, *J. Phys. Chem.* **87,** 1096 (1983).
120. G. Pacchioni, D. Plavsic, and J. Kontecky, *Ber. Bunsenges Phys. Chem.* **87,** 503 (1983).
121. P. Fantucci, J. Kontecky, and G. Pacchioni, *J. Chem Phys.* **80,** 325 (1984).

122. J. Kontecky and P. Fantucci, *Chem. Rev.* **86**, 539 (1986).
123. D. M. Lindsay, Y. Wang, and T. F. George, *J. Chem. Phys.* **86**, 3500 (1987).
124. D. M. Lindsay, Y. Wang, and T. F. George, *J. Chem. Phys.* **87**, 1685 (1987).
125. P. J. Foster, R. E. Leckenby, and E. J. Robbins, *J. Phys. B.* **2**, 478 (1969).
126. A. Hermann, E. Schumacher, and L. Woste, *J. Chem. Phys.* **68**, 2327 (1978).
127. M. M. Kappes, P. Radi, M. Shär, and E. Schumacher, *Chem. Phys. Lett.* **119**, 11 (1985).
128. M. M. Kappes, M. Shär, P. Radi, and E. Schumacher, *J. Chem. Phys.* **84**, 1863 (1986).
129. K. I. Peterson, P. D. Dao, R. W. Farley, and A. W. Castleman, Jr., *J. Chem. Phys.* **80**, 1780 (1984).
130. (a) W. A. Saunders, K. Clemenger, W. A. deHeer,and W. D. Knight, *Phys. Rev. B* **32**, 1366 (1985); (b) M. M. Kappes, M. Schaer, C. Yeretzian, U. Heiz, A. Vayloyan, and E. Schumacher, *NATO ASI Ser. B* **158**, 145 (1987).
131. W. D. Knight, K. Chemenger, W. A. deHeer, W. A. Saunders, M. Y. Chon, and M. L. Cohen, *Phys. Rev. Lett.* **52**, 2141 (1984).
132. W. D. Knight, W. A. deHeer, and K. Clemenger, *Solid State Commun.* **53**, 445 (1985).
133. (a) C. H. Wu, *J. Chem. Phys.* **65**, 3181 (1976); C. H. Wu, *J. Phys. Chem.* **87**, 1538 (1983); (b) K. Hilpert, *Ber. Bunsenges Phys. Chem.* **88**, 260 (1984).
134. (a) C. Fuchs, V. Bonocic-Koutecky, and J. Koutecky, *J. Chem. Phys.* **98**, 3121 (1993); (b) J. Uppenbrink and D. J. Whales, *J. Chem. Phys.* **98**, 5720 (1993); (c) J. Blanc, V. Bonacic-Koutecky, M. Broyer, J. Chevaleyre, P. Dugourd, J. Koutecky, C. Scheuch, J. P. Wolf, and L. Wöste, *J. Chem. Phys.* **96**, 1793 (1992).
135. (a) A. L. Roche and C. Jungen, *J. Chem. Phys.* **98**, 3637 (1993); (b) T. Kobayashi, T. Usui, T. Kumauchi, M. Baba, K. Ishikawa, and H. Kato, *J. Chem. Phys.* **98**, 2670 (1993); (c) L. Li, Q. Zhu, A. M. Lyyra, T. J. Whang, W. C. Stwalley, R. W. Field, and M. H. Alexander, *J. Chem. Phys.* **97**, 8835 (1992); (d) K. G. Dyall and A. D. McLean, *J. Chem. Phys.* **97**, 8424 (1992); (e) V. Bonacic-Koutecky, J. Pittner, C. Scheuch, M. F. Guest, and J. Koutecky, *J. Chem. Phys.* **96**, 7938 (1992).
136. V. Tarnovsky, M. Bunimovicz, L. Vuskovic, B. Stumpf, and B. Bederson, *J. Chem. Phys.* **98**, 3894 (1993).
137. W. deHeer, P. Milani, and A. Chatelain, *Z. Phys. D: At. Mol. Clusters* **19**, 241 (1991).
138. R. N. Edmonds, P. P. Edwards, S. C. Guy, and D. C. Johnson, *J. Phys. Chem.* **88**, 3764 (1984).
139. T. Ina, Y. Minowa, N. Koshirakawa, and K. Yamanishi, *Nucl. Instrum. Methods Phys. Res. Sect. B.* **B37–B38**, 779 (1989).
140. I. Yamada, H. Usui, and T. Takagi, *Nucl. Instrum. Methods Phys. Res., B* **B33**, 108 (1987).
141. S. N. Mei, S. N. Yang, J. Wong, C. H. Choi, and T. M. Lu, *J. Cryst. Growth,* **87**, 357 (1988).
142. R. M. DeGryse and J. Vennick, *in* "Proceedings, 33rd Annual Tech. Conference on Soc. Vac. Coaters,* 1990," pp. 299–302.
143. J. Gspann, *Nucl. Instrum. Methods Phys. Res. Sect. B* **B37–B38**, 775 (1989).
144. H. Abe, K. P. Charle, B. Tesche, and W. Schulze, *Chem. Phys.* **68**, 137 (1982).
145. S. Kasukabe, *J. Cryst. Growth,* **99**, 196 (1990).
146. K. Kimura and S. Bandow, *Phys. Rev. B: Condens. Matter* **37**, 4473 (1988).
147. K. Kimura, *Bull Chem. Soc. Jpn.* **60**, 3093 (1987).
148. G. W. Webb, *Mater. Res. Soc. Symp. Proc.,* 197 (1987).
149. K. Kimura and S. Bandow, *Phys. Rev. Lett.* **58**, 1359 (1987).
150. S. Kasukabe and K. Mihama, *J. Cryst. Growth* **79**, 126 (1986).
151. K. Kimura and S. Bandow, *Nippon Kagaka Kaishi,* **6**, 916 (1984).

152. K. Kimura and S. Sako, *Chem. Lett.* **6**, 973 (1984).
153. C. G. Granquist and R. A. Buhrman, *J. Appl. Phys.* **47**, 2200 (1976).
154. S. Sako, *J. Phys. Soc. Jpn.* **59**, 1366 (1990).
155. M. Allegrini, P. Bicchi, D. Datlrino, and L. Moi, *Opt. Commun.* **49**, 39 (1984).
156. P. G. Jasien and C. E. Dykstra, *J. Am. Chem. Soc.* **105**, 2089 (1983).
157. P. G. Jaisen and C. E. Dykstra, *J. Am Chem. Soc.* **107**, 1891 (1985).
158. (a) H. Imamura, T. Nobunaga, M. Kawahigashi, and S. Tsuchiya, *Inorg. Chem.* **23**, 2509 (1984); (b) M. Raimondi, E. Tornaghi, D. L. Cooper, and J. Gerratt, *J. Chem. Soc., Faraday Trans.*, 2309 (1992).
159. T. P. Martin, *Surf Sci.* **156**, 584 (1985).
160. T. Handa, "Shinku Kikai Kogyo," CA 88(8): 56906, 1977. [In Japanese]
161. (a) U. Heiz, U. Roethlisberger, A. Voyloyan, and E. Schumacher, *Israli J. Chem.* **30**, 147 (1990); (b) M. Musso, L. Windholz, F. Fuso, and M. Allegrini, *J. Chem. Phys.* **97**, 7017 (1992).
162. M. E. Rosenkrants, *in* "Proceedings, High Density Matter (HEDM) Conference, Albuqerque, NM, Feb. 24–27 1991" (M. E. Cordonnier, Ed.), p. 67, special report, Air Force Systems Command, Edward Air Force Base.
163. P. G. Carrick and C. R. Brazier, *in* "Proceedings, High Density Matter (HEDM) Conference, Albuqerque, NM, Feb. 24–27, 1991" M. E. Cordonnier, Ed.), p. 269, special report, Air Force Systems Command, Edward Air Force Base.
164. (a) A. Nakajima, K. Hoshino, J. Naganuma, Y. Sone, and K. Kaya, *J. Chem. Phys.* **95**, 7061 (1991); (b) E. Earl, R. Hernandez, and J. Simons, *J. Chem. Phys.* **97**, 8357 (1992).
165. (a) Y. C. Wang, M. Kajitani, S. Kasahara, M. Baba, K. Ishikawa, and H. Kato, *J. Chem. Phys.* **95**, 6229 (1991); (b) V. Bonacic-Koutecky, J. Gaus, M. F. Guest, and J. Koutecky, *J. Chem. Phys.* **96**, 4934 (1992).
166. C. F. Kirmzan and D. M. Lindsay, *J. Phys. Chem.* **94**, 7445 (1990).
167. (a) R. J. Van Zee, C. A. Baumann, and W. Weltner, Jr., *Chem. Phys. Lett.* **113**, 524 (1985); (b) K. Hoshino, T. Naganuma, Y. Yamada, K. Watanabe, A. Nakajima, and K. Kaya, *J. Chem. Phys.* **97**, 3803 (1992).
168. I. R. Beattie, P. J. Jones, and N. A. Young, *Inorg. Chem.* **30**, 2250 (1991).
169. R. L. DeKock, M. A. Peterson, L. K. Timmer, E. J. Baerends, and P. Vernooijs, *Polyhedron* **9**, 1919 (1990).
170. (a) I. A. Brytov, L. E. Mstibovskaya, and L. G. Rabinovich, *Izv. Akad. Nauk. SSSR, Ser. Fiz.* **49**, 1490 (1985); (b) I. A. Brytov, L. E. Mstibovskaya, and L. G. Rabinovich, *Fiz. Tverd. Tela.* (*Leningrad*) **25**, 3461 (1983).
171. C. N. Jarman, R. A. Hailey, and P. F. Bernath, *J. Chem. Phys.* **96**, 5571 (1992).
172. (a) C. Kaito and Y. Saito, *J. Cryst. Growth* **98**, 847 (1989); (b) R. L. Whetten, *Acc. Chem. Res.* **26**, 49 (1993); (c) C. Ochsenfeld and R. Ahlrichs, *J. Chem. Phys.* **97**, 3487 (1992); (d) A. Heiderreich, I. Oref, and J. Jortner, *J. Phys. Chem.* **96**, 7517 (1992).
173. E. M. Waddell, B. C. Monachan, K. L. Lewis, T. Wyatt-Davies, and A. M. Pitt, *Nist. Spec. Publ.* **756**, 309 (1988).
174. H. Oya, *Nippon Kinzoku Gakkai Kaiho* **26**, 1057 (1987).
175. A. Matsumoto, H. Sadamura, A. Inubushi, M. Okubo, S. Masuda, and K. Suzuki, *Mater. Sci. Monogr.* **38B**, 1421 (1987).
176. (a) G. N. Glavee, C. F. Kernizan, K. J. Klabunde, C. M. Sorensen, and G. C. Hadjapanayis, *Chem. Mater.* **3**, 967 (1991); (b) G. N. Glavee, K. Easom, K. J. Klabunde, C. M. Sorensen, and G. C. Hadjipanayis, *Chem. Mater.* **4**, 1360 (1992).
177. D. Zhang and K. J. Klabunde, unpublished work.
178. E. Kirkor, J. Gebicki, D. R. Phillips, and J. Michl, *J. Am. Chem. Soc.* **108**, 7107 (1986).

179. E. Kirkor, D. E. David, and J. Michl, *J. Am. Chem. Soc.* **112**, 139 (1990).
180. E. S. Kirkor, V. M. Maloney, and J. Michl, *J. Am. Chem. Soc.* **112**, 148 (1990).
181. H. Dahmane, B. Mile, H. Morris, J. A. Howard, and R. Sutcliffe, *J. Chem Soc. Chem. Commun.*, 1068 (1983).
182. R. Mulhaupt, U. Klabunde, and S. D. Ittel, *J. Chem Soc. Chem Commun.*, 1745 (1985).
183. S. Utamapanya, J. V. Ortiz, and K. J. Klabunde, *J. Am. Chem. Soc.* **111**, 799 (1989).

Early Transition-Metal Elements
(Groups 3–7)

I. Early Transition-Metal Atoms (Sc, Ti, V, Cr, Mn, Y, Zr, Nb, Mo, Hf, Ta, W, Re)

A. Occurrence and Techniques

Since these elements are so refractory their occurrence in nature has only been reported in some types of stars.[1] A variety of laboratory methods for evaporating these materials were described earlier and included electron beam, laser, arc, and induction heating.[1] In particular, electron beam and laser evaporation processes have been studied in some detail.[1]

Recent developments, such as pulsed laser evaporation has been used extensively for many refractory materials (see Chapter 2). However, more specific method development for the early transition elements has been reported. For example, liquid-phase metal vapor chemistry employing rotary reactors with electron beam capabilities has been described by Ozin *et al.*[2] Dual electron guns constructed of copper and stainless steel were fashioned and employ the feature of reverse polarity,[3,4] where the hearth (anode) is positive with an ungrounded electrostatic focusing shield. Figure 4-1 illustrates the design of this apparatus where electron beam evaporation can be employed. The atoms of the evaporating element move upward into a liquid film formed by rotation of the 5-liter Pyrex flask.

The development of this apparatus is a significant step forward for metal atom–vapor synthesis technology. It allows relatively large-scale

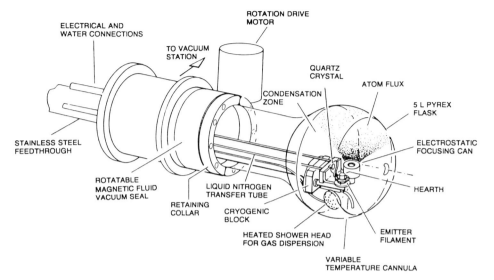

FIGURE 4-1   Perspective view of twin positive hearth electrostatically focused electron-beam metal atom reactor. Reprinted with permission from Ozin et al., Inorg. Chem. **29**, 1069 (1990). Copyright 1990 American Chemical Society.

evaporations of refractory materials, and reactants can be inlet as gases or be in the liquid phase. However, with these capabilities comes high cost and operational difficulties.

Another innovation that has served well is the development of a multisurface matrix isolation apparatus.[5]

This apparatus, illustrated in Fig. 4-2, allows cold matrices to be prepared on a large number of reflective surfaces. In this way a series of matrix reactions/depositions can be studied by reflective infrared techniques without having to break vacuum. Concentrations, matrix temperatures, and the other variables can be studied more conveniently and thoroughly using this setup.

The merging of preparative-scale and microscale metal atom chemistry is now nearly complete with the description of "preparative-scale matrix isolation"[6] as described in Chapter 2. Temperatures below 77 K can be employed and synthesis and isolation of useable amounts of compounds is possible. Additional reviews of apparatus and techniques for both preparative-scale and microscale metal atom–vapor chemistry are available.[7,8]

A new approach to separating and analyzing gas-phase metal clusters involves injection and trapping of supersonic bare metal cluster ions in a

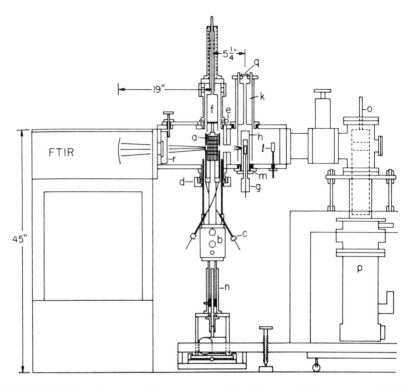

FIGURE 4-2   Multisurface matrix isolation apparatus (after Hauge et al.[5]). Figure notation: (a) Multi-sided matrix block; (b) closed-cycle helium refrigerator; (c) shutter actuator; (d) sliding O-ring seal for vertical positioning; (e) worm gear for matrix block rotation; (f) liquid nitrogen dewar; (g) high-temperature/water-cooled furnace assembly; (h) water-cooled shield; (k) liquid nitrogen-cooled de-war and shield; (l) commercial water-cooled quartz crystal microbalance; (m) rotating O-ring seal; (n) electric screwjack for vertical positioning; (o) liquid nitrogen dewar; (p) 4-in. diffusion pump; (q) window for furnace temperature measurement; (r) rotating window sets—one set of cesium iodide is used for mid-IR, the other set of polyethylene is used for the far-IR.

Fourier transform ion cyclotron resonance (FT–ICR) spectrometer.[9] Metal clusters were generated by laser evaporation in a pulsed nozzle apparatus and the cold clusters were ionized by an argon–fluoride exci-mer laser, mass selected, and then directed toward a superconducting magnet and trapped in a rectangular cell of the FT–ICR. Einzel lenses were used for directing the cluster ions. Bare Nb clusters of nuclearity 2–6 were successfully injected with very high efficiencies. See Chapter 2 for more details.

B. Physical Properties and Theoretical Studies

Spectral characteristics of transition-metal atoms imbedded in frozen rare gas matrices are intriguing. Atomic absorption spectra are significantly shifted compared with the gas phase. A simple way to rationalize these shifts is to consider that an isolated atom, for example ytttrium (Y), surrounded by immobilized Ar atoms is confined. Upon absorption of a photon, the necessary expansion of the electron cloud is impeded by the surrounding argon atoms. Therefore, more energy is needed to carry out the ground state to excited state conversion.

These matrix-induced frequency shifts have been studied for all the transition-metal atoms.[10] If the shifts to higher energy are indeed due to simple confinement as described above, it would be expected that the shifts would be largest for those rare gases that are the least polarizable, i.e., Ar > Kr > Xe. Furthermore, the frequency shifts should be predictable in passing from Ar to Kr to Xe, and related to van der Waals interaction potentials. In fact this was found to be the case for almost all the transition-metal atoms.[10] Thus, a Lennard–Jones (6–12) potential function was used to describe the interaction between trapped species and inert rare gas atoms. This is a convenient, good approximation to the nonbonding weak intermolecular potential between two atomic species "a" and "b." In this equation, σ represents the molecular size when the attractive and repulsive forces are equal, and ε represents the maximum energy of attraction of the two species, which occurs at a separation of $r = 2^{1/6}\sigma$. This approach was used successfully for predicting the experimentally observed absorption frequency shift for the optical excitation of the metal atom in question while it was surrounded by Ar, Kr, or Xe atoms:[10]

$$\phi_{ab} = 4\varepsilon_{ab}\left[\frac{(\sigma_{ab})^{12}}{r_{ab}} - \frac{(\sigma_{ab})^6}{r_{ab}}\right].$$

Interestingly, however, in a few cases, notably Y, Rh, Pd, and Pt, predictable frequency shifts were not observed. At least in the case of Y, which possesses a very diffuse electron cloud and thus a very large atomic polarizability and radius, an additional weak interaction with the most polarizable host, Xe, is indicated.

Indeed, many studies reinforce the concept that metal atoms can be significantly perturbed by rare gas matrices. For example, Re atoms trapped in Kr do not exhibit an ESR spectrum even though its ground state should be $^6S_{5/2}$.[11] Using magnetic circular dichroism and absorption spectra it was shown that the $^6S_{5/2}$ state is spin–orbit coupled with a low-lying $^4P_{5/2}$ state. In a matrix substitutional site that is tetradecahedral, this mixed state is split by crystal/matrix field forces. Thus, matrices can influence the magnitude and sign of spin–orbit coupling of excited states, can enhance spin–forbidden transitions, and can be responsible for large ground-state splittings.

Analogous studies of matrix-isolated Cr atoms showed that forbidden transitions between two allowed $^7P \leftarrow {}^7S$ bands are observable, and this may be due to spin–orbit coupling to near lying ^7P states. These studies also showed that rare gas matrices significantly enhance mixing between electronic states.[12,13]

Matrix-isolated Mn atoms and MnH molecules have been carefully studied by ESR/ENDOR (electron nuclear double resonance).[14] For the Mn atoms, which are in a ^6S electronic state, ENDOR lines were observed for ^{55}Mn ($I = \frac{5}{2}$) corresponding to hyperfine transitions within the $M_s = \pm\frac{1}{2}$ levels. Values for the hyperfine interaction constant and nuclear moment of ^{55}Mn were derived.

With a view toward obtaining atom-ion core level shifts for use in analyzing photoelectron spectroscopic data, nonrelativistic, numerical Hartree–Fock calculations have been performed on most of the common ions in the periodic table (excluding the actinides and lanthanides). Ionization potentials, core-level shifts, and relaxation energies were tabulated for the cations of the metallic and semimetallic elements and for anions of the halogens and chalcogens.[15]

C. Chemistry

1. Electrons Transfer Processes

Little work on electron transfer processes is applicable with the early transition metal atoms. However, Squires[16] has reported electron- and hydrogen-binding energies for many transition-metal atoms, as experimentally determined from gas-phase studies. Table 4-1 tabulates much of this data.

When Ti, V, Cr, Fe, Co, Ni, or Cu atoms are cocondensed with CO_2, electron transfer followed by coupling occurs, and metal oxalates are sometimes formed.[17] The efficiency of these reactions varies greatly. In the case of Ti, V, and Cr, first insertion of a metallic atom into a C–O bond leads to intermediate O–M–CO molecules. This intermediate reacted further with CO_2:

$$Ti + CO_2 \xrightarrow{15K} O{=}Ti{-}C{\equiv}O + Ti{=}O + Ti{-}C{\equiv}O + C{\equiv}O$$

Similar reaction schemes have been proposed by Almond et al.[18] and Poliakoff et al.[19] for W and Cr atom reactions with CO_2/O_2.

2. Abstraction Processes

Some attempts to deoxygenate or desulfurate organic compounds using transition-metal atoms have been reported.[20,21] Codeposition ex-

TABLE 4-1
Electron Binding Energies and D(M–H)
According to Squires[16]

Metal atom	Electron affinity (kcal/mol)	D(M–H) (kcal/mol)
Sc	4.3	47
Ti	1.8	38
V	12	38
Cr	15	41
Mn	<0	<32
Fe	3.8	30
Co	15	42
Ni	27	60
Cu	28	60
Zn	<0	20
Mo	17	46
Pd	13	<76
Ag	30	54
Cd	<0	16
Pt	49	84
Au	53	72
Li	14	55
Na	13	45
K	12	42

periments have shown that epoxides are deoxygenated by Ti, V, Cr, Co, and Ni atoms to yield alkenes. Aminoxides and DMSO also undergo deoxygenation, while nitro- and nitrosoarenes are deoxygenated to yield coupled azo- and azoxy products. Ethers generally do not undergo deoxygenation, ketones usually undergo more complex reductive reactions, and coupling and aldol products are formed.[22a] The earlier transition metals are generally more efficient for these types of reactions:

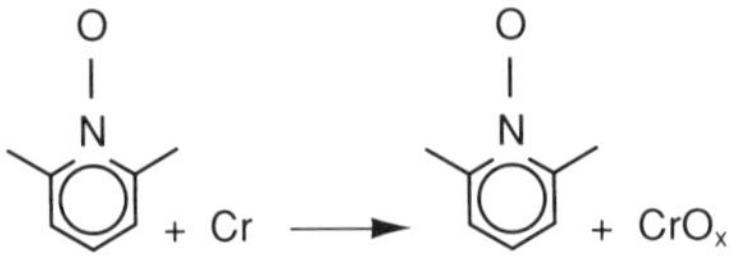

Molybdenum atoms abstract sulfur from a variety of organics in a clean and straightforward manner.[21] A comparison was made between Mo and C atom reactions:

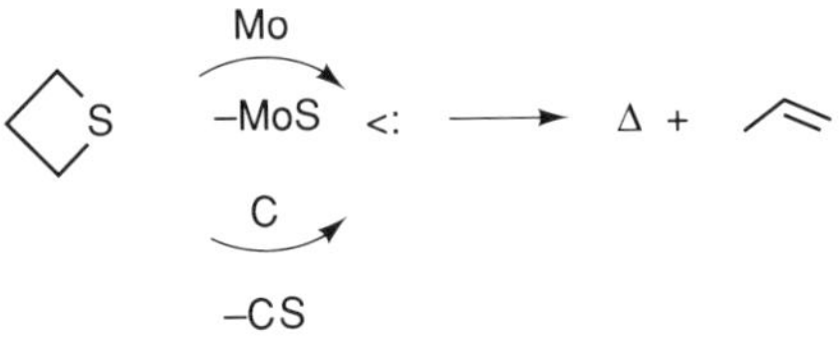

In the case of Mo atoms the ratio of cyclopropane to propene formed was 1 : 12, where for C atoms it was 10 : 1.[22b]

It seems clear that the two desulfurization processes proceed by different mechanisms, and it was proposed that Mo atoms undergo C–H insertion and formation of an allyl-intermediate.

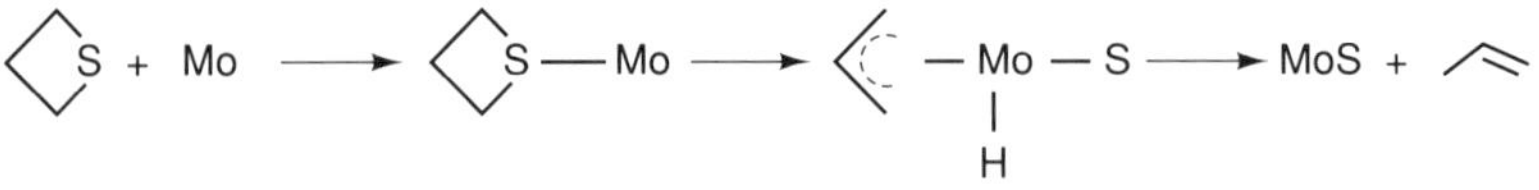

Other sulfur compounds also undergo reaction with Mo atoms, and the extent of desulfurization vs C–H insertion and reduction depends on the C–S bond strength.[21]

Codeposition of Mn or Fe atoms with dimethyl ether followed by hydrolysis yielded a variety of hydrocarbons.[23] The solid products before hydrolysis were MnO and Fe_3O_4, showing that deoxygenation was a primary reaction pathway, although multiple complex pathways are evident.

Gas-phase oxidation reactions of transition-metal atoms often proceed by abstraction processes. In these studies gas-phase metal atoms were produced by multiple photon dissociation of volatile transition-metal compounds (pulsed dye laser).[24] Ground-state atoms were produced and allowed to react with molecular oxygen in Ar buffer. Reactions (5–700 Torr, room temp) were monitored by resonance fluorescence excitation at variable time delays following the photolysis pulse. From this study emerges a trend where reactivity with respect to complex formation can be correlated with a $d^n s^1$ valence electron configuration of the metal atom. Mn and Fe atoms were unreactive; Co, Ni, Cu slightly reactive; while Cr was much more reactive. Other atoms, Sc, Ti, and V, underwent atom transfer processes.

Association and oxidation reactions of $^7S_3$ Cr atoms have also come under study.[25] The reagents O_2, NO, N_2O, and CCl_4 were studied over a buffer gas range of 1–700 Torr and between 298 and 348 K. Multiphoton dissociation of $Cr(CO)_6$ yielded Cr atoms in excited states that relaxed to ground state within 2 μsec. A summary of this work and that from other literature is given in Table 4-2 where rate constants are tabulated.

Generally, Cr atoms reacted readily with stable free radicals O_2 and NO, exhibiting third-order kinetic behavior over a wide pressure range (termolecular reactions). Reactions with CCl_4 and N_2O showed second-order behavior, consistent with Cl or O atom transfer processes. A very

TABLE 4-2

Absolute or Limiting Second-Order Rate Constants for Removal of
Ground-State ($^7S_3$) Cr Atoms by Various Molecules at 298 K

Reactant	Pressure range (Torr) Ar buffer gas	Rate constant, second-order $cm^3\ mol^{-1}\ s^{-1}$
CCl_4	0–0.6	$1.94 \pm 0.04 \times 10^{-10}$
CF_3Cl	0–50	$<8 \times 10^{-15}$
N_2O	0–150	$1.0 \pm 0.1 \times 10^{-14}$
N_2O (348 K)	0–125	$3.3 \pm 0.1 \times 10^{-14}$
CO_2	0–40	$<5 \times 10^{-14}$
OCS	0–50	$<2 \times 10^{-14}$
CH_3OH	0–50	$<3 \times 10^{-15}$
CH_3OCH_3	0–50	$<3 \times 10^{-15}$
CO	0–100	$<1 \times 10^{-14}$
C_2H_4	0–20	$<2 \times 10^{-14}$
C_6H_6	0–30	$<2 \times 10^{-15}$
CH_4	0–50	$<3 \times 10^{-15}$
H_2	0–50	$<5 \times 10^{-15}$
D_2	0–30	$<1 \times 10^{-14}$
C_6H_{14}	0–50	$<8 \times 10^{-15}$
H_2S	0–50	$<1 \times 10^{-15}$

low activation energy was found in the CCl_4 case. Many other reagents were relatively unreactive with ground-state Cr atoms, which reflects in part the effects of valence-s-orbital occupation and spin restrictions arising from the high-spin configuration of the ground ^7S state of Cr.

3. Oxidative Addition Processes

a. **Hydrogen** The interaction of H_2 with V and Ti atoms in low-temperature matrices has shown that upon codeposition at 12 K no reaction occurred.[26] However, upon UV photolysis both metals reacted by insertion to yield MH_2. Infrared spectral analysis indicated bond angles TiH_2 (145° ± 5) and VH_2 (127° ± 5). Interestingly, small Ti clusters were found to react spontaneously with H_2.

b. **Alkanes (Nonphotochemical)** Following the early work with Ni and Zr atom reactions with alkanes,[27–29] where it was clear that C–H and C–C bond activation could take place at very low temperatures, a series of reports on matrix isolation studies of M atom/CH_4 interactions have appeared. Ozin and co-workers reported that V atoms could be trapped in alkanes near 12 K with no evidence of reaction.[30] Likewise, no reactions were observed for Mg, Ca, Ti, Cr, Fe, Co, Ni, Pd, Cu, Ag, Au, Ga, In, or Sn atoms, although Al atoms appeared to yield a small amount of CH_3AlH.[31] Recently, a systematic study of all the transition metal-atoms codeposited with CH_4/Ar mixtures about 10 K was reported.[32] Interestingly, several early transition-metal atoms showed evidence of reaction under these low-temperature conditions. Thus, Nb, Ta, Mo, Re, Ru, Os, and Rh gave varying amounts of CH_3MH according to FT–IR and consumption efficiencies monitored by atomic absorption spectra (matrix-isolated atoms). In attempting to understand these results, a combination of factors were considered, including the electronic configuration of the ground state atom, diatomic M–H bond strengths, and ΔH_{vap} of these elements. A $d^n s^1$ electronic configuration appeared to be important in enhancing reactivity, and D[M–H] also showed a rough correlation. Although the results are far from conclusive, a likely mechanism is one where a $d^n s^1$ configuration reacts by H atom abstraction followed by coupling:

$$M + CH_4 \rightarrow [M\text{–}H \cdot CH_3] \rightarrow CH_3\text{–}M\text{–}H$$

$$d^n s^1.$$

However, kinetic energy of the atom incoming to the matrix may also have an effect. Since these results were obtained using laser evaporation as the source of atoms, higher kinetic energies would be expected as compared with simple oven evaporation. Further studies are needed to ascertain if laser power, wavelength, or pulse rate have an effect, and these studies are ongoing.

Theoretical approaches to this problem have been reported by Blomberg and co-workers.[33a,b] Partial agreement with the experimental results[32] was reported in that the $Rh-CH_4 \rightarrow CH_3RhH$ system showed essentially no activation energy barrier. The electronic configuration responsible for this low barrier is an efficient mixing between the $4d^9$ and $4d^8s^1$ states, which are both low lying, according to this report. Interestingly, their calculations indicated that Nb and Ni should have *high* barriers. Weisshaar and co-workers have reported kinetic analyses of gas-phase neutral metal atoms with alkanes and alkenes.[33c] Results for transition metal atoms Sc–Cu were discussed.

From the synthetic standpoint, Mo and W atoms codepositions with alkanes have yielded several new ligand stabilized metal hydrides, indicating that oxidative addition to C–H bonds must have occurred. Cyclopentane/PMe$_3$ with W yielded a $C_5H_5WH_5PMe_3$ complex,[34] while Mo with bidentate phosphines yielded a molybdenum hydride–phosphine complex:[35]

$$Mo + (Pr^i)_2 PCH_2CH_2P(Pr^i)_2 \rightarrow \rightarrow MoH_4(Pr^i_2PCH_2CH_2PPr^i_2).$$

Linear and cyclic saturated hydrocarbons can also be activated in the presence of Re atoms and benzene:[36]

c. Activated C–H Bonds   The presence of C–H bonds near an aromatic or a triene are attacked by Mo and Re atoms, and new organometallic metal hydrides have been produced.[37,38] The intermediate metal hydride species (structure unknown) reacted further in order to fill its coordination shell. For example, the Re–benzene–toluene system yielded $(\eta^6$-arene)$_2$ReH, and Re–toluene yielded a bridged hydride species:[37,38]

d. Hydrogen and Alkanes (Photochemical)   In the presence of electronically excited transition metal atoms (in frozen matrices), oxidative addition of H–H and C–H bonds is common. Billups et $al.$[39] reported that Mn, Co, Cu, Zn, Ag, and Au all reacted under photolysis with CH_4 to give CH_3MH. However, Ca, Ti, Cr, and Ni failed to do so. Ground-state adduct structures were suspected as being very important in determining if photochemical reaction occurred. That is, upon codeposition of transition metal atoms with CH_4, a weakly bound complex is formed (before photochemical oxidative addition). The structure of this complex is still not known, although three possibilities have been proposed:[40]

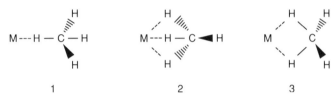

It remains to be determined which structure is correct and how the structure changes upon photochemical excitation of M.

Further progress has been reported by Ozin et $al.$[41] Both H_2 and CH_4 were found to interact with a series of metal atoms under the influence of UV. Remarkably, the reactions are reversible when longer wavelengths are utilized. Discussions of Fe, Cu, and Ag atom behavior were emphasized and the factors affecting chemical reactivity argued in favor of a $M^*(4p) \rightarrow CH_4$ (σ^*) interaction for the first-row transition-metal atoms. However, the $M^*(3d/4s) - CH_4(lt_2)$ energy separation must also be considered. In passing from Sc to Zn, the $3d$ and $4s$ orbitals are generally stabilized, becoming closer in energy to the lt_2 level of CH_4. Their interaction is thereby enhanced, and supposedly greater charge donation from CH_4 to M^* is promoted.

e. Photochemical Induced Oxidative Additions of Diazomethane and Esters   The codeposition of Cr atoms with CH_2N_2/Ar at 12 K yielded an insertion product as well as a carbene complex.[42a] A Cr–C

$$Cr + CH_2N_2 \longrightarrow N_2CrCH_2 + CrCH_2 + N_2$$

$$Cr(CH_2N_2)$$

hv 500 nm

stretching frequency was observed at 567.0 cm.$^{-1}$ It was proposed that the reaction pathway is different depending on if the Cr atom approaches the CH_2 or N_2 end of the CH_2N_2 molecule. Photochemical reactions of esters have also been observed, as described below:[42b]

$$Cr + CH_3-\overset{\overset{O}{\|}}{C}-OCH_3 \longrightarrow Cr\overset{\overset{\displaystyle CH_3}{\diagup}}{\underset{O}{\overset{O}{\diagdown\diagup}}}C-CH_3 \xrightarrow{h\nu} CH_3\overset{\overset{O}{\|}}{C}-CrOCH_3 \overset{-CH_2CO}{\longrightarrow}$$

$$CH_3-\overset{\overset{O}{\|}}{C}-Cr-H \longrightarrow CH_3\overset{\overset{O}{\|}}{C}-H + Cr$$

This sequence implies a catalytic function of Cr atoms, although this was not confirmed.

f. Organohalides   Oxidative addition of organohalides to early transition-metal atoms has not been investigated in much detail. However, an interesting synthesis of binuclear and trinuclear cluster halides of Mo, Re has been reported, as well as carbonyl halides of Re, Ir, Ru, and Pt.[43]

$$Re + BrCH_2CH_2Br \xrightarrow{THF} \longrightarrow Re_3Br_9(THF)_3$$

$$Re + Cl-\overset{\overset{O}{\|}}{C}-\overset{\overset{O}{\|}}{C}-Cl \longrightarrow \longrightarrow [Re(CO)_4Cl]_2$$

$$Ir + Cl-\overset{\overset{O}{\|}}{C}-\overset{\overset{O}{\|}}{C}-Cl \xrightarrow{P(C_6H_5)_3} \longrightarrow Ir(CO)_2(P(C_6H_5)_3)Cl_3$$

These cocondensation reactions yield solvated transition-metal halides in lower valency states and in the absence of water. By employing a large-scale apparatus with a positive hearth electron-gun furnace, 7 g of a metal could be evaporated yielding more than 12 g of product over a 4-hr period.

Chromium atoms have been codeposited with alkyl halides.[44] For example, neopentyl bromide and Cr atoms yielded the known tetrakis (neopentyl) chromium.

$$Cr + (CH_3)_3CCH_2Br \rightarrow \rightarrow [(CH_3)_3C\,CH_2]_4Cr + CrBr_x.$$

The yield of this reaction was high, but the complex mechanism leading to the product is not understood. Alkyl- and arylhalides with less steric restrictions yielded complex mixtures of products, and RCrX species could not be readily trapped and isolated.

4. Simple Orbital Mixing (Complex Formation)

a. Carbonmonoxide   The electronic states of matrix-isolated mono-carbonyl complexes of the early transition-metal atoms have come under investigation by Van Zee, Weltner, Kasai, and their coworkers.[45–49] A good indication of the strength of the M–CO bond can be taken from the magnitude of the lowering of the $\nu_{C\equiv O}$ stretch in the infrared, since d-π^* backbonding is obviously an important part of the bonding interaction, and the mass of the metal atom is relatively large.

In Table 4-3 a compilation of data is given, and the results indicate that the bonding strength roughly correlates with the energy necessary for promotion of $4s^2 3d^{n-2} \rightarrow 4s^1 3d^{n-1}$. It also appears that Mn does not bind to CO, and it is proposed that this is due to the high promotion energy involved.

In spite of the numerous works published on M atom–CO studies,[1] it appears that more data are still needed, particularly with the heavier transition metals.

b. Isocyanides   Metal atom cocondensations with cyclohexyl isocyanide have yielded a palladium trimer $Pd_3(CN\text{–}C\text{–}Hex)_6$ as the only product in 30–40% yields.[50] A crystal structure showed that the Pd_3 triangle has three bridging and three terminal (unbridged) CN–C–Hex groups.

TABLE 4-3
Matrix-Isolated M–CO Species

Metal atom	$\nu_{C\equiv O}$ (cm^{-1})	$\Delta\nu$ (2143 − $\nu_{C\equiv O}$) (cm^{-1})	Ground electronic state	Reference
Sc	1950	193	$^4\Sigma$	47
Nb	—	—	$^6\Sigma$	47
Ti	—	—	$^3\phi$	47
V	1903	240	$^6\Sigma$	47
Cr	1977	166	$^7\Sigma$	47
Mn	2140	Small	$^6\Sigma$	47
Fe	1898	245	$^3\Sigma^-$ or $^5\Sigma^-$	47
Co	1953	190	$^4\Delta$	49
Ni	1996	147	$^1\Sigma$	47
Cu	2010	133	$^2\Sigma$	47

c. Dienes and Trienes   Metal atom synthesis of new organometallics has continued for over 20 years now, and novel complexes are still being produced. Especially interesting have been diene/triene/phosphine reactions with early transition-metal atoms. A series of recent successful syntheses employing pentamethylcyclopentadiene are shown below:[51-53]

$$\text{cocond.}$$

$$W + Cp^*H \rightarrow (Cp^*)_2\,WH_2$$

$$Re + Cp^*H \rightarrow (Cp^*)_2\,ReH$$

$$Mo + Cp^*H \rightarrow (Cp^*)_2\,MoH_2.$$

Photolysis of the tungsten dihydride complex caused the elimination of H_2 and resulted in a "tucked-in" complex where oxidative addition of a ring CH_3 to the W center took place. Similarly, photolysis of Cp_2^*ReH caused $\frac{1}{2}H_2$ elimination to yield Cp_2^*Re that was easily reduced to $Cp_2^*Re^-$.[52] An extensive chemistry and photochemistry is exhibited by this interesting series of compounds.[53]

Little is known about the mechanistic details of the metal atom reaction. However, it is likely that oxidative addition of the C–H bond to M occurs stepwise ($\rightarrow Cp^*MH \rightarrow Cp_2^*MH_2$). Some mixed labeling studies where Cp^*H and Cp^*D were employed could be of interest.

Ruthenium atom codeposition with dienes at 77 K followed by warming in the presence of CO has yielded gram quantities of Ru(CO)(diene)$_2$ complexes:[54a,b]

Atoms of Ti and V react with 1, 4-*t*-butyl-1, 3-butadiene and lead to the syntheses of the

first homoleptic diene complexes of Ti and V, as well as the first examples of such "sandwich" butadiene structures.

  Cyclic dienes and trienes with transition-metal atoms have yielded a series of new complexes. Heating or ligand trapping can generally lead to stable, isolable compounds in quite good yields:[55,56a]

  d. Arenes The most important synthetic chemistry of the early transition-metal atoms continues to be the preparation of bis(arene) M(O) sandwich complexes:[56b]

The metal atom method has allowed a much extended array of such sandwich complexes, and recent work has continued unabated. In earlier years it was shown that a wide range of substituents could be tolerated on the arene ring, if metal atom cocondensation was the synthetic method employed.[1] In recent years these syntheses have become more sophisticated, and many more unusual arene-M and mixed arene/ligand-M complexes have been prepared. The best way to summarize is to show reactions and structures.

  e. Chromium and Vanadium This synthesis was carried out in diglyme solution at $-80°C$ utilizing a rotary solution reactor. An interesting aspect of the napthalene ligand is that it is comparatively labile, and arene displacements occur readily:

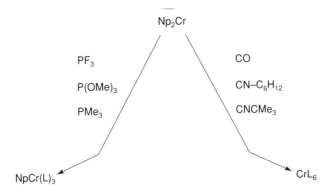

Methyl-substituted napthalenes also gave high yields of the Np_2Cr complexes. In the case of 1,4-dimethylnaphthalene the unsubstituted ring is bound to chromium. With l-methylnaphthalene, 10 different structural and geometrical isomers were detected by proton NMR.[57]

Lagowski and co-workers have described an unusual synthetic approach where solid arenes were treated with metal atoms.[58] The solid arene is simply placed in a rotary reactor and metal evaporated. Surprisingly the solid particles take up the metal atoms, and respectable yields of bis(arene) complexes of Cr were formed:

Elschenbroich *et al.* have prepared a wide series of new Cr and V analogs, for example metallocyclophanes, which show high thermal stabilities, and spectral studies have suggested a larger degree of metal → ligand spin delocalization in these compounds compared with unbridged analogs.[59]

$C_6H_5SiMe_2Cl$ $1,4\text{-}C_6H_4(SiMe_2Cl)_2$

Arene complexes containing strained ring substituents as well as large $SiMe_3$ and $Si(C_6H_5)_3$ groups have also been prepared.[60–62]

Judging from the C–C bond lengths and ^{13}C chemical shifts, the *ipso* carbon atoms of the coordinated arene in the $Si(C_6H_5)_2$ bridged compound are in a hybridization state between sp^2 and sp^3. For the vanadium paramagnetic analog, ESR measurements further showed that the ring tilting is accompanied by an increase in metal → ligand spin delocalization.[62]

(also vanadium)

Metal atom–ligand cocondensation of V with phosphabenzene has yielded bis(η^6-phosphabenzene)vanadium. This interesting derivative is reddish-brown and is less air sensitive than the ligand itself.[63] The compound exhibits good thermal stability (mp 210°C), and the phosphorus heteroatom has a much stronger influence on the electron affinity of the free arene than in the bound arene. The rotational barrier in this complex is comparable with that in the homoarene case.

$$V \ + \ 2\,C_5H_5P \longrightarrow$$

f. Niobium   The cocondensation of Nb atoms with toluene has yielded bis(η^6-toluene)niobium, which serves as a useful intermediate for the preparation of a wide series of mono- and bis-arene derivatives, as shown in Scheme 4-1.[64]

g. Scandium, Titanium, Hafnium, Zirconium, Tungsten, Manganese   For these metals, a series of new mono- and bis-arene derivatives has been prepared using metal atom reactions with arenes mixed with other ligands, particularly phosphines. Table 4-4 summarizes these findings.

Expressing a unique twist to the study of bis(arene) complexes, Elschenbroich et al. have reported a study of metal atom reactions with bis(arene) mimicks.[70] In their study of Cr and V complexes the "$C_{12}H_{12}$" derivatives were $C_6H_6 + C_6H_6$ and $C_5H_5 + C_7H_7$. Several compounds were prepared and compared:

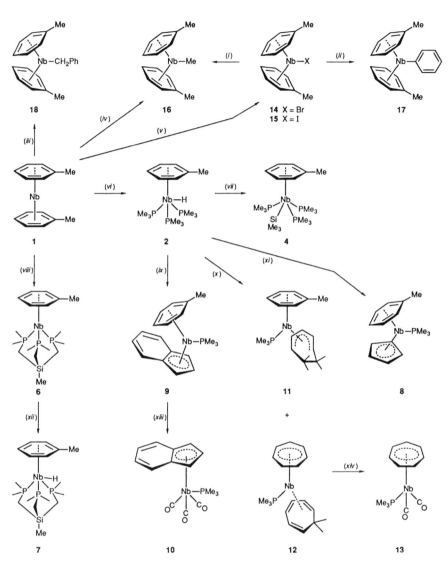

SCHEME 4-1   Some Chemistry of bis(η^6-toluene) niobium (after Green et al.[64]), (i) For X = Br, LiMe in Et$_2$O, $-80°$C with slow warm up, yield 66%; (ii) for X = Br, LiPh in Et$_2$O, $-80°$C with slow warm up, 50%; (iii) potassium film in thf then benzyl bromide, $-80°$C, 20%; (iv) MeI in light petroleum, $-80°$C; potassium film in thf, then MeI, at $-80°$C, 18%; (v) allyl bromide or iodide in light petroleum, $-80°$C, ca. 65%; (vi) neat PMe$_3$, 18 hr at RT, 74%; (vii) potassium film in thf at RT, then SiMe$_3$Cl, 33%; (viii) neat tmps, 24 hr, RT, 65%; (ix) indene in toluene, 6 hr, 60°C, 41%; (x) cycloheptatriene in toluene, 6 hr, 60°C, 38% (1 : 2 mixture of 11 and 12); (xi) cyclopentadiene in toluene, 8 hr, 60°C, 54%; (xii) potassium film in thf at RT, then water, 45%; (xiii) in toluene, CO (1.5 atm) at RT, 68%; (xiv) in toluene, CO (1.5 atm) at RT, 67%.

TABLE 4-4

Mono-Arene Derivatives of Early Transition Metals Prepared by Metal Vapor Methods

Metal	Arene	Additional ligands	Product	Reference
Zr	Toluene	PMe$_3$	(η^6-arene)$_2$ Zr(PMe$_3$)	65
Hf	Benzene	PMe$_3$	(η^6-arene)$_2$ Hf(PMe$_3$)	65
Mn	Benzene	PMe$_3$	(η^6-arene) Mn(PMe$_3$)$_2$H	66
Re	Benzene	PMe$_3$	[(η^6-arene) Re(PMe$_3$)$_2$]$_2$	66
Mn	Toluene	I$_2$	(η^6-arene)$_2$ Mn$^+$I$^-$	67
W	Toluene	—	(η^6-arene)$_2$ W^a	68,69

^{a}Several new mono-arene derivatives where the W atom is also bound to allyl, phosphino-, isocyano-, or hydrido-ligands have been derived from bis(toluene) tungsten.[69]

Redox behavior showed that the symmetrical complexes were more readily oxidized than the unsymmetrical ones.

Generally the bis(arene) complexes of these heavier metals are not isolable unless an additional ligand is available. However, Cloke *et al.* have reported the synthesis of bis (η-1, 3, 5-tri-t-butylbenzene) sandwich complexes of Ti, Zr, and Hf, and a Hf(O) carbonyl complex.[71a,b]

Cocondensation of Ti (1 g) and excess ligand followed by hexane extraction and removal of excess ligand by sublimation, and finally low-temperature crystallization from pentane yielded a 40% yield of Ti (η-Bu$_3^t$C$_6$H$_3$)$_2$ as deep-red crystals. The Zr and Hf anologs were deep green and deep purple respectively, also formed in high yield. Reactions of the Hf complex with CO resulted in immediate formation of green Hf(arene)$_2$CO with ν_{co} = 1870 cm^{-1}, as the first example of an isolable carbonyl complex of a heavy Group 4 metal in a zero-oxidation state.

Similar results were reported for scandium, and π-arene complexes of Sc(O) and Sc(I) were prepared and isolated for the first time. Again, steric hindrance in the arene ligand seems to be the key to high stability.

The hydride was believed to be formed by direct insertion of a Sc atom into a C–H bond followed by ligation. The hydride product could be favored by 195 K cocondensation to a 1 : 1 ratio, whereas a 77 K cocondensation yielded an 8 : 1 ratio favoring the Sc(O) sandwich complex.

Several matrix isolation studies of arene-M interactions have also been reported. Francis *et al.* have prepared naphthalene-Ti matrices and have attempted to spectroscopically analyze $(Np)_2Ti$, Cr, and V.[71c] The major conclusion was that the $(Np)_2Ti$ complex does exist at low temperatures, but is extremely reactive.

McCamley and Perutz have spectroscopically observed bis(benzene)V and bis(mesitylene)V in N_2 matrices at 12 K by resonance Raman.[72] The transition involved was identified as the $^2A_{1g} \rightarrow {}^2E_{1u}$ ligand to metal charge transfer transition (as determined by absorption spectra in solution and in matrices). Thus, open-shell sandwich complexes of this type exhibit structured LMCT absorption bands, and laser irradiation into these bands excites long resonance Raman progressions.

An optical, ESR, and IR study of $(\eta^6\text{-}C_6H_6)V$ and $(\eta^6\text{-}C_6F_6)V$ matrix-isolated complexes has also appeared.[73] The interest in C_6F_6 complexes stems from earlier work,[1] where it was shown that $(C_6F_6)_2Cr$ and V complexes are unstable and are actually explosive. However, $(C_6F_6)Cr(PF_3)_3$, $(C_6H_6)Cr(C_6F_6)$, and $(C_6H_6)V(C_6F_6)$ have been prepared, and spectral and theoretical analysis indicates that the central metal atom is internally oxidized by the C_6F_6 ligand with electron density of the metal being transferred to the π^*-system of C_6F_6. The matrix work lends support to this idea[73] (vanadium oxidation), and this may explain why the complex is unstable.

h. Phosphines Metal atom–vapor chemistry has led to the synthesis and isolation of the first zero-valent derivatives of Ta and a variety of new complexes of Cr, Mo, W, V, and Nb.[74] The $Me_2PCH_2CH_2PMe_2(DMPE)$ ligand was codeposited with each of these metals yielding the tris(bidentate) $(DMPE)_3M$ derivatives. Similarly, the codeposition of Cr, Mo, and W vapors with trialkyl phosphites $[P(OR)_3]$ has led to the synthesis of a series of novel zero-valent complexes, such as $Cr[P(OR)_3]_6$.[75] When R = Me the zero-valent complex can be isolated, but when R = Et steric hindrance caused thermal instability, although the complex could be observed spectroscopically at low temperature. Both complexes were reactive with H_2. In the cases of

$$Cr + P(OMe)_3 \longrightarrow Cr[P(OMe)_3]_6 \xrightarrow{H_2} CrH_2[P(OMe)_3]_5$$

$$Cr + P(OMe)_3 + toluene \longrightarrow \text{[benzene ring]}Cr[P(OMe)_3]_3$$

Mo and W the hexakis-phosphite complexes could also be prepared, although yields were very low. The W system was particularly interesting. Upon cocondensation at 77 K followed by warmup, only a $(W)_x$ slurry was formed. Apparently the WL_6 complex was photolytically unstable and decomposed under the influence of the metal evaporation source light output. However, the $W[P(OMe)_3]_6$ complex could be obtained in 14% yield by refluxing the $(W)_x$ slurry with excess $P(OMe)_3$ for a short time.[75]

i. **Polymers**   Metal atoms interact with functional groups of polymers, for example, arene or ether functionalities. Due to the physical nature of polymers, metal atoms or small clusters can be trapped and immobilized in the polymer backbone.[76-80] As with all metal atom reactions with organic substrates, a competition between trapping/complexation and metal atom aggregation is always important. With less reactive polymers, metal colloids are formed and encapsulated. This is a common feature of poly-(dimethylsiloxane) polymers where only silyl-ether functionalities are present. In contrast, when phenyl substituents are present, early transition-metal atoms invariably bond strongly to the arene moieties, and so the metal atom aggregation pathway is discouraged. Thus, Mo atoms can form bis(arene)Mo complexes:

Me — Si — Me
/
O
\
Me — Si   Me — Si — Me
/   \
O   M   O
\   /
Me — Si — Me   Si — Me
/   \
O   O
| /
Me — Si — Me
\

Cyclopentadienyl-functionalized polymers have also been treated with metal atoms. Initially a green bis(cyclopentadiene)Fe(O) complex that spontaneously coverted to the encapsulated ferrocene derivative upon warming was found at low temperature.

Naphthyl- and isocyano-functionalized polymers have also yielded interesting materials. In the latter case a $M(CNR)_6$-encapsulated moiety was spectroscopically detected (M = Cr, Mo, W, V).

Finally it should be pointed out that metal atom-induced polymerizations have yielded metal-encapsulated polymers; the metal atom itself served to catalyze polymerization of the host monomer. For example, Ge and Sn atoms induced acetylene polymerization (see Chapter 8), and in another case Na atoms initiated anionic polymerization of styrene at low temperatures (see Chapter 3).

II. Early Transition-Metal Clusters

A. Occurrence and Techniques

The refractory elements, such as several of the early transition metals, can be generated as small, gas-phase clusters in continuous beams[81] or as pulsed beams by laser vaporization.[82] These techniques are described in more detail in Chapter 2. In addition, a very useful review of clusters of many types was given by Castleman and Keesee.[83]

Lasers have also been used to photodissociate organometallic cluster compounds in the gas phase, thus forming free transition-metal clusters.[84] For example, $Co_2(CO)_8$ could be photodissociated in the gas phase with pulses of 10 ns duration from a tunable dye laser (photon flux 10^8 W/cm^2 in the sample region). Multiphoton absorption took place during a single laser pulse resulting in dissociation and ionization of the $Co_2(CO)_8$. Irradiation at 406 nm yielded Co^+ and Co_2^+ in a ratio of 10:1, and their formation sequence was proposed as

$$Co_2(CO)_8 \rightarrow 2Co + 8CO$$

$$Co_2(CO)_8 \rightarrow Co_2 + 8CO$$

$$Co \rightarrow Co^+ + e^-$$

$$Co_2 \rightarrow Co_2^+ + e^-.$$

The formation of Co_2^+ was shown not to proceed by reaction of Co^+ with $Co_2(CO)_8$ or other molecular species. In a similar way, pulsed laser dissociation of $Mo(CO)_6$ has led to Mo atom formation. Aggregation of these atoms led to the formation of molybdenum cobwebs (metallic molybdenum).[85]

Molybdenum atom aggregation in polymer media has been studied in some detail by Andrews and Ozin.[86] Kinetic analyses of Mo_2 formation from a polymer attached bis(arene)Mo complex was carried out. By increasing the cross-link density in the DC510 silcone oil polymer, the thermally activated bimolecular loss of the Mo dimer was reduced. Diffusion coefficients depended on Mo loading (concentration) and on temperature. Low values of the diffusivity were considered to be due to restricted mobility of Mo_2, as they may be bound to the polymer cross-links.

At certain temperature ranges, cluster formation is quite spontaneous.[87] Again employing the Mo atom/poly(methylphenylsiloxane) DC510 as a model system, the aggregation of polymer attached $(arene)_2Mo$ with additional atoms was investigated. Dependence on temperature was extreme. At 240–250 K, polymer attached $(arene)_2Mo$ was favored, at >270 K Mo aggregation to colloidal particles took place, while the intermediate range of 250–270 K allowed stabilization of polymer attached Mo_n clusters. At lower temperatures (<240 K) Mo atoms aggregated on the surface of the DC510 oil before good penetration could occur.[87]

Polymer-stabilized divanadium has also been prepared, both on micro- and macroscale.[88] The polymer used was a Dow Corning DC510/50, which is liquid poly(dimethylsiloxane-comethylphenylsiloxane) with the methyl:phenyl ratio = 17:1. Vanadium vapor deposition (1 g metal to 100 ml liquid polymer) yielded a clear red-brown liquid that contained -$(phenyl)_2V$ and -$(phenyl)V_2$ according to new Raman bands at 520, 430, 270, and 155 cm^{-1} with the 155 cm^{-1} band apparently characteristic of the trapped V_2 species.[88]

B. Physical Properties and Theoretical Studies

The early transition-metal dimers are of particular interest because they exhibit either extremely strong bonding (Cr_2) or extremely weak bonding (Mn_2). Due to this fascinating behavior and because the dimers are small enough that their bonding can be theoretically modeled, the literature has been rich in recent years with reports on these species.

Morse and co-workers[89] and Weltner and Van Zee[90] have given us important reviews on metal dimer molecules and the reader is referred to these for a more in-depth discussion.

Herein, some recent results are summarized (Fig. 4-3). In moving into the early transition-metal series, first the Sc_2 molecule is considered.[91–93]

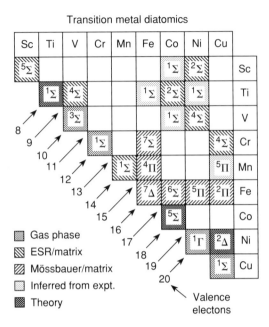

FIGURE 4-3 The ground states of possible diatomics formed from the first-row transition metals. Those in bold borders are definitely established; the others are derived as indicated. Reprinted with permission from Weltner and Van Zee, *ACS Symp. Ser.* **394**, 213 (1989). Copyright 1989 American Chemical Society.

Matrix-isolated Sc_2 exhibits a complex ESR spectrum that has been assigned by Knight *et al.*[91] It possesses a covalent bond of about 30 kcal/mol, and a ground $^5\Sigma$ state according to matrix ESR. Table 4-5 summarizes pertinent data.

The V_2 dimer has been studied in the gas phase by Smalley and co-workers.[94] This species was produced by laser evaporation of vanadium to form an expansion-cooled molecular beam. Using the technique of resonant two-photon ionization spectroscopy, a band structure near 7000 Å was rotationally resolved. These data are indicative of extensive 3d-orbital participation in the bonding and imply that the ground-state electronic configuration contains a half-filled $\pi(3d)$ or $\delta(3d)$ orbital.

The problem inherent in much of the work on these dimers, both gas phase and matrix isolated, is that mixtures are invariably present (monomers, dimers, clusters). To spectroscopically observe the pure dimer, Lindsay and co-workers[95] have generated V_2 in the gas phase, ionized it, mass selected it, neutralized it, and then deposited it in argon matrices.

TABLE 4-5
Early Transition-Metal Dimers

Molecule	$D_{(M-M)}$	ν_{M-M}	Ground state	Bond length	Ionization energy	Reference
Sc_2	1.65 eV	239 cm^{-1}	$^5\Sigma_u^-$	—	—	91
						92
						93
Ti_2	1.23 eV	408 cm^{-1}	$^1\Sigma_g^+$	—	—	89
V_2	2.49 eV	538 cm^{-1}	$^3\Sigma_g^-$	1.77 Å	—	89,94
Cr_2	1.78 eV	—	$^1\Sigma_g^+$	1.68 Å	—	89,100
Mn_2	0.79 eV	—	$^1\Sigma_g^+$	3.4 Å	6.47 eV	103
						97
						98
Y_2	1.62 eV	—	$^1\Sigma_g^+$	—	—	89
Zr_2	—	306 cm^{-1}	—	—	—	89
Nb_2	5.0 eV	426 cm^{-1}	$^3\Sigma_g^-$ or $^3\Delta_g$	—	—	110
Mo_2	4.38 eV	477 cm^{-1}	$^1\Sigma_g^+$	1.94 Å	<6.42 eV	89

Raman spectroscopy, having proved to be a powerful technique for obtaining vibrational information on clusters, was used as a probe.[96]

Several of these dimers have been studied by multiple investigators and are deserving of more detailed analysis. Of special interest has been the Cr_2 molecule because of its proposed sextuple bond. However, bonding schemes, according to theoretical work of Goodgame and Goddard,[99] do not involve a sextuple bond, but may be best described as an antiferromagnetic dimer with very low-lying electronic spin states.

It might be expected that an extremely short bond would be the consequence of sextuple bonding, and this has been examined spectroscopically for gas-phase Cr_2.[100] Resonant two-photon ionization was used to probe a band near 4600 Å that could be rotationally resolved. These studies led to the conclusion that the bond length in Cr_2 was 1.68 ± 0.01 Å from a $^1\Sigma_g^+$ ground state, the shortest metal–metal bond of any known molecule.[100] Unfortunately the agreement between theory and experiment regarding the dissociation energy of Cr_2 is not very good and experiments by Hilpert and Ruthardt[101] have reported a value of 142 ± 5 kcal/mol, which is very different from several theoretical predictions.

Matrix-isolated Cr_2 has been examined by resonance Raman.[102] Vibrational constants obtained were quite different from values obtained in the gas phase, and this discrepancy may be due to matrix effects that may be the result of the fact that the ground-state potential for Cr_2 has a double minimum.

The manganese dimer has also been investigated extensively under matrix isolation conditions.[103–107] These studies have shown Mn_2 to be a weakly bound van der Waals molecule similar to alkaline-earth metal dimers. It is an antiferromagnetic diatomic species with a $^1\Sigma_g^+$ ground state. Numerous UV–VIS absorption bands for Mn_2 have been assigned to transitions between singlet, triplet, quintet, and septet electronic spin states of the diatomic, and these assignments were based on the temperature dependence of the intensities of these adsorptions. The resonance Raman spectra in Ar, Kr, and Xe yielded vibrational frequencies of 59–68, 76.4, and 68.1 cm^{-1}, respectively, confirming that Mn_2 is indeed a weakly bound antiferromagnetic molecule.

Antiferromagnetic exchange coupling between the $3d^5$ electrons of each atom in Mn_2 has been examined by ESR of Mn_2 in frozen cyclopropane.[103] Exchange in this solid occurred such that the bond length of Mn_2 varied from about 3.2 to 3.6 Å in different spin states (S = 0 to S = 5 states). The removal of 1 electron to give Mn_2^+ resulted in the formation of one σ-bond and high spin coupling among the remaining 11 unpaired electrons to yield a $^{12}\Sigma$ ground state.

Heavier dimer molecules such as Zr_2 have also come under study. Lindsay and co-workers[108] utilized a recently developed mass selection/matrix deposition technique. They deposited Zr_2^+ that had been generated in the gas phase and then deposited electrons, forming Zr_2 in frozen argon. Using "scattering depletion" absorption spectra they found that the Zr_2 species showed absorption bands at 388 and near 630 nm, where a vibrational progression was observed. Resonance Raman spectra were obtained by exciting into the 630 nm band and were consistent with a $^1\Sigma_g^+$ or $^3\Delta_g$ ground state.[109]

Similar techniques have been employed for the studies of Nb_2[110] and W_2. The authors emphasize the importance of the Raman technique as it can provide structural information such as vibrational frequencies, force constants, and bond energies. However, the mass selection/matrix deposition technique yields very dilute samples, and so Raman must be resonantly enhanced. To do this, it is first necessary to map out the optical absorption spectra of the cluster. And the authors have found that light scattering at 90° is a somewhat more sensitive technique. For Nb_2, strong optical absorptions were observed at or near 320, 430, and 670 nm. Two absorptions were also observed in the region 580–720, including a vibrational progression.

Gas-phase spectral investigations of ultracold Mo_2 have been reported.[111] This was one of the first reports on the pulsed nozzle beam method and showed that clusters of Mo and W up to 25 atoms could be produced. Mass selected resonant two-photon ionization spectra of Mo_2

showed that its translational temperature was <6 K, rotational about 5 K, and vibrational about 325 K. Analysis of rotationally resolved spectra allowed the bond length to be determined as 1.940 ± 0.009 Å. The ionization energy of Mo_2 is <6.42 eV (compared to 7.10 eV for atomic Mo), and these data imply a substantially stronger chemical bond for Mo^{2+} than neutral Mo_2.

Clusters larger than dimers are less well understood, of course. However, some spectroscopic characterization has been reported for Cr_4 (or Cr_5).[112] A complex ESR spectrum of 16 lines between 200 and 7000 G in neon and argon matrices was found to be due to an axial molecule. Assuming that the molecule is Cr_4 or Cr_5, a unique apical atom must be bonded weakly to the base and is the locus of almost all of the s-character among the unpaired spins. Thus, the molecule exhibiting this total spin $S = 3$ is either a tetrahedral Cr_4 or square based pyramid of Cr_5. The observed hyperfine is essentially that of a weakly bound $Cr(^7S)$ atom situated on a base of Cr_3 or Cr_4, and this is quite an unexpected result.

Linear and circular dichroism studies of Mn_x clusters placed in a magnetic field have been reported by Lignières $et~al.$[113] Isolated in krypton at less than 10^{-3} parts, the Mn_x species showed distinct circular or linear dichroism signals only after the application of an external magnetic field. The clusters responsible for this behavior have an axial structure with a plane of symmetry perpendicular to the magnetic moment. These results combined with previous ESR results suggest that three of the electronic transitions observed were due to the Mn_5 cluster.

For Mn_3,[105,114] resonance Raman showed a D_{3h} trimer subject to Jahn–Teller distortions. An absorption centered at 14,750 cm^{-1} (678 nm) was assigned to Mn_3. A Mn–Mn force constant of 0.38 mdyn/Å was reported. This force constant is four times higher than that for Mn_2, and so Mn_3 is much more strongly bound than Mn_2.[105]

For Mn_5,[104] which can be prepared in concentrated matrices, it appears to be a highly oriented, axial molecule with its axis perpendicular to the substrate surface (according to ESR spectral interpretations). It contains 25 unpaired electrons ($S = 25/2$). Its structure is probably a plane pentagon.

Studies by ESR have also helped elucidate the properties of Sc_3 and Y_3.[115] The spectra establish the structure of Sc_3 as an equilateral triangle at 4–30 K with a $^2A_1'$ ground state. However, it may be a fluxional, bent molecule. The unpaired electron has little s character and is delocalized in $3d$ orbitals on three equivalent atoms. The Y_3 molecule is different, however, is not equilateral, and is most likely a distorted molecule in a $^2B_2$ ground state.

Gas-phase clusters Nb_x, Ta_x, and W_x have been subjected to multiphoton excitation ionization and dissociation.[116] Decay dynamics were elucidated, including ionization and dissociation rates for Nb_x (when $x = 5$–10). Evidence was presented for a delayed ionization process following two-photon excitation of small Nb, Ta, and W clusters at 308 and 351 nm. The dissociation rates appeared to be small for W and Ta clusters in comparison to their ionization rates. Thus, for these species it may be possible to control internal temperature, and perhaps by spectroscopically analyzing adsorbed molecules precise temperatures could be measured.

Several approaches to understanding cluster-packing schemes, minima, transition states, and rearrangement mechanisms have also been published.[117–121]

C. Chemistry of Early Transition-Metal Clusters

1. Abstraction Processes and Oxidative Addition Processes

Employing the pulsed nozzle beam apparatus for production of metal clusters, El-Sayed and co-workers have prepared a mixture of Nb_x clusters where $x = 5$–20. A series of chemical reactivity studies have been reported, including cluster size dependence on reactivity.[122] Size dependence as a mechanistic probe was the thrust of much of this work. The reaction of Nb_x with benzene showed a size threshold of

$$Nb_x + C_6H_6 \rightarrow Nb_x - C_6H_6 + Nb_xC_6$$

about $x = 4$. Minimas at $x = 8$ and 10 were rationalized due to the high stability of Nb_8 and Nb_{10}. This sudden increase in reactivity for $x = 4$ or 5 can be related to either thermodynamic or catalytic origins. A minimum number of Nb–carbide bonds may need to be formed to drive the reaction. However, a catalytic effect could be required simply to anchor the benzene ring to facilitate dehydrogenation processes.

Other organic (nonaromatic) molecules were also studied. Two conclusions were drawn: (1) a double bond must exist in the starting molecule, and (2) only the loss of an even number of hydrogen atoms takes place.[122]

Abstraction of either Br or CN from BrCN by Nb clusters took place where smaller clusters favored CN abstraction. Thus, selectivity decreased with cluster size.[123,124]

$$Nb_x + BrCN \rightarrow Nb_xBr + Nb_xCN.$$

These results were rationalized with regard to approach of the Nb cluster

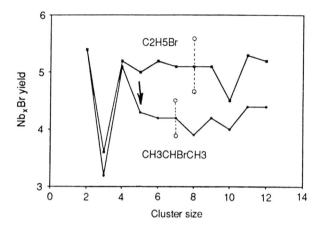

FIGURE 4-4   Mass peak intensity of Nb_xBr^+ (normalized to that of NbI produced from the reaction of a fixed amount of CH_3I) produced from the bromine extraction reaction between Nb_x and $CH_3 \cdot CH_2Br$ (top) and between Nb_x and $CH_3CHBrCH_3$ (bottom). Due to increased steric effects in reacting with the latter compound, the amount of Nb_xBr produced decreases as the size of the cluster contains five or more atoms. These clusters are too large to be able to extract the bromine from the configuration in $CH_3CHBrCH_3$, which has the bromine sterically protected (after El-Sayed[122]).

to the BrCN molecule. A change in the nature of the reactive collision from being impulsive (small clusters) to one that involves complex formation (larger clusters) was proposed.

Steric effects also change with cluster size.[122,125] Two organobromides, CH_3CH_2Br and $CH_3CHBrCH_3$, were compared in their reactions with $(Nb)_x$. The reaction of small Nb clusters ($x \leq 4$) gave the same amount of Nb_xBr^+ with both reagents. However, when $x = 5$ the yields of abstraction product decreased significantly and this was attributed to steric crowding on $CH_3CHBrCH_3$ so that approach to the Br was more hindered. This was a significant effect for Nb_5, Nb_6, Nb_7, etc. Figure 4-4 illustrates these findings.

Cluster size can also affect reaction channels. For example if a complex that can react further to give two different products, one path more energetically or sterically demanding than another, is formed, then changes in product distributions with change in cluster size might be expected. This was indeed observed as shown below.[122]

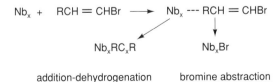

addition-dehydrogenation   bromine abstraction

It was found that Nb_xRC_2R formation was favored by larger clusters. These results imply that smaller clusters can attack Br more readily. Larger clusters probably form π-complexes more readily with the C=C bond and this sets up the addition–dehydrogenation process.

Another interesting example of Nb_x reaction is with CO_2. The products of this reaction are shown below:

$$Nb_x + CO_2 \longrightarrow [OCNb_xO] \quad \begin{matrix} \nearrow \text{ } OCNb_xO \\ x > 7 \\ \\ x < 7 \\ \searrow \text{ } Nb_xO + CO \end{matrix}$$

A common intermediate is apparently formed since the yields of $OCNb_xO$ and Nb_xO are inversely related to each other (isotopically labelled $C^{18}O_2$ was employed to clarify this process).[122] Interestingly, smaller clusters favored Nb_xO formation while larger ones favored $OCNb_xO$. This could be because larger clusters can internally stabilize the "hot" OCNbO intermediate while small cluster-CO_2 adducts are forced to decompose.

Although these are fascinating studies and much can continue to be learned, El-Sayed[122] points out the inherent problems associated with the cluster-beam/reactivity studies. First, the temperature of the clusters is not known, a very serious drawback. Also, large clusters could be fragmenting and this could be affecting conclusions:

$$Nb_x + RX \rightarrow Nb_yR + Nb_zX.$$

The uncertainty in the laser-ionization/detection step is also serious. Does ionization also cause fragmentation or unwanted ion–molecule reactions?

The interaction of Nb clusters with hydrocarbons such as cyclohexane, cyclohexene, and cyclohexadiene showed a sensitivity toward unsaturation.[126,127] Small Nb clusters did not react with the saturated hydrocarbons, but extensive dehydrogenation of the alkene and diene took place.[126] For the smaller clusters Nb_x, where $x = 1–3$, Nb_x-C_6H_6 gasphase complexes were detected. For larger clusters, where $x = 4–9$, an ion of Nb_xC_6 was dominant, especially for $x = 9$. Thus, larger clusters yield more extensive dehydrogenation, and Nb–C bond formation to give carbide like structures appears to be a strong driving force.

As briefly mentioned before, benzene itself showed a variable reaction probability depending on the size of the gaseous Nb_x species.[128] It was found that Nb_5, Nb_6, Nb_7, Nb_9, and Nb_{11} were relatively reactive toward converting C_6H_6 to C_6, while Nb_8 and Nb_{10} were not reactive in this way. This intriguing result was also supported by a report that Nb_8 and Nb_{10} were relatively unreactive with H_2 and N_2.[129]

In a closely related study, H_2 dissociative chemisorption on Nb_7^+, Nb_8^+, and Nb_9^+ was studied. In agreement with earlier studies on the neutral clusters, Nb_7^+ was found to be far more reactive than Nb_8^+ or Nb_9^+. These similar reactivity patterns for neutrals vs positive ions is not consistent with a simple electrostatic model of the dissociative chemisorption process.[130]

Some general conclusions could be derived from these studies: (1) Large Nb_x clusters seem to be necessary to energetically drive the dehydrogenation reactions to completion; (2) Nb_8 and Nb_{10} may be structurally unreactive; and (3) the number of Nb–C bonds that can be formed may be the driving and controlling force:

$$Nb_x + C_6H_6 \rightarrow (C_6H_6)Nb_x + C_6Nb_x + H_2.$$

The dependence of reactivity on cluster structure is still as mysterious as it was when Morse stated in 1986, "At this point it remains an open question why certain clusters are particularly reactive or unreactive, and why ionization potential as a function of cluster size fluctuates in the observed manner" (p. 1049).[89a]

The concept of defect sites on the surface of growing clusters remains as a viable explanation for seemingly mysterious reactivity patterns,[131] and the need for creative and probing experiments is greater than ever. Unfortunately, gas-phase studies of cluster reactivity, depending primarily on mass spectral data, suffer from lack of knowledge of the real temperature of the clusters (which could be very vibrationally hot even if translationally cold), cluster structure is unknown, and the photoionization necessary to detect the cluster and reaction products may drive the reactions.

Only a few qualitative comparative studies of early transition-metal clusters regarding chemical reactivity with hydrocarbons have been reported. In low-temperature matrices of cold hydrocarbons, Cr, Mn, Co, Ni, Pd, Cu, and Zn have been deposited, and cluster growth was in competition with reaction of the growing cluster with the alkanes.[131,132] Generally a trend was found where the early transition-metal clusters reacted with the alkanes more vigorously and completely than the later transition metals. These results are in accord with the predictions of Baetzold and co-workers,[133] where heats of adsorption and activation energies for C–H cleavage on metal surfaces were considered. This theoretical treatment dealt with the orbital energy match for C–H cleavage on certain metal ensembles. As the number of d electrons increased in passing from the early to the late transition elements, activation energy for dissociative adsorption also increased.

Another comparative study was reported by Hamrick and Morse,[134] where V, Nb, and Ta gas-phase clusters were studied in a fast flow chemical reactor. Dissociative chemisorption of D_2 and N_2 was found to be quite size selective. Reactivities of Nb and Ta were similar. Cluster sizes 4, 8, 10, and to some extent 12 were relatively unreactive for both metals. Vanadium clusters behaved differently and seemed less sensitive to size. Structural isomerism seems probable for Nb_9, Nb_{11}, Nb_{12}, and Ta_{12}, based on biexponential reaction kinetics with N_2.

III. Bimetallic and Binuclear Systems

A. Molecules

1. Heterobimetallics

Bonding and spectroscopic properties of a series of M–M' dimers have been elucidated in recent years, primarily due to the matrix isolation work of Weltner and Van Zee[135] and gas-phase work of Gingerich and Shim.[136] Atoms of two elements from opposite ends of the periodic table would be expected to form strong multiple bonds and therefore be of low spin. For example, ScNi, ScPd, YNi, YPd have $S = \frac{1}{2}$, and the unpaired spin density is largely on the light atom with about 30% s character.[137] The bonding seems to be best rationalized where d-electron-rich Pd donates electrons to d-poor Y while Y backdonates s electrons (see Table 4-6).

Weltner and co-workers[138,139] have applied the isoelectronic principle, where the number of valence electrons for M–M vs M–M' are compared, in an attempt to predict ground states. In a general way this principle holds; first-row and mixed-row diatomics containing 12, 13, 14, 15, and 18 valence electrons tend to have $^1\Sigma$, $^2\Sigma$, $^1\Sigma$, $^4\Sigma$, and $^7\Sigma$ ground states, respectively. However, the CrCu, CrAg, and CrAu species do not seem to follow this "rule."

It would be helpful to consider a few of these species in more detail. For example CrMn[140] was found by ESR to exhibit a ground-state $^4\Sigma$ and multiple bonding and in accord with that expected after comparing Mn_2 and Cr_2 (D_{Mn_2} = 42 kJ/mol, D_{CrMn} = 94, and D_{Cr_2} = 151 kJ/mol).[98] The Cr–Mn bonding can be considered as a pentuple bond $(s + d)^1\sigma^2 d\pi^4 d\delta^4$ with three unpaired spins.

The Sc–Cr molecule has also been examined.[141] It was found to have a $^6\Sigma$ ground-state, which is not like its isoelectronic counterpart TiV,

TABLE 4-6

Heterodiatomic Bimetallic Species Isolated in Low-Temperature Matrices

Species	$D_{M-M'}$	Ground state	Reference
ScCo		$^1\Sigma$	135,143
ScNi	3.28 eV	$^2\Sigma$	135,143
ScPd	3.66	$^2\Sigma$	135,143
YNi	3.06	$^2\Sigma$	135,143
YPd		$^2\Sigma$	135,143
VCo		$^1\Sigma$	135
VNi	2.82 eV	$^4\Sigma$	135
VPd	2.80 eV	$^4\Sigma$	135
VPt		$^4\Sigma$	135,142
NbNi	3.53	$^4\Sigma$	135,142
CrFe		$^7\Sigma$	135,142
CrCu		$^4\Sigma$	138
MnCu		$^4\Sigma$	138
CrFe		$^7\Sigma$	138
CrAg		$^6\Sigma$	138
CrAu		$^6\Sigma$	138
CrZn		$^7\Sigma$	135,144
MnCu			135
MnAg		$^7\Sigma$	135
MnAu			135
MnFe		$^4\Pi$	135
TiV	183 kJ	$^4\Sigma$	139
TiCo		$^2\Sigma$	135,143
TiFe		$^1\Sigma$	135,143
TiNi		$^1\Sigma$	135
CrMn	94 kJ	$^4\Sigma$	140,143
ScCr	207 kJ	$^6\Sigma$	141,143

which shows a $^4\Sigma$ ground state. According to ESR the magnetic parameters indicate a $s\sigma^2 d\sigma^2 d\sigma^{*1} d\pi^2 d\delta^2$ electronic configuration. These results imply that ScCr has less $d\pi$ bonding, and this is another case showing that the isoelectronic principle only has limited predictive power. The overall view of ScCr is that it is strongly bonded, probably involving a σ-bond and two one-electron π-bonds.

It is also interesting to compare the CrCu, CrAg, and CrAu molecules.[138] The CrCu species has a $^4\Sigma$ ground state while CrAg and CrAu have $^6\Sigma$. Thus, CrCu is believed to be triply bonded and so intermediate in properties between Cr_2 and Cu_2. Figure 4-5 demonstrates a crude MO diagram for this species.

Further discussions of these species as well as TiV, VNi, ScNi, TiCo, and others have been reported.[145–149] Bond strengths of TiV, V_2, TiCo, and VNi have also been reported by Spain and Morse,[149a] as measured by predissociation thresholds in severely conjested electronic spectra: TiV 2.068 ± 0.001 eV, V_2 2.753 ± 0.001 eV, TiCo 2.401 ± 0.001 eV, and VNi 2.100 ± 0.001 eV.

The contribution of the d-orbitals to the measured bond strength was estimated by taking into account the promotion energy required to prepare the atoms for bonding, and by comparison with the filled d-subshell coinage metal diatomics. The d-orbital contributions to the chemical bonds in TiV, V_2, Ni_2, NiPt, and Pt_2 were found to be 1.10, 1.22, 0.04, 0.46, and 0.85 eV, respectively.

2. Metallic–Nometallic Small Molecules

A few examples of hydride, carbide, and silicide diatomic molecules are of interest here. For example, MnH has been matrix isolated. The

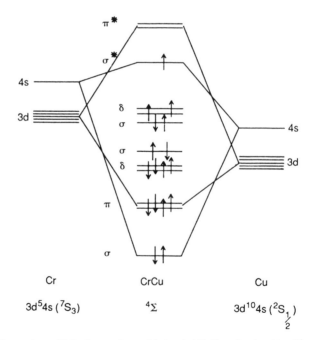

FIGURE 4-5 Approximate M.O. diagram for a triply bonded CrCu molecule with a $^4\Sigma$ ground state (after Weltner and co-workers[138]).

^{55}Mn hyperfine splitting was ~3000 MHz; however, hydrogen hyperfine could not be resolved.[150] Most of the unpaired spin is located on Mn.

The CoC and VC molecules, both of which have a $^2\Sigma$ ground state, have also been matrix isolated.[151] According to ESR the unpaired spin density resides mainly on V. However, the ESR data were difficult to rationalize due to unexpectedly large anisotropic parameters. More recent work on VC, NbC, VSi, and NbSi[152a] has shown that metal hyperfine splittings can vary with matrix medium. This is rationalized by attributing a strong unsymmetrical interaction of these ionic molecules with the matrix. Discussion of the large dipole moments of these molecules as well as ScO and TiN is also given, and electronic structure and spectroscopic properties of SeO, TiN, and VC have been considered in some detail,[152b] as well as reactions of ScO$^+$, TiO$^+$, and VO$^+$ with Δ_2.[152c]

Electronic ground states of RhC($^2\Sigma$), PtC($^1\Sigma$), IrC($^2\Delta_i$), VC($^2\Sigma$), and CoC($^2\Sigma$) have been summarized and discussed.[153] In continuing work, ScC$_2$($^2\Sigma$), VC$_2$($^4\Sigma$), and CoC$_2$($^2\Sigma$) have been observed in solid argon matrices.[153] The ground states observed correspond to analogous MO$_2$ species. It was proposed that these MC$_2$ molecules exist as unsymmetrical MCC species.

Some other unusual species have been matrix deposited. For example, K$_2$MoO$_4$ and K$_2$WO$_4$ have been spectroscopically ellucidated.[154] These molecular species were prepared by vaporization of the parent solid substance and the molecules matrix trapped. Isotopic labeling experiments established a D$_{2d}$ geometry for K$_2$WO$_4$, K$_2$CrO$_4$, and K$_2$MoO$_4$. Bond angles were determined by high-resolution IR spectra.

$$
\begin{array}{ccc}
\text{O} & & \text{O} \\
/ \; \backslash & / & \backslash \\
\text{K} & \text{Mo} & \text{K} \\
\backslash \; / & \backslash & / \\
\text{O} & & \text{O}
\end{array}
$$

IR and UV–VIS spectra of MoOF$_4$, MoOCl$_4$, WOCl$_4$, and WOBr$_4$ have also been obtained.[155] The spectra indicated multiple trapping sites in frozen argon, but only one site in frozen nitrogen. Matrix electronic spectra were of excellent quality with vibrational fine structure. The IR spectra showed that quite pure monomers were matrix isolated. All structures contain a distorted six-coordinate arrangement about the metal.

A few reports on the low-temperature chemistry of metal oxide vapors have also appeared. The vapors of MoO$_3$ and WO$_3$ have been co-deposited with water, methanol, THF, and other chemicals at 77 K.[156,157] The free molecules of these species were shown to be considerably more reactive than the bulk solids, as would be expected. Complexes were formed with 2,4-pentanedione, acetone, formic acid, and methanol. However, polymeric structures were formed, and so synthetic utility was limited. Oxygen–halogen exchange occurred

$$WO_{3\,vap} + CH_3C\overset{O}{\overset{\|}{-}}CH_2 - C\overset{O}{\overset{\|}{-}}CH_3 \xrightarrow{\text{cocond.}} \longrightarrow$$

with BCl_3, HCl, and $SiCl_4$, while PCl_3 was oxidized by MoO_3 or WO_3 molecules:

$$2WO_3 + 3BCl_3 \xrightarrow{\text{cocond.}} \longrightarrow WOCl_4 + WO_2Cl_2 + (BOCl)_3.$$

With CH_3OH a methoxy adduct was formed, while THF yielded an amorphous solid:[157]

$$MoO_3 + CH_3OH \xrightarrow{\text{cocond.}} \longrightarrow Mo_2O_5\,(OCH_3)_2 \cdot 2CH_3OH$$

Other high-temperature refractory metal oxides exhibit somewhat more unusual chemistry. Thus, VO and TiO molecules have been prepared by evaporation of MO or M_2O_3 solids. The evaporation of these oxides from open tungsten boats yielded M atoms, MO, and MO_2 molecules. Vapor compositions were similar to Knudsen cell experiments in the case of VO and V_2O_3, but somewhat different for TiO and Ti_2O_3.[158] The best sources of molecular VO and TiO were the solid M_2O_3 oxides. The monoxide molecules reacted with chlorine to produce oxometal chlorides in high yields. They reacted with 2,4-pentanedione and other organics containing relatively acidic hydrogens to form water as a by-product via protonation of the oxo-moiety; coordination compounds of titanium and vanadium as acac complexes were produced in high yields. The facilty of loss of the oxo-ligand was surprising.[158]

$$2\,TiO(g) + 3/2\,Cl_2 \longrightarrow ClTiO + Cl_2TiO$$

$$2\,VO(g) + 5/2\,Cl_2 \longrightarrow OVCl_2 + OVCl_3$$

$$VO + 3\,CH_3C\overset{O}{\overset{\|}{-}}CH_2C\overset{O}{\overset{\|}{-}}CH_3 \longrightarrow V(acac)_3 + H_2O + 1/2\,H_2$$

B. Heterobimetallic Clusters

The characterization of mixed-metal clusters is an extremely difficult but important undertaking. In the field of catalysis, bimetallic small particles are causing a revolution in synthesis and interpretation. Herein some new approaches to the study of heterobimetallic clusters are introduced.

The advent of the gas-phase cluster-beam method is beginning to have an effect on this important field. The synthesis of small amounts of bimetallic clusters can be carried out by laser evaporation of alloys, physical mixtures of the two metals,[159–164] or by evaporating a metal in the presence of another volatile metal-containing complex.[160] It is particularly intriguing that mixed-metal bonding can be forced to take place in systems that do not form stable alloys. Overall, these studies have shown that the mixed-metal clusters form in statistical distribution for clusters with more than about six atoms. For smaller clusters, departure from statistical behavior is sometimes observed, and this can be attributed to chemical bonding differences or ionization probability (for MS detection) differences. Some of the bimetal combinations studies to date are listed:

Cr–Bi	Cu–In	Cr–Ge
Ag–Mn	Cu–Ga	Co–Mn
Cr–Sn	Ni–Cr	Co–V
Fe–Bi	Ni–Al	Nb–Al
Fe–Sb	Ta–Mn	V–Al

In some specific cases, certain pure clusters seem to be building blocks for the mixed clusters, for example, Co_4 in the production of $Co_n Mn_m$ clusters. Sometimes phase diagrams of the bulk material can be used to explain gas-phase cluster distributions.[162]

Gas-phase reactivity of metal "alloy clusters" can also be studied.[164,165] Unpredictable results are the norm in this field of study. In the case of gas-phase $Co_n V_m^+$ cluster ions, reactivity toward H_2 was studied in a fast flow reactor.[164] Compared with pure Co_n^+, the reactivity of $Co_{n-1} V^+$ cluster ions showed a sharp decrease in activity, similar to that of analogous neutral clusters. Geometrical structure was proposed as a rationale, but much more data are necessary before some true understanding exists.

The reaction of O_2 with $Nb_n Al_m$ and $V_n Al_m$ clusters has also been investigated. The Jellium model, which treats the nuclei (and core electrons) of the constituent atoms as smeared out in a specified potential well, and all valence electrons are allowed to fill electronic shells, was used to rationalize the unreactivity of certain species (such as Al_{13}^-, Al_{23}^-, and $NbAl_4^-$).

The high reactivity of certain metal clusters can be used to advantage to prepare ultrafine particles of refractory materials. For example, the high reactivity of Ti clusters is demonstrated by experiments where Ti metal was evaporated into a gaseous medium of He and N_2. The Ti clusters reacted to form TiN particles, which were collected as a powder made up of crystals of 5–20 nm (fcc NaCl crystal structure).[166] Actually, a similar method can be employed for the production of ultrafine particles of several metal nitrides.[167]

C. Film Formation from Clusters (and Atoms)

Ion cluster beams (ICB) in a reactive environment have been used to prepare smooth metal-oxide films.[168] For example, evaporation of Ti into an inert gas yielded Ti clusters that were oxidized by O_2 present. Cluster ionization and acceleration allowed film deposition on silicon and glass substrates. Acceleration voltage affected refractive indexes of the resultant TiO_2 films and was observed to change the crystal structure from anatase to rutile. Smooth, crystalline films were produced by this "reactive ICB" technique. Further development of this technique for practical purposes included the low-energy ion bombardment of the growing ICB produced films.[169]

A review of the ICB technique and its applications to formation of epitaxial Al films, GaAs, CdMnTe, and CdTe/PbTe multilayer expitaxial films has been presented by Yamada et al.[170]

The properties of Ti thin films (about 500 Å) produced by ICB techniques have also been studied. The substrate was p-type silicon. The films were deposited at 1–2 Å/sec at a pressure of 3×10^{-7} Torr. Varying beam conditions, such as the fraction of ionized clusters and acceleration potential, were explored. Very uniform films were produced with good interfacial properties.[171] Likewise, ICB depositions of Cr, Cu, Ag, and Au have been reported. Nozzle design is important, and a conical shape improves performance.[172]

The method has been developed to the point that large-area deposition of films is now possible, such as on a 250-mm-diameter substrate, and depositions of $TiSi_2$, $PbTiO_3$, and $YBa_2Cu_3O_7$ have been carried out.[173] The characteristics of Au films on copper, Al films on silicon, and the crystallizing of FeO_x films were also described. An additional improvement of the technique for production of oxide and nitride films is by using dual beams, one an ICB of metal clusters and the other an ionized gas beam. As before, the ICB is produced by vaporizing metals and ejecting the vapor through a multinozzle into high vacuum where the atoms are clustered by cooling due to adiabatic expansion. The clusters are partially ionized by an electron shower and accelerated toward the substrate. The gas beam is produced by ionizing neutral gas molecules by an electron shower and likewise accelerated toward the substrate. These two beams collide and combine together on their way to the substrate. This technique for preparing, for example, SiO_2 and TiN films allows high deposition rates at relatively low temperatures over large-area substrates. It is evident that the cluster–gas reaction rate is enhanced by this treatment.[174] ICB techniques allow less film damage and high deposition rates at low temperatures. (see Chapter 2 for a further description of ICB). Additional applications of ICB are for producing CdMnTe films,[175] the manufacture of high-X-ray reflection films, high-damage-threshold Al mirrors, and high-density TiO_2 and Al_2O_3 films,[176] as well as superconducting $YBa_2Cu_3O_7$ films.[177]

Vapor-phase reactions of metal atoms and clusters can lead to ultrafine particles of metal sulfides. Thus, reaction of Mo vapor with sulfur vapor (or Pb with S) yielded Mo_3S_4 (or PbS) ultrafine powders.[178]

Sputtering techniques for producing clusters and fine particles have also advanced in recent years. For example, W, Cu, and Ag ultrafine particles were formed. In addition, tungsten carbide (deficient in carbon) was formed by reactive sputtering where gas-phase clustering of W atoms is done in the presence of a reactive gas such as a hydrocarbon.[179]

The ICB method as a technique for production of large metal clusters and use of these clusters to prepare metallurgical coatings continues to come under intense study.[180] Deposition, epitaxy, crystallography, mechanical, electrical, optical, and magnetooptical properties of metal and intermetallic films can be controlled by acceleration voltage and electron current for ionization. Films with high density and strong adhesion were reported for Cu–Ni, CdTe–PbTe, and CdMnTe.[180] Analogously, the ICB method has been used to prepare MnBi films for magnetooptical memory applications.[181] Both Mn and Bi were evaporated simultaneously and allowed to cluster via adiabatic expansion, ionized, accelerated, and deposited on glass substrate. Annealing of the films caused crystallization as low as 100°C, which is far below the temperature needed for conventional vacuum evaporation/deposition.

Reactive ICB techniques, where Ti clusters have been formed in the presence of oxygen gas, have yielded TiO and TiO_2 films.[182] Control of stoichiometry and film properties was possible by ionization current, acceleration voltages, and oxygen pressure.[183]

In addition to the ICB method,[183] there are several other vaporization/deposition processes for producing films. Table 4-7 serves as a brief summary of electron beam, crucible evaporations, and others.

TABLE 4-7
Electron Beam and Crucible Evaporations and Reactive Evaporation for Production of Particles and Films

Material evaporation	Technique	Gas	Particles	Films	Comments	Reference
Al	Crucible	H_2	Al			184
Ti			TiH_2			
Mo	Electron beam	Argon	Mo		3–20 nm, various crystal habits	185
W			W			
Mo/Si	Electron beam	Vacuum	—	Mo_5Si_3	Hexagonal	186
In	Plasma-assisted electron beam	O_2		Metal oxides, nitrides, sulfides, fluorides		187
Sn		N_2				
Cd		NH_3				
Zn		CH_4				
Ti		H_2S				
Ge		N_2O				
		CF_4				
Ni	Tungsten arc	H_2	Ni			188
Fe		Ar	Fe			
Ti			Ti			
Ti	Crucible	H_2	TiH_2	—	fcc structure	189

Nb	Electron beam	Vacuum		Nb	Superconductivity at 9.2 K	190
Ti	Electron beam	N_2 (0.1 Torr)	TiN(10 nm)		Reactive gas evaporation technique	191
Zr			ZrN(5 nm)			192
Hf		NH_3	HfN(5 nm)			
Al			AlN(10 nm)			
Si,C	Arc. evaporation	Inert gas	SiC		Arc struck between C and metal electrodes. Ultrafine particles formed of the carbides	193
Ti,C			TiC			
V,C			VC			
Zr,C			ZrC			
Ni,Cr,Al	Electron beam	Vacuum	Ni–Cr–Al		Films	194
Co			Co–Cr–Al			
Fe			Fe–Cr–Al			
Ti			also with			
Y			Ti,Y,Si,			
Si			Hf,Zr			
Hf						
Zr						

References

1. K. J. Klabunde, "Chemistry of Free Atoms and Particles," Academic Press, New York, 1980.
2. G. A. Ozin, M. P. Andrews, C. G. Francis, and K. Molnar, *Inorg. Chem.* **29**, 1068 (1990).
3. M. L. H. Green, *J. Organomet. Chem.* **200**, 119 (1980).
4. M. L. H. Green and D. O'Hare, *in* "High Energy Processes in Organometallic Chemistry" (K. Suslick, Ed.), ACS Symposium Series 333, American Chemical Society, Washington, DC, 1987.
5. R. H. Hauge, L. Fredin, Z. H. Kafafi, and J. L. Margrave, *Appl. Spetrosc.* **40**, 588 (1986).
6. J. Godber, H. Huber, and G. A. Ozin, *Inorg. Chem.* **25**, 2909 (1986).
7. M. P. Andrews, "Encyclopedia Polymer Science Engineering," Vol. 8, 2nd ed., Wiley, New York, 1987.
8. M. P. Andrews, "Experimental Organometallic Chemistry," ACS Sym. Series 357 (A. L. Wada and M. Y. Darensbourg, Eds.) Chap. 7, p. 158, Am. Chem. Soc., Washington, DC, 1987.
9. J. M. Alford, P. E. Williams, D. J. Trevor, and R. E. Smalley, *Int. J. Mass Spectrom. Ion Process.* **72**, 33 (1986).
10. G. H. Jeong and K. J. Klabunde, *J. Chem. Phys.* **91**, 1958 (1989).
11. R. Pellow, M. Eyring, and M. Vala, *J. Chem. Phys.* **90**, 1440 (1989).
12. M. Vala, R. Pyzalski, J. Shakhsemampour, M. Eyring, J. Pyka, T. Tipton, and J. C. Rivoal, *J. Chem. Phys.* **86**, 5951 (1987).
13. J. C. Rivoal and M. Vala, *J. Chem. Phys.* **86**, 5958 (1987).
14. R. J. Van Zee, D. A. Garland, and W. Weltner, Jr., *J. Chem. Phys.* **85**, 3237 (1986).
15. J. Q. Broughton and P. S. Bagus, *J. Electron Spectrosc. Relat. Phenom.* **20**, 127 (1980).
16. R. R. Squires, *J. Am. Chem. Soc.* **107**, 4385 (1985).
17. J. Mascetti and M. Tranquille, *J. Phys. Chem.* **92**, 2177 (1988).
18. M. Almond, A. J. Dows, and R. N. Perutz, *Inorg. Chem.* **24**, 275 (1985).
19. M. Poliakoff, K. P. Smith, J. J. Turner, and A. J. Wilkinson, *J. Chem. Soc., Dalton Trans.*, 651 (1982).
20. S. Togashi, J. G. Fulcher, B. R. Cho, M. Hasegawa, and J. A. Gladysz, *J. Org. Chem.* **45**, 3044 (1980).
21. A. H. Reid, P. B. Shevlin, T. R. Webb, and S. S. Yun, *J. Org. Chem.* **49**, 4728 (1984).
22. (a) J. T. Miller and C. W. DeKock, *J. Org. Chem.* **46**, 516 (1981); (b) P. S. Skell, K. J. Klabunde, J. H. Plonka, J. S. Roberts, and D. L. Williams-Smith, *J. Am. Chem. Soc.* **95**, 1547 (1973).
23. W. E. Billups, M. M. Konarski, R. H. Hauge, and J. L. Margrave, *Tetrahedron Lett.* **21**, 3861 (1980).
24. C. E. Broun, S. A. Mitchell, and P. A. Hackett, *J. Phys. Chem.* **95**, 1062 (1991).
25. J. M. Parnis, S. A. Mitchell, and P. A. Hackett, *J. Phys. Chem.* **94**, 8152 (1990).
26. Z. L. Xiao, R. H. Hauge, and J. L. Margrave, *J. Phys. Chem.* **95**, 2696 (1991).
27. S. C. Davis and K. J. Klabunde, *J. Am. Chem. Soc.* **100**, 5973 (1978).
28. R. J. Remick, T. Asunta, and P. S. Skell, *J. Am. Chem. Soc.* **101**, 1320 (1979).
29. S. C. Davis, S. J. Severson, and K. J. Klabunde, *J. Am. Chem. Soc.* **103**, 3024 (1981).
30. W. E. Klotzbucher, S. A. Mitchell, and G. A. Ozin, *Inorg. Chem.* **16**, 3063 (1977).
31. K. J. Klabunde and Y. Tanaka, *J. Am. Chem. Soc.* **105**, 3544 (1983).
32. K. J. Klabunde, G. H. Jeong, and A. W. Olsen, *in* "Selective Hydrocarbon Activation: Principles and Progress" (J. A. Davies, P. L. Watson, A. Greenberg, and J. F. Liebman, Eds.), Chap. 13, p. 433, VCH, New York, 1990.

33. (a) M. R. A. Blomberg, P. E. M. Siegbahn, and M. Svensson, *J. Phys. Chem.* **95**, 4313 (1991); (b) M. R. A. Blomberg, P. E. M. Siegbahn, and M. Svensson, *J. Am. Chem. Soc.* **114**, 6095 (1992); (c) D. Ritter, J. J. Carroll, and J. C. Weisshar, *J. Phys. Chem.* **96**, 10636 (1992).
34. M. L. H. Green and G. Parkin, *J. Chem. Soc., Chem. Commun.*, 1467 (1984).
35. F. G. N. Cloke, V. C. Gibson, M. L. H. Green, V. S. B. Mtetwa, and K. Prout, *J. Chem. Soc., Dalton Trans.*, 2227 (1988).
36. J. A. Bandy, F. G. N. Cloke, M. L. H. Green, D. O'Hare, and K. Prout, *J. Chem. Soc., Chem. Commun.*, 240 (1984).
37. M. L. H. Green and D. O'Hare, *J. Chem. Soc., Dalton Trans.*, 403 (1987).
38. M. L. H. Green and D. O'Hare, *J. Chem. Soc., Dalton Trans.*, 2409 (1986).
39. W. E. Billups, M. M. Konarski, R. H. Hauge, and J. L. Margrave, *J. Am. Chem. Soc.* **102**, 7393 (1980).
40. W. E. Billups, private communication.
41. G. A. Ozin, J. G. McCaffrey, and J. M. Parnis, *Angew. Chem. Int. Ed. Engl.* **25**, 1072 (1986).
42. (a) W. E. Billups, S. C. Chang, R. H. Hauge, and J. L. Margrave, *Inorg. Chem.* **32**, 1529 (1993); (b) W. E. Billups, J. P. Bell, R. H. Hauge, E. S. Kline, A. W. Moorehead, J. L. Margrave, and F. B. McCormick, *Organometallics* **5**, 1917 (1986).
43. P. R. Brown, F. G. N. Cloke, M. L. H. Green, and R. C. Tovey, *J. Chem. Soc., Chem. Commun.*, 519 (1982).
44. T. J. Groshens, Ph.D. thesis, Kansas State University, 1988.
45. R. J. Van Zee and W. Weltner, Jr., *J. Am. Chem. Soc.* **111**, 4519 (1989).
46. S. B. H. Bach, C. A. Taylor, R. J. Van Zee, M. T. Vala, and W. Weltner, Jr., *J. Am. Chem. Soc.* **108**, 7104 (1986).
47. W. Weltner, Jr. and R. J. Van Zee, *in* "Computational Chemistry: The Challenge of *d*- and *f*-Electrons" (D. R. Salahub and M. C. Zerner, Eds.), ACS Sym. Ser., 1989.
48. R. J. Van Zee, J. B. H. Bach, and W. Weltner, Jr., *J. Phys. Chem.* **90**, 583 (1986).
49. P. H. Kasai and P. M. Jones, *J. Am. Chem. Soc.* **107**, 813 (1985).
50. C. G. Francis, S. I. Khan, and P. R. Morton, *Inorg. Chem.* **23**, 3680 (1984).
51. F. G. N. Cloke, J. C. Green, M. L. H. Green, and C. P. Morley, *J. Chem. Soc. Chem. Commun.*, 945 (1985).
52. F. G. N. Cloke and J. M. Day, *J. Chem. Soc. Chem. Commun.*, 967 (1985).
53. F. G. N. Cloke, J. P. Day, J. C. Green, C. P. Morley, and A. C. Swain, *J. Chem. Soc., Dalton Trans.* 789 (1991).
54. (a) D. N. Cox and R. Roulet, *Helv. Chim. Acta* **67**, 1365 (1984); (b) F. G. N. Cloke and A. McCamley, *J. Chem. Soc., Chem. Commun.*, 1470 (1991).
55. F. G. N. Cloke, M. L. H. Green, and P. J. Lennon, *J. Organomet. Chem.* **188**, C25 (1980).
56. (a) E. P. Kundig, P. L. Timms, B. A. Kelly, and P. Woodward, *J. Chem. Soc., Dalton Trans.*, 901 (1983); (b) P. Treichel, G. P. Essenmacher, H. F. Efner, and K. J. Klabunde, *Inorg. Chim. Acta* **48**, 41 (1981).
57. E. P. Kuendig and P. L. Timms, *J. Chem. Soc., Dalton Trans.*, 991 (1980).
58. R. M. Markle, T. M. Pettijohn, and J. J. Lagowski, *Organometallics* **4**, 1529 (1985).
59. C. Elschenbroich, J. Hurley, W. Massa, and G. Baum, *Angew. Chem. Int. Ed. Engl.* **27**, 684 (1988).
60. C. Elschenbroich, J. Koch, J. Schneider, B. Spangenberg, and P. Schiess, *J. Organomet. Chem.* **317**, 41 (1986).
61. C. Elschenbroich and J. Koch, *J. Organomet. Chem.* **229**, 139 (1982).
62. C. Elschenbroich, J. Hurley, B. Metz, W. Massa, and G. Baum, *Organometallics* **9**, 889 (1990).

63. C. Elschenbroich, M. Nowotny, B. Metz, W. Massa, J. Graulich, K. Biehler, and W. Sauer, *Angew. Chem. Int. Ed. Engl.* **30**, 547 (1991).
64. M. L. H. Green, D. O'Hare, P. Mountford, and J. G. Watkin, *J. Chem. Soc., Dalton Trans.*, 1705 (1991).
65. F. G. N. Cloke and M. L. H. Green, *J. Chem. Soc.*, 127 (1979).
66. M. L. H. Green, D. O'Hare, and J. M. Wallis, *J. Chem. Soc., Chem. Commun.*, 233 (1984).
67. W. E. Billups, A. W. Moorehead, P. J. Ko, J. L. Margrave, J. P. Bell, and F. B. McCormick, *Organometallics*, **7**, 2230 (1988).
68. F. G. N. Cloke and M. L. H. Green, *J. Chem. Soc., Dalton Trans.*, 1938 (1981).
69. P. R. Brown, F. G. N. Cloke, M. L. H. Green, and N. J. Hazel, *J. Chem. Soc., Dalton Trans.*, 1075 (1983).
70. C. Elschenbroich, E. Bilger, and B. Metz, *Organometallics*, **10**, 2823 (1991).
71. (a) F. G. N. Cloke, M. F. Lappert, G. A. Lawless, and A. C. Swain, *J. Chem. Soc., Chem. Commun.*, 1167 (1987); (b) F. G. N. Cloke, K. Klan, and R. N. Perutz, *J. Chem. Soc., Chem. Commun.*, 1372 (1991); (c) C. G. Francis and P. D. Morand, *Inorg. Chem.* **24**, 56 (1985).
72. A. McCamley and R. N. Perutz, *J. Phys. Chem.* **95**, 2738 (1991).
73. M. Andrews, S. Matlar, and G. Ozin, *J. Phys. Chem.* **90**, 744 (1986).
74. F. G. N. Cloke, P. J. Fyne, M. L. H. Green, M. J. Ledoux, A. Gourdon, and C. K. Prout, *J. Organomet. Chem.* **198**, C69 (1980).
75. S. D. Ittel, F. A. Van-Catledge, and C. A. Tolman, *Inorg. Chem.* **24**, 62 (1985).
76. C. G. Francis, H. Huber, and G. A. Ozin, *J. Am. Chem. Soc.* **101**, 6250 (1979).
77. G. A. Ozin, M. P. Andrews, and R. West, *Inorg. Chem.* **25**, 580 (1986).
78. M. P. Andrews and G. A. Ozin, *Chem. Mater.* **1**, 174 (1989).
79. M. P. Andrews, *in* "Metallization of Polymers" (E. Sacher and J. J. Piraux, Eds.), ACS Symp. Series 440, Chap. 18, p. 242, Am. Chem. Soc., Washington, DC, 1990.
80. M. P. Andrew, M. E. Galvin, and S. A. Heffner, *Mater. Res. Soc. Proc.* **131**, 21 (1989).
81. S. J. Riley, E. K. Parks, C. R. Mao, L. G. Pobo, and S. Wexler, *J. Phys. Chem.* **86**, 3911 (1982).
82. L. Heinbrook, M. Rasanen, and V. E. Bondybey, *J. Phys. Chem.* **91**, 2468 (1987).
83. A. W. Castleman and R. G. Keesee, *Acc. Chem. Res.* **19**, 413 (1986).
84. V. Vaida, N. J. Cooper, R. J. Hemley, and D. G. Leopold, *J. Am. Chem. Soc.* **103**, 7022 (1981).
85. C. W. Draper, *J. Phys. Chem.* **84**, 2089 (1980).
86. M. P. Andrews and G. A. Ozin, *J. Phys. Chem.* **90**, 3353 (1986); and references therein.
87. C. G. Francis, H. Huber, and G. A. Ozin, *Inorg. Chem.* **19**, 219 (1980).
88. M. Andrews, G. A. Ozin, and C. G. Francis, *Inorg. Synth.* **22**, 116 (1982).
89. (a) M. D. Morse, *Chem. Rev.* **86**, 1049 (1986); (b) M. Doverstäl, B. Lindgren, U. Sassenberg, C. A. Arrington, and M. D. Morse, *J. Chem. Phys.* **97**, 7087 (1992).
90. W. Weltner, Jr. and R. J. Van Zee, *ACS Symp. Ser.* **394**, 213 (1989).
91. L. B. Knight, Jr., R. J. Van Zee, and W. Weltner, Jr., *Chem. Phys. Lett.* **94**, 296 (1983).
92. M. A. Douglas, R. H. Hauge, and J. L. Margrave, *J. Phys. Chem.* **87**, 2945 (1983).
93. (a) M. Moskovits, D. P. DiLella, and W. Limm, *J. Chem. Phys.* **80**, 626 (1984); (b) H. Akeby, L. G. M. Pettersson, and P. E. M. Siegbahn, *J. Chem. Phys.* **97**, 1850 (1992).
94. P. R. R. Langridge-Smith, M. D. Morse, G. P. Hansen, R. E. Smalley, and A. J. Merer, *J. Chem. Phys.* **80**, 593 (1984).
95. Z. Hu, B. Shen, Q. Zhou, S. Deosaran, J. R. Lombardi, D. M. Lindsay, and W. Harbich, *J. Chem. Phys.* **95**, 2206 (1991).

96. M. Moskovits, Ed., "Metal Clusters," Wiley, New York, 1986.
97. M. F. Jarrold, A. J. Illies, and M. T. Bowers, *J. Am. Chem. Soc.* **107**, 7339 (1985).
98. A. R. Miedema, *Faraday Symp. Chem. Soc.* **14**, 136 (1980).
99. M. M. Goodgame and W. A. Goddard, III, *J. Phys. Chem.* **85**, 215 (1981).
100. (a) M. L. Michalopoules, M. E. Geusic, S. G. Hansen, D. E. Powers, and R. E. Smalley, *J. Phys. Chem.* **86**, 3914 (1982); (b) S. M. Casey and D. G. Leopold, *J. Phys. Chem.* **97**, 816 (1993).
101. K. Hilpert and K. Ruthardt, *Ber. Bunsen Gesellschaft Phys. Chem.* **91**, 724 (1987).
102. M. Moskovits, W. Limm, and T. Mejean, *J. Phys. Chem.* **89**, 3886 (1985).
103. M. Cheeseman, R. J. Van Zee, H. L. Flanagan, and W. Weltner, Jr., *J. Chem. Phys.* **92**, 1553 (1990).
104. C. A. Baumann, R. J. Van Zee, S. V. Bhat, and W. Weltner, Jr., *J. Chem. Phys.* **78**, 190 (1983).
105. K. D. Bier, T. L. Haslett, A. D. Kirkwood, and M. Moskovits, *J. Chem. Phys.* **89**, 6 (1988).
106. A. D. Kirkwood, K. D. Bier, J. K. Thompson, T. L. Haslett, A. S. Huber, and M. Moskovits, *J. Phys. Chem.* **95**, 2644 (1991).
107. J. C. Rivoal, J. S. Emampour, K. J. Zeringue, and M. Vala, *Chem. Phys. Lett.* **92**, 197 (1982).
108. Z. Hu, B. Shen, Q. Zhou, S. Deosaran, J. R. Lombardi, D. M. Lindsay, and W. Harbich, *J. Chem. Phys.* **95**, 2206 (1991).
109. (a) Z. Hu, Q. Zhou, J. R. Lombardi, and D. M. Lindsay, in press; (b) Z. Hu, B. Shen, J. R. Lombardi, and D. M. Lindsay, *J. Chem. Phys.* **96**, 8757 (1992).
110. (a) Z. Hu, B. Shen, Q. Zhou, S. Deosaran, J. R. Lombardi, and D. M. Lindsay, in press; (b) Z. Hu, J.-G. Dong, J. R. Lombardi, and D. M. Lindsay, *J. Chem. Phys.* **97**, 8811 (1992).
111. J. B. Hopkins, P. R. R. Langridge-Smith, M. D. Morse, and R. E. Smalley, *J. Chem. Phys.* **78**, 1627 (1983).
112. R. J. Van Zee, C. A. Baumann, and W. Weltner, Jr., *J. Chem. Phys.* **82**, 3912 (1985).
113. J. Lignières, B. d'Humieres, and J. C. Rivoal, *Z. Phys. D: At., Mol. Clusters* **19**, 207 (1991).
114. M. Moskovits, D. P. DiLella, and W. Limm, *J. Chem. Phys.* **80**, 626 (1984).
115. L. B. Knight, Jr., R. W. Woodward, R. J. Van Zee, and W. Weltner, Jr., *J. Chem. Phys.* **79**, 5820 (1983).
116. A. Amrein, R. Simpson, and P. Hackett, *J. Chem. Phys.* **95**, 1781 (1991).
117. J. Uppenbrink and D. J. Wales, *J. Chem. Soc. Faraday Trans.*, 215 (1991).
118. D. M. P. Mingos, *Acc. Chem. Res.* **23**, 17 (1990).
119. G. A. Ozin and S. A. Mitchell, *Angew. Chem. Int. Ed. Engl.* **22**, 674 (1983).
120. G. A. Ozin and D. F. McIntosh, *J. Phys. Chem.* **90**, 5756 (1986).
121. (a) S. M. Owen, *Polyhedron* **7**, 253 (1988); (b) L. Lian, C.-X. Su, and P. B. Armentrout, *J. Chem. Phys.* **97**, 4075 (1992).
122. M. A. El-Sayed, *J. Phys. Chem.* **95**, 3898 (1991).
123. L. Song, A. Eychmuller, and M. A. El-Sayed, *J. Phys. Chem.* **92**, 1005 (1988).
124. A. Eychmuller, L. Song, and M. A. El-Sayed, *J. Phys. Chem.* **93**, 404 (1989).
125. L. Song and M. A. El-Sayed, *Chem. Phys. Lett.* **152**, 281 (1988).
126. R. J. St. Pierre, E. L. Chronister, L. Song, and M. A. El-Sayed, *J. Phys. Chem.* **91**, 4648 (1987); also see p. 5228.
127. L. Song and M. A. El-Sayed, *J. Phys. Chem.* **94**, 7907 (1990).
128. R. J. St. Pierre and M. A. El-Sayed, *J. Phys. Chem.* **91**, 763 (1987).
129. M. E. Geusic, M. D. Morse, and R. E. Smalley, *J. Chem. Phys.* **82**, 590 (1985).

130. J. M. Alford, F. D. Weiss, R. T. Laaksonen, and R. E. Smalley, *J. Phys. Chem.* **90,** 4480 (1986).

131. K. J. Klabunde, G. H. Jeong, and A. W. Olsen, "Selective Hydrocarbon Activation Principles and Progress" (J. A. Davies, P. L. Watson, A. Greenberg, and J. F. Liebman, Eds.), Chap 13, pp. 433–467, VCH, New York, 1987.

132. A. W. Olsen, Ph.D. thesis, Kansas State University, 1989.

133. (a) R. C. Baetzold, *Solid State Commun.* **44,** 781 (1982); (b) E. Shustorovich, R. C. Baetzold, and E. L. Muetterties, *J. Phys. Chem.* **87,** 1100 (1983).

134. Y. M. Hamrick and M. D. Morse, *J. Phys. Chem.* **93,** 6494 (1989).

135. W. Weltner, Jr. and R. J. Van Zee, "The Challenge of *d* and *f* Electrons: Theory and Computation" (D. R. Salahub and M. C. Zerner, Eds.), ACS Sym. Soc., Vol. 394, ACS, Washington, DC, 1989.

136. (a) K. A. Gingerich, *Faraday Symp. Chem. Soc.* **14,** 109 (1980); (b) I. Shim and K. A. Gingerich, *Chem. Phys. Lett.* **101,** 528 (1983).

137. R. J. Van Zee and W. Weltner, Jr., *Chem. Phys. Lett.,* in press.

138. C. A. Baumann, R. J. Van Zee, and W. Weltner, Jr., *J. Chem. Phys.* **88,** 5272 (1983).

139. C. A. Baumann, R. J. Van Zee, and W. Weltner, Jr., *J. Chem. Phys.* **88,** 1815 (1984).

140. M. Cheeseman, R. J. Van Zee, H. L. Flanagan, and W. Weltner, Jr., *J. Chem. Phys.* **92,** 1553 (1990).

141. W. Weltner, Jr., *J. Phys. Chem.* **94,** 7808 (1990).

142. M. Cheeseman, R. J. Van Zee, and W. Weltner, Jr., High Temp. Sci. **25,** 143 (1988).

143. A. R. Miedema, *Faraday Symp. Chem. Soc.* **14,** 136 (1980).

144. J. C. Rivoal, C. Grisola, and M. Vala, Phys. *Chem. Small Clusters B* **158,** 617 (1987).

145. K. A. Gingerich, *Faraday Symp. Chem. Soc.* **14,** 109 (1980); and references therein.

146. G. A. Ozin and W. E. Klotzbucher, *Inorg. Chem.* **18,** 2101 (1979).

147. R. J. Van Zee and W. Weltner, Jr., *Chem. Phys. Lett.* **107,** 173 (1984).

148. R. J. Van Zee and W. Weltner, Jr., *High Temp. Sci.* **17,** 181 (1984).

149. (a) E. M. Spain and M. D. Morse, *J. Phys. Chem.* **96,** 2479 (1992); (b) E. M. Spain, J. E. Behm, M. D. Morse, *J. Chem. Phys.* **96,** 2511 (1992).

150. R. J. Van Zee, D. A. Garland, and W. Weltner, Jr., *J. Chem. Phys.* **84,** 5968 (1986).

151. R. J. Van Zee, J. J. Bianchini, and W. Weltner, Jr., *Chem. Phys. Lett.* **127,** 314 (1986).

152. (a) Y. M. Hamrick and W. Weltner, Jr., *J. Chem. Phys.* **94,** 3371 (1991); (b) S. M. Mattar, *J. Phys. Chem.* **97,** 3171 (1993); (c) D. E. Clemmer, N. Aristov, and P. B. Armentrout, *J. Phys. Chem.* **97,** 544 (1993).

153. J. J. Biachini, R. J. Van Zee, and W. Weltner, Jr., *J. Mol. Struct.* **157,** 93 (1987).

154. I. R. Beattie, J. S. Ogden, and D. D. Price, *J. Chem. Soc. Dalton Trans.,* 505 (1982).

155. W. Levason, R. Narayanaswamy, J. S. Ogden, A. J. Rest, and J. W. Turff, *J. Chem. Soc. Dalton Trans.,* 2501 (1981).

156. N. D. Cook and P. L. Timms, *J. Chem. Soc. Dalton Trans.,* 239 (1983).

157. C. W. Dekock and L. V. McAfee, *Inorg. Chem.* **24,** 4293 (1985).

158. T. J. Groshen and K. J. Klabunde, *Inorg. Chem.* **29,** 2979 (1990).

159. V. E. Bondybey, G. P. Schwartz, and J. H. English, *J. Chem. Phys.* **78,** 11 (1983).

160. K. Laitting, P. Y. Cheng, and M. A. Duncan, *J. Phys. Chem.* **91,** 6521 (1987).

161. E. A. Rohlfing, D. M. Cox, R. Petkovic-Luton, and A. Kaldor, *J. Phys. Chem.* **88,** 6227 (1984).

162. Y. Sone, K. Hoshino, T. Naganuma, A. Nakajima, and K. Kaya, *J. Phys. Chem.* **95,** 6830 (1991).

163. T. G. Taylor, K. F. Willey, M. B. Bishop, and M. A. Duncan, *J. Phys. Chem.* **94,** 8016 (1990).

164. A. Nakajima, T. Kishi, T. Sugioka, Y. Sone, and K. Kaya, *J. Phys. Chem.* **95**, 6833 (1991).
165. A. C. Harms, R. E. Leuchtner, S. W. Sigsworth, and A. W. Castleman, Jr., *J. Am. Chem. Soc.* **112**, 5673 (1990).
166. M. D. Bentzon and F. Kragh, *Z. Phys. D: At., Mol. Clusters* **19**, 299 (1991).
167. H. Fuchida, M. Tsuneizumi, and M. Nagase, *Jpn Koka: Tokkyo Koho* **A2** (1990).
168. K. Fukushima and I. Yamada, Appl. Surf. Sci. **43**, 32 (1989); also see *J. Appl. Phys.* **65**, 619 (1989).
169. H. Ito, Y. Minowa, T. Ina, K. Yamanishi, and S. Yasunaga, *Rev. Sci. Instrum.* **61**, 604 (1990).
170. I. Yamada, H. Takaoka, H. Usui, and T. Takagi, *J. Vac. Sci. Technol.* A **4**, 722 (1986).
171. S. Huq, V. K. Raman, R. A. McMahon, and H. Ahmed, *in* "Proceedings, AIP Conference, 1988, Vol. 167, p. 291.
172. S. Yasunaga, K. Yamanishi, H. Tsukazaki, and Y. Kawagoe, *Ion. Plasma Assisted Tech. Int. Conf.* **6**, 13 (1987).
173. T. Ina, Y. Minowa, N. Koshirakawa, and K. Yamanishi, *Nucl. Instrum. Methods Phys. Res. Sect.* B **B37–B38**, 779 (1989).
174. H. Ito, Y. Minowa, and T. Ina, *Nucl. Instrum. Methods Phys. Res. Sect.* B **B37–B38**, 814 (1989).
175. K. Yamano, T. Sota, T. Koyanagi, K. Nakamura, and K. Matsubara, *Nippon Oyo Jiki Gakkaishi* **13**, 175 (1989) CA 112 (10):86962f; see also *Jpn. J. Appl. Phys.* **28**, L669 (1989); *J. Appl. Phys.* **65**, 1381 (1989).
176. I. Yamada, *Mater. Res. Soc. Symp. Proc.* **152**, 183 (1989).
177. K. Yamanishi, Y. Kawagoe, S. Yasunaga, K. Imada, and O. Wada, *Oyo Butsuri* **58**, 1227 (1989) (CA 111(26):245797d.
178. C. Kaito, Y. Saito, and K. Fujita, *Jpn. J. Appl. Phys.* **26**, L1973 (1987).
179. S. Yatsuya, T. Kamakura, K. Yamauchi, and K. Mihama, *Jpn. J. Appl. Phys.* **25**, L42 (1986).
180. H. Takaoka, I. Yamada, and T. Takagi, *J. Vac. Sci. Technol.* A **36**, 2665 (1985).
181. T. Takagi, K. Matsubara, N. Kondo, K. Fujii, and H. Takaoka, *in* "Proceedings, 11th Solid State Devices, 1979/1980," p. 507.
182. K. Fukushima, I. Yamada, and T. Takagi, *J. Appl. Phys.* **58**, 4146 (1985).
183. T. Takagi, K. Matsubara, H. Takaoka, and I. Yamada, *Thin Solid Films* **63**, 41 (1979).
184. S. Yalsuya, K. Yamauchi, T. Kamakura, A. Yanagida, H. Wakayama, and K. Mihama, *Surf. Sci.* **156**, 1011 (1985).
185. S. Iwama and K. Hayakawa, *Surf. Sci.* **156**, 85 (1985).
186. C. Kaufmann, *Wiss. Z. Tech. Hochsch* **26**, 513 (1984).
187. P. Nath, U. S. Patent 4514437, 1985.
188. R. Okada, Y. Ibaraki, T. Araya, and S. Hioki, *Gakka Ronbunshu,* **3**, 68 (1985) CA 102(26) 224349h.
189. S. Yatsuya, A. Yanagida, K. Yamauchi, and K. Mihama, *J. Cryst. Growth* **70**, 536 (1984).
190. X. Meng, G. Cui, J. Li, W. Guo, S. Wang, X. Wu, T. Zhang, and Y. Wei, *Diwen Wuli* **6**, 23 (1984) CA 101(8):64340x.
191. K. Hayakawa and S. Iwama, *Shinku* **25**, 320 (1982) CA97(10):77010p.
192. S. Iwama and C. Asada, *Seimitsu Kikai* **48**, 248 (1982) CA97(10):76954n.
193. Y. Ando and R. Uyeda, *J. Cryst. Growth* **52**, 178 (1981).
194. D. H. Boone, S. Shen, and R. McKoon, *Thin Solid Films,* **64**, 299 (1979).

5

Late Transition Metals (Groups 8–10)

I. Late Transition-Metal Atoms (Fe, Co, Ni, Ru, Rh, Pd, Os, Ir, Pt)

A. Occurrence and Techniques

The occurrence of these atoms in nature is only possible in the high-temperature atmosphere of stars and the sun.[1] In the laboratory, every technique known has been applied to produce vapors of these important metals, and these studies were summarized earlier.[1] Preferred vaporization techniques, vapor compositions, temperatures, etc., were also summarized earlier.

More recent advances in techniques applicable for these metals are described in Chapter 2 herein. In addition, a more advanced rotary metal atom–vapor reactor has been described recently.[2]

A new technique particularly applicable for these metals has been described by Mitchell and Hackett.[3] Several of the metal carbonyls of these metals are volatile, for example, $Ni(CO)_4$ and $Fe(CO)_5$. Thus, gaseous $Fe(CO)_5$ has been completely dissociated in the gas phase by multiphoton absorption of visible light (1–2 mJ/pulse at 552 nm from an excimer-pumped dye laser). Iron atoms can be generated in this way and their reactivity was studied near room temperature in the gas phase. Ground- and excited-state atoms were detected by resonance fluorescence excitation using a second dye laser. An interesting result was that quenching gases such as N_2O, C_2H_4O, and O_2 do not remove a significant portion of excited-state atoms by chemical reaction processes. Thus, chemical reaction rates are minor or insignificant for all *quenching* molecules investigated.

B. Physical Properties and Theoretical Studies

In Chapter 4, a summary of the optical spectral shifts that occur when transition-metal atoms are guests in frozen rare gas hosts was given.[4] It was learned that most metal atoms interact with rare-gas atoms simply by dispersion forces (van der Waals), but that in a few instances more significant "bonding" occurred. For the late transition metals Rh was found to interact more strongly than by just van der Waals forces. Nonlinear correlations between matrix-induced frequency shifts and polarizabilities of rare-gas matrices were observed.[4,5]

Palladium atoms in low-temperature matrices have come under particularly intense study. Absorption and emission spectra and decay kinetics of the $^3D_3$ metastable state of Pd in argon have been reported by Ozin and Garcia-Prieto.[6] The matrices Ar, N_2, and N_2/Ar were investigated, and it was determined that the optical absorption bands of $Pd(N_2)_x$ ($x = 1$, 2, 3) were also in the Pd atom absorption region. Broadband and laser-induced fluorescence spectra of Pd atoms in neat Ar revealed the coexistence of Pd ($4d^{10}$, $^1S_0$) and Pd ($4d^9$, $^3D_3$) atoms; the stabilization of the latter state appears to be the result of interaction of Pd atoms in a special trapping site with the matrix cage. In addition, two electronic states of $Pd(N_2)$ were identified, namely $^1\Sigma^+$ and $^3\Sigma^+$.

As an aid to optical spectra, magnetic circular dichroism (MCD) has been helpful for lending understanding of matrix-isolated atoms. For Fe atoms in Ar, Kr, and Xe, absorption and MCD spectra have been recorded as a function of temperature (12–30 K) in the visible and UV regions.[7] Experimental results and theoretical calculations have supported the concept that spin–orbit coupling is dominant (as in the gas-phase free atom). Thus, the influence of the matrix on the Fe atom is minimal (no vibronic coupling). However, two matrix sites exist in Ar but only one in Kr (and Xe). That is, neither matrix field effects on matrix-induced Jahn–Teller effects were important in this system. These studies also aided the assignment of various optical transitions for Fe atoms. MCD in conjunction with absorption spectra and magnetization studies of Fe atoms in frozen Xe have also aided the assignment of electronic transitions.[8] Actually, magnetization measurements indicate considerable Fe–Xe interaction, but little evidence for such interaction was observed in the spectra. The magnetization data suggest a $^5D_4$ ground state with large axial crystal field splitting.

A closely related spectral technique, magnetic linear dichroism (MLD), has been brought to bear as well.[11] MLD is the difference in absorbance of light polarized parallel from that polarized perpendicular to a magnetic field whose direction is perpendicular to the propagation direction of the incident radiation.[11] Theoretical results for MLD and MCD suggest that for $\Delta J = 0$ positive signs for both MLD and MCD bands are expected, while for $\Delta J = 1$ bands for both MLD and MCD should be negative. But for $\Delta J = -1$, MLD should be negative and MCD positive. Thus, using MLD and MCD together on the same sample, different allowed excited state J values can be determined.

Using these combined techniques, essentially all absorption bands have been assigned for Fe and Ni atoms.[9,10] For Ni, most bands in the 270- to 360-nm region originate from atoms in sites whose ground state is $^3D_3$ ($3d^94s^1$), whereas bands in the 230- to 240-nm region arise from atoms whose matrix ground state is $^3F_4(3d^84s^2)$. The $^3F_4$ ground-state atoms disappeared on annealing of the cold matrix. Thus, the gas-phase ground-state ($^3F_4$) is not the favored state in certain frozen matrices, although it can be trapped in certain (unstable) matrix sites. Both states of Ni are observable and depend on the matrix material. (The $^3D_3$ state lies 205 cm^{-1} in energy above the $^3F_4$ ground state). Both states can be observed in Ar, Kr, and Xe. However, Ne matrices seem to mimic the gas phase more closely and only the $^3F_4$ state was observed.[11]

The electronic ground state of matrix-isolated Ni atoms has also been discussed by Barrett et al.[12] Isolated in krypton, the ground state is primarily $^3D_3$ but the $^3F_4$ state can coexist. In SF_6 the ground state is $^3F_4$. Thus, it is clear that matrix interaction and matrix trapping sites can affect ground states.

A long-lived excited state of matrix-isolated Pd atoms has been detected by MCD.[13] This transient species shows a very strong MCD signal but no detectable absorption spectrum. The most likely ground state for the transient is $^3F_5$ ($4d^85s^2$).

Ground-state, free atoms have been discussed in terms of a new electronegativity concept. The average one-electron energy of the valence-shell electrons of such atoms could be termed their electronegativity,[14] and it is argued that electronegativity is the third dimension of the periodic table. A thorough discussion of average one-electron energies and other definitions of electronegativity is given by Allen.[14]

C. Chemistry

1. Oxidative Addition

a. Hydrogen (H_2) and Alkanes (RH) (Thermal and Photochemical)   Palladium atoms generated by laser evaporation react with H_2 and D_2. A variety of products were detected by ESR, such as PdH, PdD, PdH_2^+, and

PdD_2^+ (all ESR active). It is likely that ESR-inactive species PdH_2 and PdD_2 were also present.[15] The reactions of Pd and H_2 (D_2) probably took place in the gas phase since the laser evaporation was carried out in the presence of gaseous H_2 and products trapped in low-temperature matrices. The ESR results prove that the H atoms are equivalent in PdH_2^+ (probably a $^2A_1$ ground state).

Laser evaporation of Pd in the presence of gaseous CH_4 led to the generation of $PdCH_3$ radicals that have been trapped in low-temperature matrices.[16] Isotope labeling and ESR analysis allowed experimental description of the electronic structure. Comparison of PdH with $PdCH_3$ indicates that a significant amount of s/d hybridization on Pd is apparent for $PdCH_3$, which agrees with earlier calculations on the bonding in $Pd(CH_3)_2$.

A systematic study of *all* the transition metal atoms codeposited with CH_4/Ar mixtures about 10 K was recently reported.[17] These results are discussed in Chapter 4. To summarize findings pertinent to the late transition metals: a $d^n s^1$ configuration appears to be important, and the oxidative addition may proceed by H-atom abstraction followed by coupling:

$$M + CH_4 \rightarrow [M\text{--}H\text{---}\cdot CH_3] \rightarrow CH_3MH$$

$$d^n s^1 \quad M = Ru, Rh, Os.$$

Ruthenium, osmium, and rhodium atoms were all categorized as highly reactive with CH_4 (along with Nb, Ta, Mo, Re, Cu, and Ag). Other important features were that the D[M–H] value needs to be reasonably high (50–60 kcal/mol) to favor reaction, again suggesting an H-atom abstraction process. And finally, the heat of vaporization correlated very roughly with reactivity, which suggests that kinetic energy of the atoms does play a role in the reaction mechanism.

Theoretical approaches to understanding the M + $CH_4 \rightarrow CH_3MH$ reaction have been reported.[18] Partial agreement with the experimental results[17] was reported in that the Rh atom showed no activation energy barrier. The electronic configuration responsible for this low barrier is due to mixing between the $4d^9$ and $4d^8 5s^1$ states, which are both low lying. Interestingly, their calculations[18] indicate that Nb should have a high barrier, and this is not in accord with the experimental results.[17]

In the presence of *electronically excited* transition-metal atoms (in frozen matrices), oxidative addition of H–H and C–H bonds is common. Billups *et al.* reported that Mn, Co, Cu, Zn, Ag, and Au all reacted under photolysis to yield CH_3MH.[19] However, Ca, Ti, Cr, and Ni failed to do so under the same conditions. Ground-state atom-CH_4 adducts are believed to be important. The exact geometric structure of these adducts may determine if photochemical reaction occurs (see Chapter 4 for further discussion).

Ozin and co-workers have also investigated photochemical reactions.[20,21] A startling finding was that the reactions could be reversed using different wavelength light:

$$Fe + CH_4 \underset{h\nu\ 420\ nm}{\overset{h\nu\ 300\ nm}{\rightleftharpoons}} CH_3FeH$$

The iron atom transition to an excited state allowed oxidative addition to occur, and the $4p^1$ configuration may be important in determining this reactivity.[22,23] Also, it is interesting that the reverse reaction using 420-nm light is extremely efficient, and the two-way reaction is reversible over several cycles. It was surmised that the 420-nm photoexcitation of nonlinear CH_3FeH populates a low-lying electronic state having antibonding character that leads to a facile and concerted reductive elimination of CH_4:

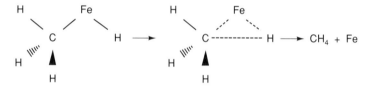

A discussion of the factors affecting metal atom reactivity argued in favor of a $M^*(4p) \rightarrow CH_4(\sigma^*)$ interaction for the first-row transition-metal atoms. However, the $M^*(3d/4s) - CH_4(lt_2)$ energy separation must also be considered. In passing from Sc to Zn, the $3d$ and $4s$ orbitals are generally stabilized and become closer in energy to the lt_2 level of CH_4. Their interaction is thereby enhanced, yielding greater electron density donation from CH_4 to M^*.

Nickel atoms seemed to be a special case with regard to photochemical activity.[24a] However, closer study revealed that photoexcited Ni atoms did react with CH_4, but the IR absorption bands for CH_3NiH were very weak. The reverse reaction was also observed with wavelength ≥ 400 nm. With these results it can be concluded that all the metal first-row atoms from Mn to Zn when photoexcited react with CH_4 in low-temperature matrices to form CH_3MH. In the case of the early transition metals reaction may occur, but the CH_3MH species may be unstable and automatically revert back to starting materials (this may be due to photolytic processes as well). Also σ-complexes may be important intermediates in these processes, as recent investigations with Co/CH_4 have suggested.[24b]

$$Co \text{---} H - C - H \quad (\text{with } H \text{ above and } H \text{ below on } C)$$

In support of this idea, Carroll and Weisshaar have shown that gas-phase, room-temperature Pd atoms form σ-complexes with alkanes.[24c]

Ethane, propane, and cyclopropane also oxidatively add to photoexcited Fe atoms. The normal alkanes reacted by C–H oxidative addition, and a series of CH_3FeH, C_2H_5FeH, and C_3H_7FeH species was produced.[25] Interestingly, the yield of RFeH with ethane and propane was less sensitive to alkane concentration in the matrix, suggesting that they react somewhat more efficiently than CH_4. The ν_{Fe-H} bands were assigned as CH_3FeH (1684 cm^{-1}), C_2H_5FeH (1665 cm^{-1}), and C_3H_7FeH (1659 cm^{-1}) in argon.

When cyclopropane was employed, no ν_{Fe-H} band was produced, new bands at 535 and 556 cm^{-1} appeared, and the C–C bond apparently was attacked instead of the C–H bond.

$$Fe^* + \underset{CH_2}{\overset{CH_2}{\diagdown}}\bigg|_{CH_2} \longrightarrow Fe \underset{CH_2}{\overset{CH_2}{\diagup\diagdown}} CH_2$$

Interestingly, this reaction only occurred in dilute matrices. In pure cyclopropane no reaction took place, and this implies that cyclopropane is an extremely efficient heat sink, thus allowing rapid Fe* → Fe conversion and loss of energy into the matrix. Indeed, this was confirmed by an earlier study of the Ni/cyclopropane system.[17]

b. Hydrogen Halides   Parker et al. have reported on the matrix reactions (nonphotochemical) of Fe atoms with hydrogen halides.[26a] In Ar and Kr the ν_{Fe-H} band was observed at 1735–1755 cm^{-1} when the halogens were Cl, Br, and I, and Mössbauer spectra indicated that the iron species was in an Fe^{2+} oxidation state in XFeH. However, in the case of HF, a Fe–HF complex was formed and upon warming FeF_2 and FeF_3 were formed. Thus, oxidative addition to HF is much less efficient than with other HX molecules.

c. Unsaturated Hydrocarbons and Ethers   Blomberg et al. have used theoretical methods to probe ethene-M atom interactions.[26b] The largest binding energies were found for the later transition elements, which have an optimized mixing between purely covalent bonding and donation–back-donation bonding. Purely covalent bonding generally requires promotion to an excited state of the atom, and this type of bonding yields metallacyclopropanes.

Spontaneous as well as photochemical reactions of Fe and Fe_2 in the presence of ethylene and ethylene oxide have taken place to yield C–H and C–O oxidative addition products.[27,28] FT–IR spectra showed the presence of C_2H_3FeH and C_2H_3FeOH from these reactions. In the case of ethylene oxide a spontaneous C–O insertion took place to form the first unligated metallaoxetane. Upon photolysis of this product with visible wavelengths, decomposition to an ethene complex of FeO occurred. Further photolysis in the UV range caused C–H oxidative insertion:

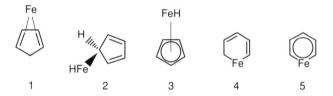

Dienes such as cyclopentadiene are particularly susceptible to attack by metal atoms. Photochemical activation is not usually necessary. For example, Fe atoms and cyclopentadiene yield Cp–Fe–H species in low temperature matrices.[29] In such a reaction several possible intermediate products enroute to ferrocene can be envisioned:

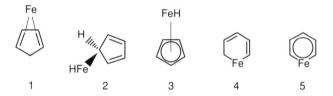

Careful IR studies of Fe/CpH/argon matrices suggested that the most likely structure is **3**. This conclusion was based on IR bands that resembled a ferrocene-like ring, coupled with the fact that ν_{Fe-H} bands were detected at 1749, 1745, and 1731 cm^{-1}. Perdeuterocyclopentadiene yielded ν_{Fe-D} bands at 1264, 1262, and 1257 cm^{-1} in accord with expected isotope shifts.

Iron atoms with benzene and cyclohexadienes in argon matrices yielded Fe–benzene complexes and photolytic dehydrogenation of cyclohexadiene.[30] With benzene alone, IR absorptions suggested the presence of $Fe(C_6H_6)$, $Fe(C_6H_6)_2$, and possibly $Fe_2(C_6H_6)$. With 1,4-cyclohexadiene simple adducts were formed, but on photolysis with UV light these adducts rearranged to FeH_2 and benzene.

Fe + [benzene] $\xrightarrow[12K]{\circ \text{ Argon}}$ [benzene]–Fe $\xrightarrow{\text{UV}}$ [benzene]–Fe + H$_2$

Alkyl-substituted arenes are also likely to undergo oxidative addition to metal atoms. For example, interesting new chemistry of Os atoms with mesitylene and other arenes has been reported by Green and co-workers.[31] The CH_3 group was attacked and new bridging η^6-arene complexes formed:

Os + [mesitylene] $\longrightarrow$ [complex structure]

A second minor product was the dehydro-analog possessing an Os≡Os triple bond. Benzene also undergoes oxidative addition; first a (η^6-benzene)(η^4-benzene)Os(O) complex was formed, and upon addition of PMe$_3$ (η^6-benzene)Os(H)(C$_6$H$_5$)(PMe$_3$) was produced. Further studies of the bis-(benzene) adduct demonstrated the presence of an intramolecular ring exchange.[32] Also, the reaction of Os atoms with a mixture of benzene and 2-methylpropane showed that activation of a saturated alkane was possible, and a remarkable product that required three C–H activation processes was formed:

Os + C$_6$H$_6$ + CH$_3$CHCH$_3$ (with CH$_3$) $\longrightarrow$ [complex structure]

These results again demonstrate that no organic "solvent" is really inert to metal atoms or clusters forming from metal atoms, even at low temperatures.

Metal atom synthesis of Mn, Ru, and Os complexes with mixed ligands also can occur incorporating an oxidative addition step.[33]

$$Ru + \bigcirc + PMe_3 \longrightarrow \underset{\underset{(Os)}{PMe_3}}{\overset{\bigcirc}{\underset{|}{Ru}}}\overset{}{\underset{}{C_6H_5}}$$

(Os)

The chemistry of these new materials has been elucidated, including studies of added arenes. Both the η^6-benzene and the σ-C_6H_5 ligand were exchanged, for example, when $RuH(\eta^6$-$C_6H_6)(PMe_3)(\sigma$-$C_6H_5)$ was heated with C_6D_6, and the perdeuterocomplex $RuD(\eta^6$-$C_6D_6)(PMe_3)(\sigma$-$C_6D_5)$ was formed, demonstrating the surprising lability of C–H and η^6-arene bonds in these materials.

d. Organohalides   Although an active area in the past,[1] oxidative addition of organohalides to the late transition metals has not received as much attention in recent years. However, a few interesting reports have appeared, for example, reactions of Re, Ru, Rh, Ir, and Pt with oxalyl chloride have yielded carbonyl-halide complexes in good yields.[34]

$$Rh + Cl\overset{\overset{O}{\|}}{-}C\overset{\overset{O}{\|}}{-}C-Cl \xrightarrow[77K]{cocond.} \xrightarrow{CO} Rh(CO)_2(\mu-Cl)_2$$

Similar experiments gave high yields of $Ru(PMe_3)_3(CO)Cl_2$, $[Ru(CO)_3 Cl(\mu$-$Cl)]_2$, $Ru(PPh_3)_3(CO)_2Cl_2$, $Rh(PPh_3)_2COCl$, $Ir(PPh_3)(CO)_2Cl_3$, and cis-$Pt(CO)_2Cl_2$.

The reaction of perfluoroalkyl and -aryl halides with metal atoms has continued to be of considerable synthetic utility.[1,35,36]

$$2\,Ni + 2C_6F_5Br \xrightarrow[77K]{cocond.} 2C_6F_5NiBr \xrightarrow[toluene]{warm} \bigcirc\!\!\!-Ni(C_6F_5)_2 + NiBr_2$$

$$Pd + CF_3(CF_2)_6CF_2Br \xrightarrow[77K]{cocond.} CF_3(CF_2)_6CF_2PdBr \xrightarrow{CH_3CN} CF_3(CF_2)_6CF_2PdBr(CH_3CN)_2$$

The R_fMBr intermediates are surprisingly stable even though coordinatively unsaturated. Upon warming in the presence of appropriate ligands, disproportionation occurred in the case of C_6F_5NiBr and C_6F_5CoBr, and toluene was taken up as an η^6-arene ligand. Then studies led to the synthesis of a variety of new π-arene complexes that have exhibited a rich organometallic and homogeneous catalytic chemistry.[37–43]

Platinum atoms also react with organohalides and several examples are shown:[44]

$$2\ Pt + 2C_6F_5Br \xrightarrow[77K]{codep} 2\ C_6F_5PtBr \xrightarrow{acetone} (C_6F_5)_2Pt(acetone)_2 + PtBr_2$$

$$\downarrow PEt_3$$

$$(C_6F_5)_2Pt(PEt_3)_2$$
cis and trans

$$Pt + CF_3\overset{O}{\underset{\|}{C}} - Cl \longrightarrow CF_3\overset{O}{\underset{\|}{C}} - Pt - Cl \xrightarrow{warm} CF_3PtCl$$

$$\underset{CO}{|}$$

$$\xrightarrow{PEt_3} CF_3\overset{O}{\underset{\|}{C}}Pt(PEt_3)_2Cl$$

$$CF_3 - \underset{\underset{CO}{|}}{Pt} - Cl + CF_3\overset{O}{\underset{\|}{C}} - Pt - Cl \longrightarrow CF_3\overset{O}{\underset{\|}{C}} - \underset{\underset{CO}{|}}{Pt}\overset{Cl}{\underset{Cl}{\diagup\diagdown}}\underset{CO}{Pt} - \overset{O}{\underset{\|}{C}}CF_3$$

$$Pt + C_6H_5CH_2Cl \longrightarrow C_6H_5CH_2PtCl \xrightarrow{warm} \text{bibenzyl and polymers}$$

$$\xrightarrow{P(OEt)_3} C_6H_5CH_2Pt(Cl)[P(OEt)_3]_2$$

In spite of the many relatively stable species prepared, quite a few examples of unstable RMX species have been encountered, and in some cases only weakly interacting RX–M adducts were formed at 77 K with the late transition metals. On warming often these RX–M adducts fall apart allowing metal atom aggregation.[45]

The (η^6-toluene)Ni(C_6F_5)$_2$ complex discussed earlier has also yielded a series of new diene complexes that have been structurally characterized.[46]

Comparisons of bond parameters suggest that the Ni–C_6F_5 bond length is quite sensitive to small changes in the degree of π-back bonding of the dienes and arenes.

e. Silicon–Silicon, Si–H, and Si–Cl   The Si–Si bond is very susceptible to attack by Ni atoms, and oxidative addition to form the Si–Ni–Si moiety takes place. This reaction was useful for the preparation of a new series of (η^6-arene)Ni(SiX$_3$)$_2$ complexes.[47]

$$X = F, Cl$$

The reactions of Cl_3SiH and $SiCl_4$ yielded the same materials by complex routes and in lower yields.

The discovery of these compounds led to their eventual preparation by more conventional approaches, including extensions to additional σ-bonding ligands.[48,49]

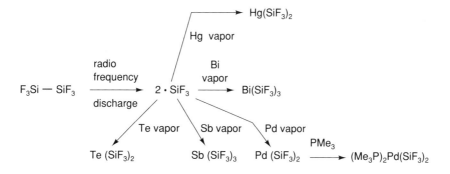

Prior homolytic dissociation of the Si–Si bond by use of a high-energy discharge, followed by reaction with metal atoms, also yields Si–M–Si complexes.[50] Although prior formation of SiF_3 radicals is not necessary with metals such as Ni (the atoms react spontaneously with the Si–Si bond), this procedure allows extension to less reactive atoms such as Te, Bi, Sb, Hg, Cd, Zn, and Pd.

2. Ligand Displacements by Metal Atoms

One of the most interesting developments in metal atom chemistry in recent years has been reaction with diazomethane to yield $M{=}CH_2$ species or N_2MCH_2.[51] Iron atoms cocondensed with CH_2N_2/Ar yielded both $Fe{=}CH_2$ and $N_2Fe{=}CH_2$. Photolysis of the matrix caused the unexpected reductive elimination of Fe from $N_2Fe{=}CH_2$ at wavelengths ≥ 500 nm. Shorter wavelengths caused $Fe{=}CH_2$ to be converted to HFeCH. It was also possible to add H_2 or H_2O to the $Fe{=}CH_2$ species.

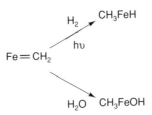

Nickel atoms behaved similarly to iron.[52] Both $Ni{=}CH_2$ and $N_2Ni{=}CH_2$ were formed spontaneously upon cocondensation at 10 K. Photolysis of the matrix ($\lambda \approx 400\text{–}500$ nm) enhanced the yield of $N_2Ni{=}CH_2$, suggesting that photoinduced diffusion of reagents may be an important step. Hydrogen reacted with $Ni{=}CH_2$ to yield CH_3NiH, which could also be formed by photolysis of $CH_4{-}Ni$ matrices[53] (see earlier discussion).

Although this is very interesting chemistry of metal atoms, perhaps the most important contribution is the spectroscopic data derived. Table 5-1 summarizes some of the more interesting IR data for $M{=}CH_2$, $N_2M{=}CH_2$, and CH_3MH species.[54] From these data it is evident that the nature of bonding is significantly different for Fe vs Ni. For example the high ν_{Ni-C} frequency indicates that there is significant double bond character. Also note the high frequency for the $\nu_{N{\equiv}N}$ for Ni vs Fe in $N_2M{=}CH_2$. It appears that Ni binds more strongly to CH_2 but more weakly with N_2 as compared to Fe. Similar trends are found when comparing CH_3FeH and CH_3NiH with ν_{Ni-C} and ν_{Ni-H} both higher in energy.

3. Addition to Boranes and Carboranes

Beginning in 1980, Sneddon and co-workers reported numerous metal atom reactions with boranes and carboranes.[55–64] These metal atom reactions have yielded numerous new metallaboron clusters and metallacarborane complexes. Often complex mixtures and low individual product yields result. However, the discovery of so many new metal atom reac-

TABLE 5-1

IR Data for Matrix Isolated Metal Carbene and Metal Alkyls:
$M = CH_2$, $N_2M = CH_2$, and CH_3MH (in Argon, cm^{-1}), where
$M = Fe$, Ni[52-54]

Species	ν_{M-C}	ν_{M-H}	$\nu_{N \equiv N}$
CH_3FeH	523.5	1683, 1675	—
CH_3NiH	554.9	1945	—
$Fe = CH_2$	623.6	—	—
$Ni = CH_2$	696.2	—	—
$N_2Fe = CH_2$	618.8	—	1812
$N_2Ni = CH_2$	688.9	—	2180

tions and new compounds has made this work important and interesting. These studies have shown that metal atoms can affect the direct conversion of boron hydrides to metallaboron clusters and boron hydrides and alkynes into metallacarborane clusters. These processes can be carried out at low temperature and often the metal atoms can serve as sites for cage construction (see Fig. 5-1).

It was also discovered that η^6-arene–metal complexes can be formed as part of the structure:[59]

$$Fe + C_6H_3(CH_3)_3 + B_{10}H_{10} \rightarrow 1\text{-}[\eta^6\text{-}C_6H_3(CH_3)_3]FeB_{10}H_{10}$$
mesitylene

Presumably hydrogen evolved is taken up by other reactive species.

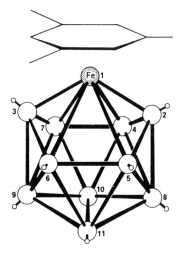

FIGURE 5-1   Metal atom reaction product with a borane and arene.[59]

Another interesting aspect is the finding that toluene-solvated Fe atoms react with carboranes:[57]

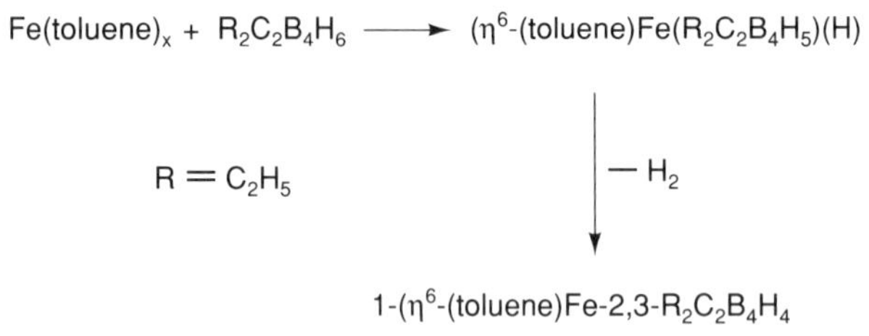

$$Fe(toluene)_x + R_2C_2B_4H_6 \longrightarrow (\eta^6\text{-(toluene)}Fe(R_2C_2B_4H_5)(H)$$

$$R = C_2H_5 \qquad\qquad \Big| -H_2$$

$$1\text{-}(\eta^6\text{-(toluene)}Fe\text{-}2,3\text{-}R_2C_2B_4H_4$$

This can be viewed as an initial oxidative addition of B–H at the metal, followed by H_2 elimination.

Additions of secondary ligands such as cyclopentadiene, pentamethylcyclopentadiene, or alkynes has allowed access to a wide variety of capped carborane structures and thiaborane clusters. Some representative examples are listed below (the list is not exhaustive):

closo-7-I-2,3-$(\eta\text{-}C_5H_5)_2Co_2$-1-SB$_9H_8$
nido-8-$(\eta\text{-}C_5(CH_3)_5)Co$-7-SB$_9H_{11}$
closo-2-$[\eta^6\text{-}C_6H_5CH_3]$-2,1-FeSB$_{10}H_{10}$
iso-8-$(\eta\text{-}C_5H_5)CoB_{17}H_{21}$
2-$[\eta^6\text{-}C_6H_3(CH_3)_3]Fe$-1,6-$C_2B_7H_9$
nido-$(\eta\text{-}C_5H_5)_2Co_2S_2B_2H_2$
1-$(\eta\text{-}C_5H_5)Co$-4,5-$((CH_3)_3Si)C_2B_6H_6$
2-$[\eta^6\text{-}C_6H_5CH_3]Fe$-6,7,9,10-$(CH_3)_4C_4B_5H_5$
6,8,7,9-$(\eta\text{-}C_5H_5)_2Co_2S_2B_5H_5$
6-$(\eta\text{-}C_5H_5)CoB_9H_{13}$

4. Simple Orbital Mixing

Metal atom complexation to donor–acceptor ligands continues to be studied extensively. Since the Group 8–10 transition-metal atoms are rich in d electrons, which can be donated to the ligand, π-acid ligands generally yield the most stable complexes. However, essentially all molecules processing π- or nonbonding electrons form discrete complexes at low temperatures with transition-metal atoms.

a. Alkenes Matrix isolation of M-atom–ethene complexes has been reported earlier.[1] Studies of M-alkene complexes of C_2F_4, $CF_3CF{=}CFCF_3$, norbornene, and others have also been reported.[1] In more resent years the reactivity of coordinatively unsaturated Co-alkene complexes has been evaluated.[65] The product of Co atom and ethene cocondensation was designated as $(C_2H_4)_mCo_n$, and this material exhibited high reactivity with acetaldehyde, formaldehyde, and CO/H_2 mixtures, and proposed reaction schemes are shown below in order to rationalize organic products obtained. Deuterium labeling experiments helped substantiate these stepwise reactions, and demonstrated the high reactivity of the Co–ethene complex. However, reaction with added substrates was only successful if their melting points were low enough to bring about mobility in the Co–C_2H_4 matrix before the $(C_2H_4)_mCo_n$ complex decomposed. Also, it should be pointed out that these reactions are not very selective and a number of products were formed simultaneously.

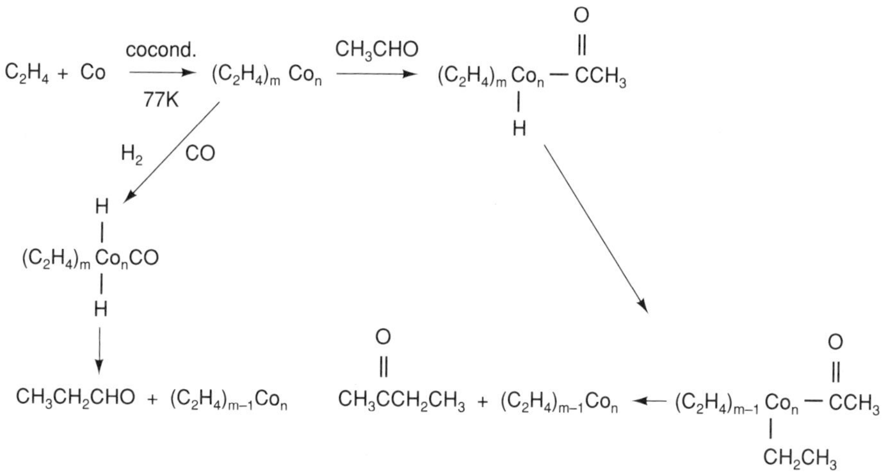

$$C_2H_4 + Co \xrightarrow[77K]{cocond.} (C_2H_4)_m Co_n \xrightarrow{CH_3CHO} (C_2H_4)_m Co_n \overset{\displaystyle O}{\underset{\displaystyle H}{\overset{\|}{-}} CCH_3}$$

$$H_2 \diagdown CO$$

$$(C_2H_4)_m \overset{H}{\underset{H}{|}} Co_n CO$$

$$CH_3CH_2CHO + (C_2H_4)_{m-1}Co_n \qquad CH_3\overset{O}{\overset{\|}{C}}CH_2CH_3 + (C_2H_4)_{m-1}Co_n \longleftarrow (C_2H_4)_{m-1} \underset{CH_2CH_3}{\overset{O}{\overset{\|}{Co_n}}} - CCH_3$$

Nickel atoms have been codeposited with several substituted ethene molecules.[66] The desire to synthesize new organometallics capable of decomposing to yield nickel films was the driving force for this work. Several homoleptic nickel-alkene complexes of Ni(alkene)₃ stoichiometry were formed by cocondensation at 77 K, but not isolated. Reaction of these complexes with PF_3 yielded a series of new $(PF_3)_3Ni$-alkene compounds, and these materials were found to decompose between -16 and $+6°C$ to yield $Ni(PF_3)_4$, alkene, and Ni metal.

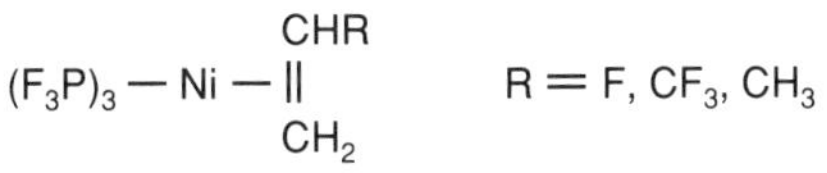

$$(F_3P)_3 - Ni - \overset{CHR}{\underset{CH_2}{\|}} \qquad R = F, CF_3, CH_3$$

Ethene itself undergoes C–H insertion, hydrogen exchange, and dimerization upon cocondensation with Fe atoms at 77 K and subsequent warming.[67] It is possible that very labile ethene–Fe_2 species were involved:

$$(C_2H_4)_2Fe_2 \; \rightleftharpoons \;
\begin{array}{c}
CH_2 = CH \quad H \\
| \qquad\quad | \\
Fe - Fe \\
| \\
C_2H_4
\end{array}
\; \rightleftharpoons \;
CH_2 = CH
\begin{array}{c}
\\
| \\
Fe - Fe \\
| \\
CH_2CH_3
\end{array}$$

Iron atom–alkene reactions in cold matrices[68] have been analyzed by IR and Mössbauer spectroscopies. Only at low Fe concentration and <18 K was $Fe(C_2H_4)$ observed. Dimers and higher clusters of $(C_2H_4)_2Fe_2$ and $(C_2H_4)_mFe_n$ were very readily formed. Mössbauer data indicated that the electronic configuration of the Fe in the alkene complexes is approximately $4s^1 3d^7$. These data are also consistent with the idea that C_2H_4 (and C_3H_6) are primarily σ-donors with only minimal $d_\pi \rightarrow \pi^*$ back bonding.

An interesting study of gas-phase Fe, Co, Ni, and Cu atoms reacting with linear alkanes and alkenes has been reported.[69] A hollow cathode discharge was used to produce metal atoms in a fast-flow reactor with He buffer gas. From the linear decay of M atom concentration (laser-induced fluorescence detection) vs hydrocarbon concentration, effective bimolecular rate constants were obtainable at 300 K and 0.8-Torr helium. The atoms $Fe(d^6s^2, {}^5D)$, $Co(d^7s^2, {}^4F)$, $Ni(d^8s^2, {}^3F)$, $Ni(d^9s, {}^3D)$, and $Cu(d^{10}s, {}^2S)$ were allowed to interact with propane, n-butane, ethene, propene, and 1-butene. The results were unexpected and striking. For one thing, these room-temperature, neutral atoms were much less reactive than corresponding cations. No reactions at all were observed with the alkanes. Also, Fe, Co, and Cu atoms did not react with alkenes, although a slow reaction of Co with 1-butene ($k = 9 \times 10^{-14}$ cm^3/sec) was observed. In contrast, Ni did react with ethene ($k = 5.0 \times 10^{-13}$ cm^3/sec), and moderately rapidly with other alkenes. It was estimated that Ni reacted one time in 500 collisions with C_2H_4, 1 of 25 with C_3H_6, and 1 of 2 with 1-butene, and these imply activation energies not larger than 3.5, 1.9, and 0.4 kcal/mol, respectively. It was proposed that the initial adduct of $Ni–C_2H_4$ is more strongly bound than with the other metals, and this allows entry into other reaction channels. Indeed, matrix IR work supports this stronger binding adduct formation as do theoretical calculations.

Chemical reactivity of gas-phase Fe atoms near room temperature with respect to abstraction or adduct formation has been studied in Ar buffer gas.[70] In this study the Fe atoms were produced by multiphoton dissociation of $Fe(CO)_5$ or ferrocene. The rate of consumption of Fe atoms was followed by resonance fluorescence excitation at variable time delay following the initial photolysis pulse. Upper limits for second-order rate constants were estimated for O_2, CO, H_2O, $(CH_3)_2CO$, C_2H_4, C_2H_2, N_2O, C_2H_4O, and CF_3Cl and fell within the range of $2–10 \times 10^{-15}$ cm³/molecule/sec, and these correspond to less than one reaction per 10^5 hard sphere collisions. Adducts with C_6H_6, 1,3-butadiene, NH_3, and NO formed with rates of $2–60 \times 10^{-14}$ cm³/molecule/sec, and binding energies for these adducts were less than 7 kcal/mol. These results again demonstrate the remarkably low reactivity of Fe atoms under such conditions and are in stark contrast to low-temperature matrix results. Apparently these gas-phase results are due to repulsive ground-state interactions and low probabilities for transitions from repulsive to attractive potential surfaces.

b. Dienes Derivatives of 1,3-butadiene, when cocondensed with nickel atoms at 10 K, exhibit optical spectra that are rather insensitive to substituents on the diene, but sensitive to metal nuclearity. IR and optical data support an η^2-bonding scheme rather than the expected η^4-scheme for the 1 : 1 adduct.[71]

A number of other interesting reactions with dienes have been observed. Several are summarized in Scheme 5-1. 1,3-Butadiene is an excellent ligand for Fe and Ru if one additional two-electron ligand is added. Another interesting point is the facility with which hydrogen transfers occur. This is quite striking with the arene/cyclopentadiene experiments,[73,77] where cyclopentadiene donates one hydrogen to a coordinated arene to produce a hexadienyl radical coordinated to Fe. Such hydrogen migrations are common in metal atom–diene chemistry.[1]

c. Alkynes The chemistry of alkynes with metal atoms is really quite remarkable. Acetylene itself complexes with Fe and Ni atoms in frozen argon.[79,80] Codeposition at very low temperatures (15 K) yields an $Fe(C_2H_2)$ complex when concentrations of Fe and acetylene are kept low (in Ar). This is believed to be a H-bonded complex rather than a

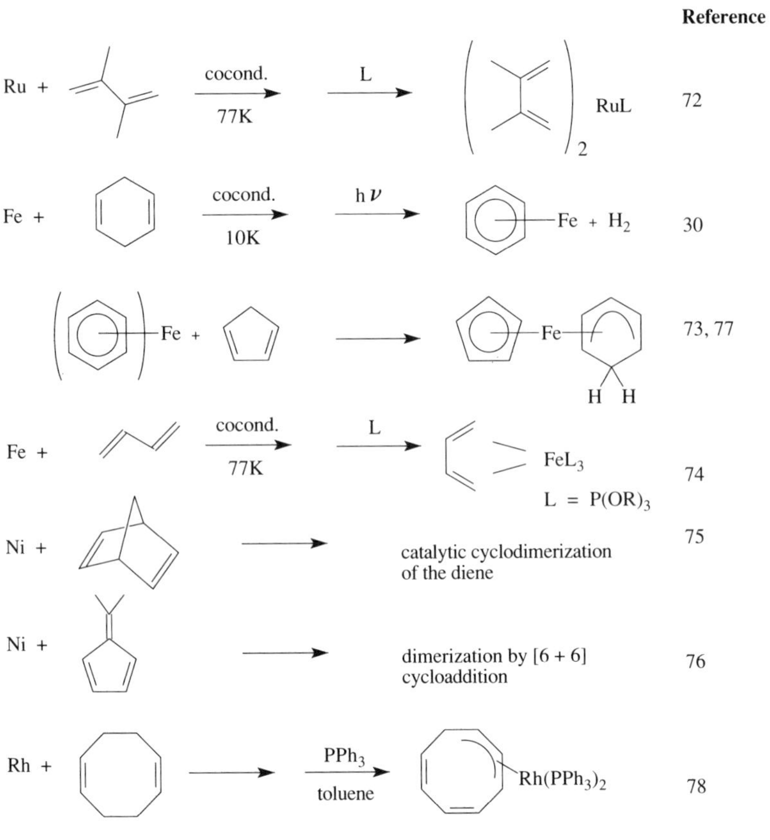

SCHEME 5-1   Diene reactions with metal atoms.

π-complex, since no perturbed $\nu_{C=C}$ band was observed in the IR, but a perturbed ν_{C-H} band at 3270 cm^{-1} was observed.

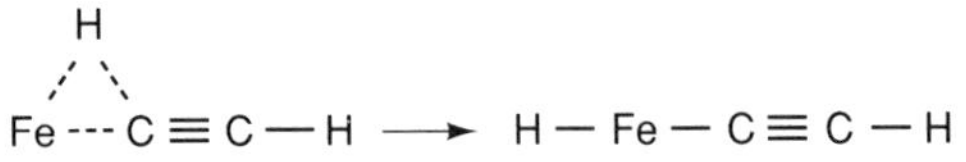

Upon photolysis in the near-UV oxidative addition occurred to yield HFeC$_2$H with ν_{C-H} 3276, $\nu_{C=C}$ 1975, and ν_{Fe-H} at 1764 cm^{-1}. In the case of Ni, a different initial adduct formed, a π-complex formed, and upon photolysis in the visible a vinylidene-Ni species was produced, a reaction that can be reversed upon UV photolysis:

$$\underset{\substack{| \\ H}}{\overset{\substack{H \\ |}}{Ni-\underset{C}{\overset{C}{\underset{|||}{C}}}}} \quad \underset{h\upsilon,\ UV}{\overset{h\upsilon,\ visible}{\underset{\longleftarrow}{\longrightarrow}}} \quad Ni-C=CH_2$$

The differences between Fe and Ni are striking, and these results demonstrate the extreme sensitivity of light-driven rearrangement reactions in the presence of metal atoms.

A most unusual synthesis of substituted ferrocenes can be carried out by cocondensing Fe atoms with alkynes.[81]

$$Fe + RC\equiv CR \xrightarrow[77\ K]{cocond.} \xrightarrow{warm}$$

$$(C_5R_5)_2Fe + \text{trimers and tetramers of the alkyne}$$

This remarkable transformation requires the cleavage of at least one $C\equiv C$ bond. Although the mechanism is unknown, it may be that a cyclobutadiene complex and a carbyne complex react with each other to form the five-membered rings necessary.

Zenneck and co-workers have made use of the high reactivity of arene solvated metal atoms with alkenes and alkynes.[82,83]

$$Fe + toluene \longrightarrow Fe(tol)_x \xrightarrow{C_2H_4} \text{[arene]}Fe(C_2H_4)_2$$

$$R = SiMe_3 \quad RC\equiv CR$$

First the toluene solvate was prepared by cocondensation (a very useful intermediate in many organometallic and catalysis preparations).[1] One of the arene ligands was then displaced by ethene, and this labile system was allowed to react with a sterically hindered, electron-rich alkyne, forming a series of novel, alkyne complexes. A wide variety of (η^6-arene)Fe(diene) were also prepared similarly.

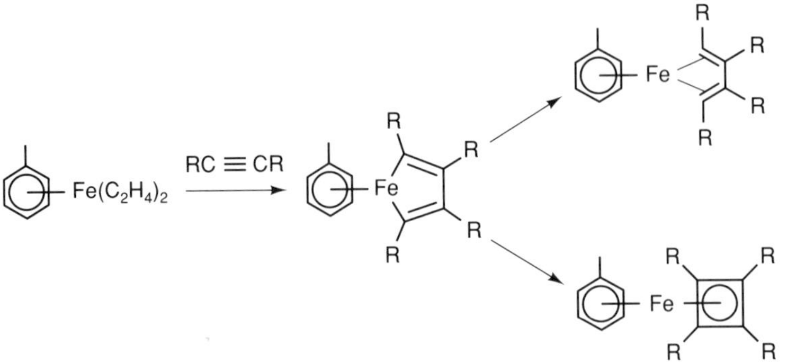

Zenneck and co-workers have explored the chemistry of arene/alkene-solvated atoms in depth. Arenes, boron heterocycles, alkynes, phospha-cetylenes, with solvated Fe, Co, and Ni have led to a rich new field of organometallic synthetic chemistry.[82]

The codepositon of cyclopentadiene and alkynes with Co atoms has also led to a series of new alkyne complexes.[84] For example, Co atoms with CpH and $CF_3C{\equiv}CCF_3$ yielded several new compounds often structurally unique, although the mechanism of formation is not well understood. More than 10 compounds were formed in this reaction, of which 9 were structurally characterized. Examples are shown below.

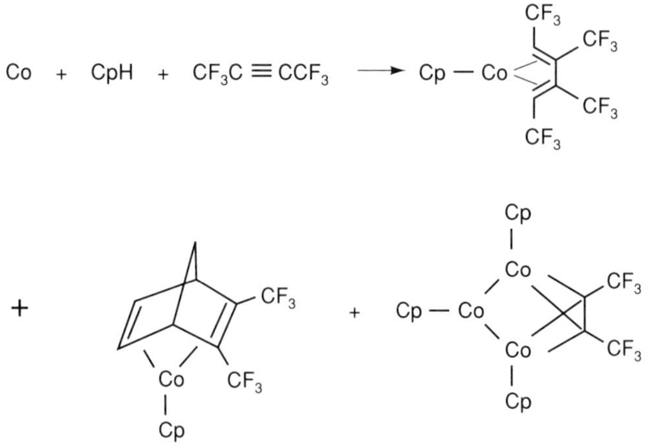

A reaction intermediate thought to be important is CpCoH and possibly CpCo(H)Co(H)Cp.[84]

The proposal of the intermediacy of a Co–Co-bonded species was the prelude to an especially exciting discovery by Schneider and co-workers.[85,86a] The reaction of Co atoms with pentamethylcyclopentadiene (Cp*) led to a triple-decker compound and a $Cp_2^*Co_2H_2$ complex (at first thought to be a compound possessing a Co=Co double bond).[86]

$$Co \; + \; C_5HMe_5 \xrightarrow[-120°C]{\text{methycyclohexane}}$$

20%   0-5%   5%

These are rather unique structures and have been examined by x-ray diffraction. For $Cp_2^*Co_2H_2$ a structural determination showed a linear arrangement of two Cp*–Co units. The Co–Cp*-ring distance is 1.689(1) Å compared with 1.724(1) Å in the triple-decker compound, showing a stronger interaction. The Co–Co length is 2.253(1) Å.

The formation of these unique compounds can be explained by the intermediary of Cp*CoH units. As Cp*H is depleted and sufficient Co atoms are deposited, Cp*CoH may dimerize to form the $Cp_2^*Co_2H_2$ or react with the Cp*Co (Cp*H) complex to form the triple decker with hydrogen release.

d. Arenes   Codeposition of the later transition-metal atoms with arenes yields weakly bound arene–M complexes. Stoichiometry can vary; 1:1 complexes in dilute matrices can form, but 2:1 is the most likely stoichiometry when excess arene is present. The best description based on the chemistry observed is a 2:1 complex with one very labile π-arene, perhaps bound as η^2- or η^4-arene:[1]

Arene-complexed metal atoms have been termed "solvated metal atoms" and these materials have proven to be one of the most useful and versatile organometallic reagents of these metals in their zero-valent state.[87]

Examples of the use of solvated metal atoms have already been given in this chapter, and as we deal with arene–metal atom reactions, many more become apparent.

Zennick and Frank report a beautiful example of the use and reactivity of toluene-solvated Fe atoms.[82,83,88] In the reaction with ethene, one toluene ligand was easily replaced, thus yielding another useful reagent, the chemistry of which with alkynes is described in the previous section.

The benzene–Fe matrix has been investigated by Mössbauer spectroscopy at both 7 and 77 K.[89] On the basis of a benzene concentration study, it was concluded that $Fe(C_6H_6)_2$ complexes and $Fe(C_6H_6)$ were formed. The species with two C_6H_6 ligands existed as 20-electron bis(η^6-C_6H_6)Fe and (η^6-C_6H_6)Fe(η^4-C_6H_6). In addition, UV–VIS and IR spectra of arene–Fe matrices have been reported.[90] In this case naphthalene and 1-methylnaphthalene with Fe in the 30–290 K temperature range exhibited UV–VIS spectra that were assigned to two products, (η^6-naphthalene)$_2$Fe and (η^4-naphthalene)$_2$Fe. Assignments were based on a series of model arenes and cyclohexadiene as a model for η^4-binding.

Further reactions of solvated metal atoms with organometallic reagents, both microscale and macroscale, have been reported.[91] Electron transfer from the solvated metal atom to dimeric organometallics was a dominant reaction pathway in the 140–180 K temperature range. In this way a dimer was reduced to monomeric anions. According to low-temperature IR and UV–VIS studies, monomeric anions and bimetallic anions were formed, depending on reaction conditions; the reactions seemed to be quite sensitive to solvent.

$$(CO)_5Mn-Mn(CO)_5 + Fe(THF)_x \rightarrow 2[Mn(CO)_5]^- + Fe^{2+}(THF)_x$$

$$(CO)_5Mn-Mn(CO)_5 + Fe(toluene)_x \xrightarrow{\text{THF}} \longrightarrow$$

$$[(toluene)_yFe_2Mn(CO)_5]^- \xrightarrow{\text{CO}} [Fe_2Mn(CO)_{12}]^-$$

Many of the unique reactions of toluene solvated Fe, Co, and Ni atoms are illustrated in Scheme 5-2.[30,77,82,83,88,92–94]

Toluene solvated Rh atoms are also unique systems, and there is evidence that hydridorhodium species are formed.[95] Codeposition of Rh atoms with toluene at 77 K followed by warming formed a red-brown solution stable to $-50°C$, and NMR studies indicate the presence of a Rh–H species. Upon treatment with P(OME)$_3$ the toluene ligands are displaced and a HRh[P(OMe)$_3$]$_4$ complex was formed. It was proposed

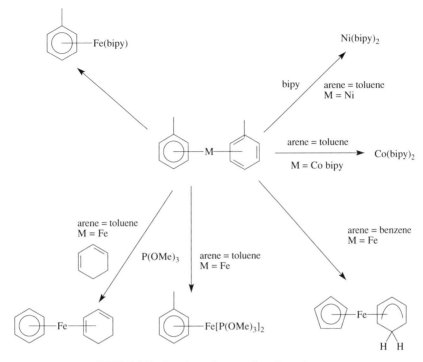

SCHEME 5-2   Reactions of arene-solvated metal atoms.

that oxidative additions of $-CH_3$ to Rh was an intermediate step followed by homolysis of the Rh-alkyl bond. $C_6H_5CH_3 + Rh \rightarrow [C_6H_5CH_2RhH] \rightarrow$ (η^6-toluene)RhH with loss of benzyl species.

e. Phosphines, Phosphites and Amines   Earlier work with metal atoms and phosphines led to a long list of new phosphine complexes, often with additional ligands.[1] Some examples are $Ni(PF_3)_3PH_3$, $C_6F_5PdBr(PEt_3)_2$, and (η^6-toluene)$Fe(PF_3)_2$. More recent work has added a few new systems, for example, (η^6-toluene)$Fe[P(OMe)_3]_2$.[94] Also, the stabilization of PF_2H by complexation to Ni atoms was reported in detail.[96]

As mentioned earlier, toluene-solvated Co and Ni atoms react with bipyridine (bipy) by complete displacement. However, toluene-solvated Fe atoms allow only one bipy ligand to add.[97] The results demonstrate that the weakly bound arene ligand on Co and Ni are more readily displaced, and this indicates that the toluene–iron complex is more strongly bound, which is also indicated by its somewhat better thermal stability (about $-30°C$, vs -40 and $-60°C$ for Fe, Co, Ni). The bonding in this (arene)Fe(bipy) complex is quite unique. An unusually short Fe–N bond distance of 1.902(1) Å was observed, and distances within the bipy ligand deviate from free bipy. These structural aspects and supporting IR data provide convincing evidence that a substantial amount of electron density is transferred from the Fe atom to the π^*-orbitals of bipy. This, in turn, causes a tighter binding with the toluene ligand.

In a similar way, diphos reacted with toluene-solvated Fe atoms to displace one arene ligand, thus yielding a very stable (η^6-toluene) Fe(diphos) complex.[97]

The reactions of Ni atoms with NH_3 in argon and krypton matrices were carried out as a follow-up to the work with CH_4/Ni and H_2O/Ni. In the NH_3 case initial Ni–NH_3 and Ni(NH_3)$_2$ complexes were formed, and upon photolysis oxidative addition took place to yield $HNiNH_2$ and $HNiNH_2(NH_3)$. The H–Ni–NH_2 species exhibited IR bands (in Ar) at 619.2 for NH_2, 676.5 for $\nu_{Ni—N}$, 1918 for $\nu_{Ni—H}$, and 3368 for $\nu_{N—H}$.[98] Similar results were reported for Fe–NH_3 matrices.[99]

f. Carbon Monoxide   Extensive earlier work reported essentially on all metal atom elements codeposited with CO.[1] A few additional studies have been carried out in recent years, for example NO and CO with Pd and Pt atoms.[100]

Rhodium atoms deposited with CO/hydrocarbon matrices in a rotating cryostat yield $Rh(CO)_4$, $Rh_2(CO)_8$, and $Rh_4(CO)_{12}$, all normally unstable species. However, they could be detected by ESR and FTIR methods.[101] There was no evidence of $Rh(CO)_{1-3}$ species. The ESR parameters for $Rh(CO)_4$ are consistent with a $^2B_2$ ground state in D_{2d} symmetry with the unpaired electron on Rh occupying the $d_{x^2-y^2}$ orbital. It exists in a slightly distorted tetrahedral geometry.

g. S_2N_2 Reactions   Nickel atoms codeposited with S_2N_2 allowed the synthesis of a very unusual material $Ni(S_2N_2H)_2$.[102] In this case the hydrogen was provided by methanol upon workup:

$$Ni + S_2N_2 \longrightarrow [Ni(S_2N_2)_x] \xrightarrow{\text{MeOH}} Ni(S_2N_2H)_2.$$

5. Catalytic Processes with Metal Atoms

The catalysis of isoprene hydrosilylation is possible with certain metal atoms,[103] especially Ni, Co, or Fe with Ni as the most active catalyst, giving quantitative yields below 0°C. A possible mechanistic scheme is shown below:

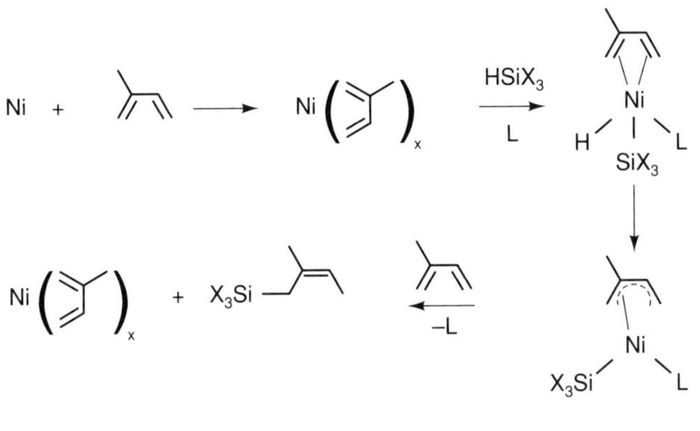

6. Metal Atoms and Surfaces

Condensation of Ir, Re, W, and Pd atoms onto an atomically smooth surface of Ir(111) has been studied to clarify if or how rather large amounts of condensation energy are transferred to the lattice.[104] Would condensing atoms persist in an excited state during which they hop around the surface slowly losing kinetic energy, or is energy transfer so efficient that thermalization is accomplished after a few oscillations close to the original point of impact? Based on studies of diffusion of condensing atoms, it was concluded that rapid localization at the first point of impact is fairly common for strongly bound metal atoms, even on smooth surfaces. It still is unclear as to how to visualize this rapid energy transfer, and this appears to be a rich area for further experimental and theoretical studies.

7. Metal Atoms with Organometallics

The reaction of Co and Ni atoms with $Fe(CO)_5$, $CpCo(CO)_2$ in the presence of mesitylene or pentamethycyclopendiene (Cp^*H) results in the formation of new trinuclear compounds; examples are shown below.[105]

$$Co + C_6H_3Me_3 + Fe(CO)_5 \xrightarrow[\text{77 K}]{\text{cocond.}} (C_6H_3Me_3)_2\ FeCo_2(CO)_5$$

$$Ni + Cp^*H + Fe(CO)_5 \xrightarrow[-120°C]{\text{methylcyclohexane}}$$

$$(Cp^*)_2\ FeNi_2(CO)_5 + (Cp^*)_2Ni_2(CO)_2$$

In each case the metal atoms follow the 18-VE rule. It appeared that reactive fragments such as Co–mesitylene and Ni–Cp play crucial roles in the buildup of the clusters.

II. Late Transition-Metal Clusters

A. Occurrence and Techniques

Free metal clusters of the Group 8 metals have not been observed anywhere in nature, such as the cooler regions of stars, due to their relative fragility. However, in the past 12 years many extremely useful methods have been developed for their production and study. Probably the most exciting has been the laser evaporation–cluster beam method that was described in Chapter 2. Tremendous advances have now been made due to this method, as described in this chapter. A recent report on analysis of the cluster beams by photoionization is of value.[106] Quantitative analysis of the neutral products of free jet expansions was possible based on the principle of measuring the yield of an ion characteristic of each component cluster at a photon energy just below that at which production of the same ion from larger clusters (photofragmentation) can be detected. So far this method has only been applied to cluster beams of ammonia–chlorobenzene, but it should in principle be applicable to metal clusters as well.

Of course the well-established gas-evaporation method (GEM), where metals are evaporated from crucibles in moderate pressures of inert gases, has also been of great use for the Group 8–10 metals. This method is also described in Chapter 2. Indeed, one of the earliest reports on this procedure was made by Kimoto et al.[107] Ultrafine metal particles were produced by evaporation of the elements in argon gas at relatively low pressures, and they were studied by electron microscopy.

An additional method that should be mentioned is the laser multiphoton dissociation (and ionization) of polynuclear metal carbonyls in supersonic beams.[108] For example, pulsed dye-laser irradiation of supersonically cooled $Mn_2(CO)_{10}$ and $Fe_3(CO)_{12}$ in the 3800- to 5000-Å wavelength range causes collision-free multiphoton dissociation/ionization. The ionic fragments M^+, M_2^+, and M_3^+ were dominant. The lack of carbonyl-containing fragment ions is reasonable since the fragmentation energy for such species (1 eV) is much lower than the ionization energy (8 eV).

Another widely used method of preparing nanoscale metallic and bimetallic particles is by metal atom aggregation in low-temperature matrices.[1] This approach is very convenient and entails cocondensation of metal atom vapors with excess solvent/diluent at 4–77 K. Synthetic scale depositions are generally carried out at 77 K. Upon warming, atom migration takes place and clustering begins. Cluster growth competes with interaction of the growing cluster with the host matrix material. Although some control of final particle size and crystallinity can be achieved by choice of matrix material (xenon, alkanes, aromatics, ethers) and warming procedure, uncontrollable growth is often a problem.[1,109–111] This procedure is often referred to as the SMAD method for producing small metal particles and catalysts (solvated metal atom dispersion; see Chapter 2).[87] It has been an especially useful tool for producing new heterogeneous catalysts (SMAD catalysts)[112,113] and bimetallic particles and these studies are discussed later in this chapter (see Fig. 5-2).

B. Physical Properties and Theoretical Studies

1. Ionization Energies

Measurements on the energies needed to ionize Fe_2, Fe_3 ..., Fe_{25} have been reported.[114] Actually the IEs were bracketed by use of various ionizing lasers. It was found that the IE (Fe atom, 7.870 eV) decreased to the work function of the metal (4.4 eV), but the trend downward with

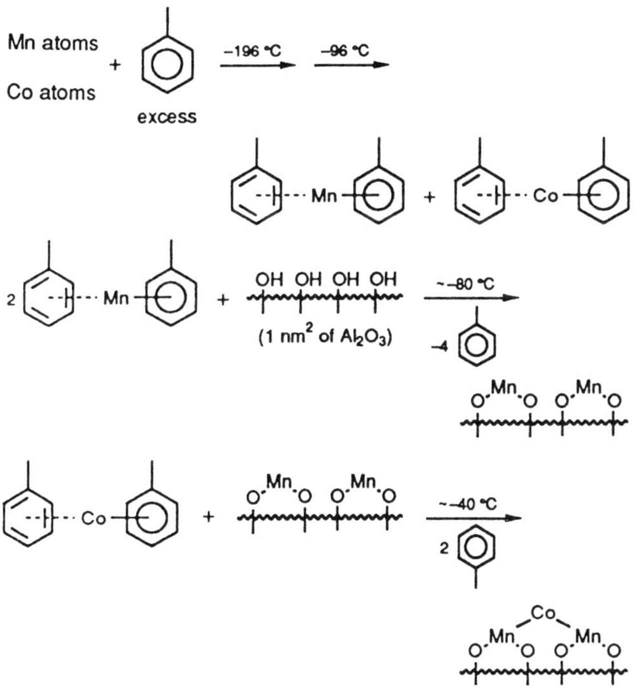

FIGURE 5-2 Growth of Co–Mn particles by the SMAD method.[111b]

increase in cluster size was by no means linear. Thus, Fe_2 was about 5.9 eV and Fe_3 and Fe_4 about 6.4 eV. However, clusters Fe_{9-12} fell below 5.6 eV while Fe_{13-18} were above 5.6 eV. The ionization energy of the Fe_{25} cluster still exceeded the work function by 0.3 eV.

Nickel clusters of Ni_{3-90} have also been analyzed.[115,116] Near-threshold photoionization measurements showed that the ionization energies of $(Ni)_n$ did not monotonically approach the work function as n increases, but exhibited an oscillation as a function of cluster size, and such behavior is still not easily explained. Indeed many transition-metal clusters display this behavior: dramatic and nonmontonic variations with cluster size up to about 10–20 atoms followed by a relatively steady decrease in IE for larger clusters. The relatively poor agreement with theoretical predictions (based on the conducting spherical drop model) suggest that the electronic and structural properties of Ni clusters in the Ni_3–Ni_{30} size range have not yet evolved to the corresponding properties of bulk *fcc* metal. Likewise, Fe, Co, and Nb clusters do not obey the conducting spherical drop model.

A useful technique for lowering IE values and as an aid in structural elucidation is to allow NH_3 to adsorb on the gas-phase clusters.[117,118] The interesting feature is that the adsorption of NH_3 on cluster surfaces lowers IEs and this IE lowering is *linear* in the number of adsorbed NH_3 molecules. Extrapolation of this dependence to zero coverage allows estimation of the IE of the bare cluster. In using these results, comparison of the bare cluster IEs with predictions of the simple spherical drop model suggests that for transition-metal clusters the Fermi level can change significantly with cluster size.

2. Structure and Electronic States

As mentioned before, adsorption of NH_3 on gas-phase metal clusters can lend information about cluster structure.[119] By determining the number of binding sites for a simple adsorbate such as NH_3, and variations in these values with cluster size, systematics observed can help elucidate cluster structure. In addition, specific adsorption sites can be probed with weaker adsorbate molecules. For Ni clusters, the number of NH_3 molecules adsorbed (as determined by photoionization/MS) show pronounced minima in the 50- to 116-atom size range for many clusters that are particularly stable as bare clusters (show "magic number" phenomenon and are thus present in larger amounts than expected). It is assumed that these magic numbers arise from closing of shells and subshells of the MacKay icosahedra, so the NH_3 adsorption data support the MacKay icosahedra structures. However, for room-temperature clusters in the size vicinity of a third shell closing (~147 atoms), the NH_3 results *do not* suggest a spherical nearly closed shell structure. In addition, for clusters of greater than 50 atoms NH_3 prolonged adsorption can change cluster structure such that more NH_3 can be adsorbed (shell opening?).

Another important point mentioned by Riley and co-workers is that the tendency to form icosahedral closed-packed structures with Ni_n does not carry over for Fe_n.[119] Thus, we see again evidence that the d-electron-rich metal atom clusters tend to easily form close-packed structures, but d-electron-poor metal atom clusters are more controlled by kinetic factors, and many unusual metastable structures are probably possible. Hydrogenation of Co and Ni clusters can also help elucidate structure[120] and has shown that Co_n and Ni_n readily form icosahedral species. In this study the clusters were exposed to H_2 and then NH_3 or H_2O. Adsorption amounts allowed further clarification of structure through correlations of predicted vs observed adsorption amounts (see Fig. 5-3).

The structures of Ni_n and Pd_n ($n = 4$–23) have also been discussed by

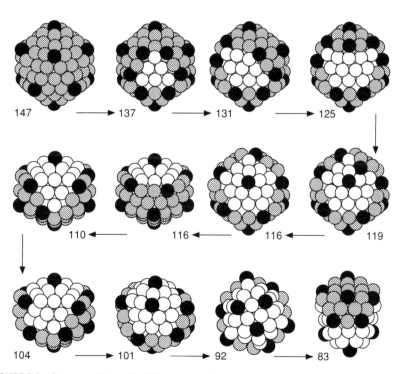

FIGURE 5-3 Drawings of closed (sub)shell clusters in the proposed growth sequence from $n = 83$–147. Black denotes primary NH_3 adsorption sites; gray, third shell atoms; and white, second shell atoms (after Riley and co-workers[120a,b]).

Stave and DePristo.[120c] Using a corrected effective medium theory, structural optimization was accomplished through simulated annealing. Two general conclusions were: (1) transition-metal clusters generally maximize the minimum coordination of any atom on the surface of the cluster, usually dictating a spherical shape; (2) in contrast, rare gas clusters maximize the number of interatomic distances close to the optimal distance for the pairwise interaction between rare gas atoms. In other words, the structure of transition-metal clusters cannot be characterized as similar to fragments of the bulk crystal structure or as similar to structures assumed for rare gas clusters.

Ferromagnetic resonance (FMR), a method very similar to ESR and using the same instrumentation, has been employed to deduce atomic structure, size, and shape of Ni particles in thin MgF_2 films.[121] The samples were prepared by sequential deposition of Ni vapor, then MgF_2 vapor (or AlF_3), with several cycles. The Ni clusters thus formed were protected by the MgF_2 matrix. Transmission high-energy electron diffraction (THEED) and FMR showed that the lattice parameters were equal to bulk Ni for these small particles. The size of the particles were in the 2–4 nm range and the shape believed to be plate-like structures lying parallel to the MgF_2 film.

For pure, ultrafine Co particles a new crystal structure has been reported.[122] A long-period stacking structure with the stacking order ABC/BCA/CAB was found in particles 20–500 nm in diameter. It is believed that transformation of *fcc* Co on cooling formed this structure. Heavily faulted/defective particles were observed in the electron microscope, and many appear to have this new form.

Although the goal of this section is to review the physical properties of nonligand-stabilized clusters, the work of Edwards, Johnson, Lewis and their co-workers[123–126] on the emergence of paramagnetism in osmium cluster compounds should be mentioned here, since it does relate to cluster structure. A series of osmium carbonyl cluster compounds has been analyzed for intrinsic paramagnetism at low temperature. For example, $H_2Os_{10}C(CO)_{24}$ exhibits such paramagnetism (according to high-sensitivity magnetic susceptibility measurements), below 70 K.[123] In contrast two low nuclearity clusters, $Os_3(CO)_{12}$ and $Os_6(CO)_{18}$, were diamagnetic over the whole temperature range. But the decanuclear cluster exhibited paramagnetism conforming almost exactly with the simple Curie law $X_m = C_m/T$, giving a nominal magnetic moment of 0.62 μ_B per cluster molecule.

The behavior of $H_2Os_{10}C(CO)_{24}$ is in contrast to bulk metal (where the electronic energy levels form a continuum). In particulate metals or cluster compounds, the energy differences between successive levels are no longer small compared with other important energies, such as thermal and electronic Zeeman energies. In terms of magnetic properties, $H_2Os_{10}C(CO_{24})$ actually does behave as a small metallic particle. ESR studies confirm this behavior and broad absorptions are observed at low temperature, but above 100 K the absorption disappears. This ESR behavior is analogous to conduction–electron spin resonance.

Extensions of this interesting work to other cluster compounds have shown that enhanced paramagnetism can sometimes be encountered when compared with localized and itinerant-electron counterparts. With regard to magnetic properties, cluster size does not have to be very large for this "metallic property" to be exhibited.[124] In the studies of 21 osmium cluster compounds between Os_n where $n = 3$ to 40, clusters of nuclearity less than 10 are diamagnetic showing temperature-independent magnetic susceptibilities and no ESR spectra. All those clusters where $n > 10$ are paramagnetic, show ESR spectra and temperature-dependent susceptibility following Curie law behavior. Indeed, this temperature-dependent paramagnetism is thought to be of widespread occurrence in metal cluster compounds.[125] Edwards and co-workers set out a good working molecular orbital scheme to try to explain this behavior.[125,126]

Several reports on the preparation, trapping, and spectral analysis of metal dimers and trimers have appeared. Some preliminary work on gas-phase cluster production, ionization, mass separation, condensation, and neutralization into cold rare gas matrices has been reported. Attempts to isolate Ni_3, Ag_3, and Ag_5 were carried out and absorption spectroscopy attempted, but cluster concentrations were so low that good spectra were not obtained.[127]

Iron atoms and clusters Fe_1–Fe_6 have been analyzed with regard to band structure and excited states[128,129] by using a theoretical approach of symmetry-adapted and symmetry-broken SCF calculations. In the SCF approximation, the atomic $3d$ ionization energy was considerably smaller (2.4 eV) than the experimental value. For Fe_4–Fe_6, near the Fermi level the d-electron ionized states that accompanied another d-electron excitation to the $4s$-like orbitals were found. These results suggest the existence of many low-lying d-electron-excited states, which could open channels for chemical reactions in a facile manner. Such ionized states were not found for Cu_n or Zn_n, and the effect in Ni_n was present but less striking compared with Fe_n.

The Fe_2 molecules was studied by ab initio SCF and CI calculations. A ground state of $^7\Delta_u$ was predicted and at this state, formally, Fe_2 should be bound by bonds approximated as $1\frac{1}{2}$ $3d$ and 1 $4s$. The $3d$ bonds were predominant and the $4s$ electrons were dispersed to the outer regions due to piling up of $3d$ electrons in the internal regions.

Potential energy surfaces of eight low-lying electronic states of Rh_3[130]

have also been probed by theoretical methods (complete active space multiconfigurational self-consistent field CASSCF). An isosceles triangle structure (D_{3h}) is most stable, and $^2A_1$ and $^2A_2$ states are probable candidates for the ground state of Rh_3. All electronic states are very densely packed. The Rh_3 species is predicted to be considerably more stable than Rh_2.

The stabilities of certain metal dimers can be analyzed by mass spectral methods, for example, Knudsen cell-generated Pt_2 and Pt–Y.[131] The Pt–Y bond is quite strong. For example, $D_0^0(Pt_2) = 358 \pm 15$ kJ/mol while Pt–Y = 470 ± 12 kJ/mol, $Pd_2 = 104 \pm 21$, and $Au_2 = 221 \pm 2$. A valence bond approach was used to rationalize bonding and bond strengths:

Pt=Pt	$6s$–$6s$ and $5d$–$5d$ bonding (Pt in a $5d^96s$ state)
Pd—Pd	A large amount of energy needed to promote the Pd atom to a $4d^96s$ state. Also, $4d$–$4d$ bond less effective.
Au—Au	Only a $6s$–$6s$ bond is possible (Au in a $5d^{10}6s$ state).
Pt=Y	Pt is d^9s and Y is d^2s giving d–d and s–s bonding.

3. Magnetic Properties

The magnetic properties of Fe, Co, and Ni ultrafine particles have been of interest for many years. Since such particles can be prepared in size ranges below the single domain size of 220 Å (in bulk α-Fe metal), the saturation magnetization M_s and coercivity H_c would be expected to vary considerably from bulk iron. However, complicating features of surface oxide effects on magnetism are still not well understood. Herein some of the recent studies are reviewed, but an exhaustive review is not appropriate at this time.

Magnetic properties of free Co clusters in a gaseous beam were studied as a function of cluster size Co_n ($n = 20$–200) and cluster temperature. The magnetic moments per atom *increase* with applied magnetic field and cluster size but decrease with increasing temperature consistent with superparamagnetic behavior. The true magnetic moments were 2.08 ± 0.20 μ_B per atom and exceed the bulk value. In addition magnetically resolved isomers were noted for $n = 55$–66.[132]

Magnetic properties of Fe clusters have been of interest and Stern–Gerlach deflections of cold Fe clusters ($n = 15$–650 atoms) have been measured. The clusters deflected uniquely in the direction of the increasing field, and this indicates that spin relaxation had taken place within the isolated clusters.[133,134] The average magnetic moments *increased* with increasing cluster temperature and with increasing strength of the external magnetic field. In all cases the magnetic moments were below the value of bulk α-Fe.

Iron particles prepared by gas evaporation in Ar (1–6 Torr) and passivated by exposure to low pressures of O_2 were isolated and studied by SQUID magnetometry and Mössbauer spectroscopy.[135,136] Particles with size below 10 nm showed a small coercivity (about 100 Oe at room temperature), which increased drastically upon cooling to 10 K (1000–1500 Oe). The larger particles of 10–20 nm had higher coercivities (approximately 1000 Oe), which increased by 100% at 10 K. Mössbauer studies showed that the Fe was surrounded by ultrafine Fe_3O_4 or γ-Fe_2O_3 grains. At room temperature, the oxide component gave a very broad absorption superimposed on the α-Fe spectrum. The enhanced coercivity at room temperature can be attributed to anisotropy induced by exchange interaction between the ferromagnetic α-Fe core and the ferri- or antiferromagnetic oxide layer.

Table 5-2 summarizes a series of additional reports on the magnetic properties of ultrafine particles of Fe, Co, Ni, FeCo, and iron/cobalt

TABLE 5-2
Magnetic Properties of Fe, Co, Ni Ultrafine Particles

Metal	Prep. method	Comments	Reference
Fe	$Fe(CO)_5$ decomposition in nitrogen plasma	α-Fe core with γ-Fe_2O_3 coating. No Fe_3O_4 detected. Magnetic ordering below 77 K.	138
Fe	Gas evaporation method	Saturation magnetization (M_s) decreases with particle size. The temperature dependence of M_s was irreversible, probably due to oxide coating causing a pinning of spins in the oxide layer.	139
Fe–Co Ni	Sputter deposited onto sputter etched polyimide substrates	Magnetic prop. were sensitive in Co rich alloys. Large particles of 250 nm showed negative anisotropy (K_u); as size decreased K_u became positive (see text for further discussion).	137
$(FeCo)_{80}B_{20}$ $Fe_{80}B_{20}$	Radiofrequency sputter deposition onto sputter etched polyimide substrate	Elongated finger-like projections of amorphous ferromagnetic material (diam <1000 Å). H_c and anisotropy are strongly dependent on Fe/Co ratio and on annealing conditions.	140

boride particles. These are examples (not exhaustive). A particularly interesting report dealt with Fe–Co and Ni particles deposited on etched polyamide substrates.[137] The magnetic properties were particularly sensitive to Co composition, and larger particles (250 nm) showed negative anisotropy (K_u). As size decreased K_u became positive. The coercive field H_c increased from a few hundred Oe to a maximum of over 1000 Oe. Below a particle diameter of 100 nm the M_s, K_u, and H_c fell, and XRD patterns sharpened, and this was due to stress relaxation, although other morphological changes may also be important.[137] All dropped rapidly toward zero, presumably due to the onset of paramagnetism. Heat treatment of the particles led to increases in H_c, K_u and M_s.

Some very interesting work on the effects of magnetic fields on the gas-evaporation growth process has been carried out. Varying external field strengths were used and Fe, Co, and Ni particles prepared. The ultrafine particles were formed and arranged themselves in zigzag and straight chains. The zigzag chain was rationalized such that the connection of particles occurred where each particle had a closed magnetic domain.[141] Conversely, the straight chain was explained by connection of particles with each particle having a single domain. These differences are due to coalescence growth in a particle smoke and saturation magnetization of the substance. A summary of several other preparative methods of ultrafine particles is given in Table 5-3, along with properties of product particles.

C. Chemistry

1. Gas-Phase Metal Clusters

Clusters produced by laser evaporation/cluster growth in cold inert gases can now be studied by chemical reaction methods. Under these clusters growth conditions, kinetic control of cluster growth may be possible depending on gas pressure and exact temperature. However, this gas-phase cluster growth process is far from well understood, and reactivity behavior is often difficult to rationalize. Riley and co-workers[162–165] observed cluster size dependence and annealing temperature dependence for the reactions of Fe_n with H_2. These authors were the first to report a very interesting phenomenon: raising the temperature of the cluster-helium stream sometimes caused a *lowering* in cluster reactivity. This finding would suggest that the initial cluster growth is kinetically controlled, perhaps leading to more reactive "defective" clusters. Upon an-

TABLE 5-3
Atom-Clustering Methods for Preparing Ultrafine Particles

Material	Method	Comments	Reference
Fe, Ag, Si	Arc-plasma sputtering	380 mg/min production of Fe particles.	142
Ni	Multisource electron beam	High rate of particle production.	143
Fe	Microwave plasma	Fe particles prepared by gas evaporation and then processing through a microwave plasma. The proportion of γ-Fe to α-Fe could be controlled.	144
Fe	Gas evaporation	Convection of inert gas makes growth zones for hexagonal plates and multiply twinned particles. Growth model proposed.	145
Fe	CO_2 laser irradiation of SF_6-$Fe(CO)_5$	γ-Fe particles, 76-Å spheres, fcc structure, paramagnetic at 4.2 K. Lasers for evaporation, induced reactions/decompositions are reviewed. The best method is IR laser photosensitized decomposition of $Fe(CO)_5$ to give high purity, small diameter, and narrow distribution of γ-Fe particles.	146 147 154
Fe, Co Fe–Co	Evaporation into oil with surfactant, Ar gas pressure	30- to 35-Å particles formed to produce a magnetic fluid. Ar gas pressure could be used to control particle sizes.	148
Fe–Ni Fe–Cu Fe–Si	Arc melting, vaporization, and clustering in H_2/Ar atmosphere at 0.1 MPa	Fe–Ni composition remained same as bulk alloy, Fe/Cu and Fe/Si ratios varied	149
Fe	Gas evaporation, trapping on TiO_2 with potassium promotor	Catalytic activities for CO reduction to light olefins were very high.	150
Ni	Arc plasma	135 g/hr produced at electrode substrate angle of 45° in Ar–H_2 atm.; arc current 400 A.	151
Ni	Gas evaporation, parti-	30-nm particles, used in	152

(*continued*)

TABLE 5-3 (*continued*)

Material	Method	Comments	Reference
Cu–Zn	cles heated in H_2 in a fluidized bed	selective hyd. of dienes to monoenes and enantio-selective hyd. of methyl acetoacetate to opt. active 3-hydroxyl butyrate.	
Ni Fe$_3$O$_4$ Cu–Zn	Gas evaporation	A review of the advantages of the gas evaporation method (purity, alloy compositions, size control). Chained particles produced and studied.	153
Co	Gas evaporation	10- to 200-nm particles were studied by TEM. A very regular twinning pattern consisted of parallel slices, and these were resolved into finer twins along different (111) planes. Martensitic transformation discussed.	155
Ni	Gas evaporation	Defect structure of ultrafine Ni particles (15–20 nm) studied by position annihi-lation method. Bulk defects were vacancies or vacancy clusters with occluded gases. Low-temperature annealing studied.	156
Fe, Se, C Si/B	Laser ablation/evapora-tion/compound decom-position	Use of laser in producing ultrafine powders is dis-cussed.	157
Ni	Gas evaporation	Superior catalytic properties for hydrogenation were carefully studied; compari-sons made with Raney nickel.	158
Ni, Cu	Isothermal annealing	Sintering of ultrafine parti-cles in H_2 at 300–500°C. Initial growth by factor of 10 for Ni. Liquid-like coalescence of solid-phase particles.	159
Fe	Rf plasma on Fe powder	Median size of 10 nm parti-cles was prepared. Growth	160

(*continued*)

TABLE 5-3 (*continued*)

Material	Method	Comments	Reference
Pd	Vacuum evaporation	took place in a fog state and a Brownian collision–coalescence mechanism governed the process. Ultrafine Pd with mean size of 2.0 nm were produced by vacuum evaporation into a running oil. The structure of the particles was *fcc* (same as bulk), but lattice parameter was larger by 1.3%.	161

nealing the cluster shape may change/collapse to a more thermodynamically stable structure which would probably be less reactive.

Furthermore, chemical reactivity of the Fe_n clusters seemed to change abruptly with cluster size. Reactivity with H_2 and binding energies of adsorbed NH_3 and H_2O seemed to follow similar patterns, pointing to a common origin; fundamental changes in the structure of the bare clusters. Overall, for Fe_n ($n = 2-165$), NH_3 chemisorption patterns gave evidence for metastable structures.

Free iron clusters have also been allowed to react with O_2, H_2O, and CH_4.[166] In this case reactivity with O_2 and H_2O was evident, but the clusters were inert toward CH_4.

$$Fe_n + O_2 \rightarrow Fe_n - O_2 \quad (n = 2-15)$$

$$Fe_n + H_2S \rightarrow Fe_nS + H_2$$

$$Fe_n + CH_4 \rightarrow \text{no reaction}$$

It was noted that Fe atoms did not react with any of these reagents under the conditions employed. As cluster size increased reactivities increased and at Fe_6 and larger, reactivities had leveled off. This result was tentatively explained by changes in Fe_n ionization thresholds (implying that $Fe_n \rightarrow Fe_n^+$ may be a rate determining process).

Dependence of metal cluster reaction kinetics on charge state has also been probed.[167] Chemisorption of H_2 by neutral and positively charged Fe clusters was studied under identical reactor conditions. Both Fe_n and Fe_n^+ showed nonmonotomic reaction rate dependences on n, which varied by 10^4 between $n = 1$ and 31. Reaction rates were affected by the positive charge on Fe_n^+; clusters $n = 4$–6 and >17 were more reactive as Fe_n^+, while for $n = 3$ and 9–14 Fe_n was more reactive than Fe_n^+. These results were rationalized in terms of cluster valence electronic structure. In addition, the geometric structure of the clusters in turn determines the number, energy, and spacial orientation of the valence orbitals which can interact effectively with H_2. Size selective reactivity patterns are dependent on electronic and geometric effects, which are intertwined and inseparable.

Theoretical studies[168] of H_2 and CH_4 with Co, Pd, and Cu clusters have helped explain why Co_3 and other Co_n clusters are reactive toward H_2 while Co, Co_2, Cu, Cu_2, Cu_3, and Cu_n are not under the same conditions.[169a] Using the ASED theoretical approach, it was found that the stability of the first formed dihydride ($M_n + H_2 \rightarrow M_nH_2$) may be a strong factor in determining reactivity. Purely kinetic effects seem unlikely for clusters since so many degrees of freedom are available. In the case of Pd_n the states of lowest repulsion of CH_4 were found to be d^0 and d^{10} (fewest s and p electrons).[169b]

Some elegant studies of gas-phase Ni clusters interacting with CO have been carried out.[170] Nickel cluster ions, size selected by quadrupole MS, were treated with CO to yield gas-phase $Ni_n(CO)_k^+$ and $Ni_nC(CO)_l$ species, where n ranged from 1 to 13. The value of k and l varied with cluster size n. The number of CO molecules attached correlated very well with electron counting rules of Lauher (who calculated most favorable molecular geometries of any given transition-metal cluster as well as the bonding capabilities of such clusters).[171] Table 5-4 shows the predicted and experimentally observed number of CO ligands that binded to the Ni clusters, based on geometries shown in Fig. 5-4.

It should be noted, however, that the Ni_9–Ni_{12} clusters do not bind as many CO molecules as would be predicted. It seems that for these larger clusters their behavior is more like a metal crystallite with a coating of

TABLE 5-4

Predicted and Observed Number of CO Ligands that Attached to Ni_n^+ Clusters

	Ni_2^+	Ni_3^+	Ni_4^+	Ni_5^+	Ni_6^+	Ni_7^+	Ni_8^+	Ni_9^+	Ni_{10}^+	Ni_{11}^+	Ni_{12}^+	Ni_{13}^+
Maximum number predicted	7	9	10	11	13	15	17	19	21	23	25	20
Maximum number observed	9	8	10	12	13	15	16	17	18	19	20	20

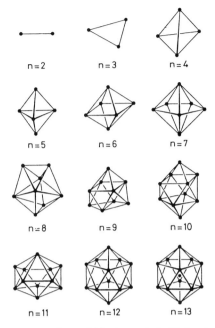

FIGURE 5-4 Shapes of polyhedral clusters Ni_n^+ where n is 2–13. Reprinted with permission from Fayet et al., J. Am. Chem. Soc. 109, 1737 (1987). Copyright 1987 American Chemical Society.

ligands, and their structures may not be perfectly closed. Once the Ni_{13}-sized cluster is reached, the first one to possess an interstitial Ni atom, a more perfect and predictable cluster is formed. Also, there is some evidence that clusters possibly derived from fragmentation processes lead to open clusters.

Mingos and Wales reported a reinterpretation of cluster growth processes, based in part on the Ni_n/CO gas-phase data.[172] Circumstantial evidence for a face-capping cluster growth sequence is presented, and this is based in terms of the *polyhedral skeletal electron pair theory*, which provides the most complete account of the stoichiometries. The concept presented is that a pentagonal bipyramidal Ni_7 cluster is a key structure in growth sequences (Fig. 5-5).

Further analysis of CO chemisorption on gas-phase metal clusters was possible by using a pulsed fast flow reactor to study clusters of V, Fe, Co, Ni, Cu, Nb, Mo, Ru, Pd, W, Ir, Pt, and Al for M_n ($n = 1$–14).[173a] The clusters were produced by pulsed laser evaporation and were allowed to enter a zone containing CO. The $M_n(CO)_x$ species formed were detected by photoionization time-of-flight MS. Most transition metals where

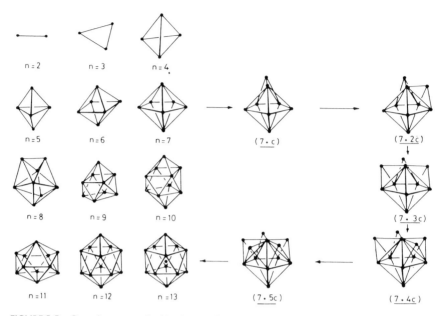

FIGURE 5-5   Growth sequence for Ni₇ clusters. On the left-hand side the deltahedral structures with *n* = 4–12 are illustrated together with the centered icosahedron with *n* = 13. On the right-hand side capped structures connecting the pentagonal bipyramidal and the centered icosahedron are illustrated. Reprinted with permission from Mingos and Wales, *J. Am. Chem. Soc.* **112**, 931 (1990). Copyright 1990 American Chemical Society.

$n = \geq 5$ readily chemisorbed CO, and reactivity toward larger clusters increased by a factor of 2 or 3. Usually M, M_2, M_3, and M_4 were relatively unreactive with CO, and this observation was explained in terms of a competition between unimolecular decomposition and collisional stabilization. Qualitative reactivity comparisons with CO were made and it was found that Fe, Mo, Cu, and Al clusters were least reactive toward CO. Theoretical methods have also been applied; chemisorption of CO on model clusters of Rh and Pd allowed a comparative analysis of site selection.[173b]

A series of interesting results with gas-phase Pd and Pt clusters have been reported as well.[174] Reactions with H_2, D_2, N_2, CH_4, CD_4, C_2H_4, and C_2H_6 have demonstrated again that strong reactivity variations exist depending on cluster size. Small Pd clusters could bind up to three deuterium atoms per Pd atom in the cluster. Strong discontinuities in D_2 uptake may be an indication of cluster structure differences. General reactivity trends were D_2, $H_2 > N_2 > C_2H_4 > CD_4$, CH_4, C_2H_6. Activated dissociative chemisorption was explained in terms of simple frontier orbital considerations involving the electronic structure of both the cluster and the adsorbate. Reactivities with H_2 or D_2 can be a sensitive probe of relative differences in cluster ionization energy.

Cox, Kaldor, Trevor and their co-workers have also studied Pt clusters intereacting with a series of alkanes and aromatics.[175-177] In the case of CH_4, activation was realized for the first time on an unsupported metal cluster, and the reaction had a distinct cluster size dependence, with Pt_2–Pt_5 most reactive and Pt_6–Pt_{24} less reactive. A mild correlation of reactivity with the availability of low-coordination metal atoms was found. Thus, low-coordination metal atoms activate CH_4 more readily than closely packed metal surface atoms. Indeed, this explanation correlates with earlier explanations of reactivity of growing clusters in low-temperature matrices[17] and is based on the defect model of reactivity (see later discussion).

The CH_4 reactions yielded $Pt_nC_{1,2}H_y$ species, but larger clusters retained more hydrogen. However, platinum clusters reacting with larger alkanes caused extensive dehydrogenation down to a C/H ratio of nearly one, indicating that aromatization of certain alkanes was facile and cluster size dependent. Mono- and diadducts of alkanes with Pt_2–Pt_4 were extensively dehydrogenated. For Pt_2–Pt_8 clusters reacting with n-hexane and 2,3-dimethylbutane, the degree of dehydrogenation increased with cluster size. In the case of cyclohexane, dehydrogenation to benzene occurred, and a diadduct was prevalent.

$$\left(\bigcirc \right)_2 \!\!-\!\! Pt_n + 6H_2$$

Benzene itself was dehydrogenated on Pt_3 and larger clusters. Qualitative comparisons of rates indicated that these clusters have at least the same reactivity for reversible C–H bond breaking in benzene as a clean surface of Pt(111). However, the analogy with crystal surfaces broke down when considering the seemingly excessive number of adsorbed aromatic moieties, more than could be accommodated considering steric factors.

Kaldor and co-workers estimated that the temperature of the Pt clusters was about 300–600 K and that hydrocarbon-cluster reaction time was about 100 μsec. Studies of lower temperatures and longer reaction times would be of interest. Unfortunately, the pulsed cluster-beam technique does not really accommodate such studies, and a continuous-flow system would be better (see Chapter 2 for a description of both of these cluster-beam methods).

Finally, the interesting work of Schnabel and co-workers indicated that low-pressure catalytic processes can take place in the gas phase where benzene precursors were observed to grow on Fe_4^+. Methane oxidative addition to Fe_n ions has also been studied by qualitative molecular orbital methods.[177b]

2. Cluster Chemistry in Matrices and Cold Liquids

The growth of clusters in low-temperature matrices should allow kinetic control of atom by atom growth. Depending on the matrix environment, cluster structure may be affected.

Of course most low-temperature matrix work has been concerned with spectroscopic analysis of the clusters themselves and their adduct formation with reactive molecules such as CO and C_2H_4.[1,178–180]

A series of papers dealing with "Clustering of Metal Atoms in Organic Media" describes growth and reactivity patterns.[87,181–183] The metal atom solvation–desolvation process proceeds in stages.

Initially, a weak M-solvent complex forms at 77 K, and upon warming solvent liquifaction, cluster growth occurs. However, in competition with cluster growth is reaction with the host solvent. Eventual cluster size and crystallinity are dependent on metal concentration, solvent reactivity, and warm-up procedure. In fact during the cluster growth, unusual and unexpected reactions with the host solvent can take place at surprisingly low temperatures.[17] By control of these parameters, Ni particles of

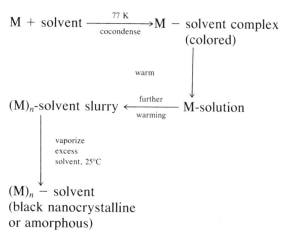

SCHEME 5-3   Metal cluster growth from metal atoms in organic media.[87,131,182]

pseudo-organometallic character were produced, with crystallite sizes so small that they were nonferromagnetic. Numerous characterization techniques and chemical reactions were carried out on these materials, especially Ni particles derived from pentane. It was concluded from these studies that the growing Ni clusters reacted with the host alkane by C–H and C–C bond activation in the cold liquid phase during the warm-up procedure (about 150 K). The reactions that took place were extensive, and multiple C–H and C–C activation processes occurred.

The novelty of these findings is that such reactions can take place at low temperatures and implies that the growing clusters are intrinsically more reactive than clean metal surfaces. This further implies that reactive defect sites must be produced during the growth process.

Comparisons of various transition metals in this low-temperature chemistry indicated that the early transition metals were somewhat more reactive (see Chapter 4 for more discussion). Basically this trend is supported by theoretical work of Baetzold and co-workers,[184,185] where metal surfaces were considered. However, the unusual nature of the growing clusters in organic media adds a novel feature. Kinetically controlled growth must allow defects and cavities to form. Such sites allow for a better orbital energy match for C–H cleavage on the $(M)_n$ ensembles.[186]

Analogous uses of toluene-solvated metal atoms for preparation of ferrofluids and magnetic fine particles should be mentioned.[187,188] Solvated Co atoms were allowed to interact with surfactant molecules, and the Co atoms nucleated to form surfactant-stabilized particles that remained in a fluid state. The particle size was about 50 Å when a Co/toluene/Manoxol OT combination was employed. Somewhat smaller particle diameters of about 30 Å were stabilized when a liquid crystal surfactant, sodium bis (2-ethylhexyl)sulfosuccinate (AOT) was used. These individual particle sizes are comparable to the micelle cavity shape and dimensions; thus, the Co particles tended to assemble into rods about 110 Å long by 30–40 Å wide[189] (Scheme 5-4).

Stable magnetic fluids have also been prepared by evaporating the metals onto the surface of an oil mixed with surfactant. Atom clustering occurred and the magnetic fluids that result were superparamagnetic and had saturation magnetization values of 165, 270, and 84 G for the Fe, Co, and Ni fluids, respectively.[190]

Nanometer-scale iron particles have been prepared by iron atom clustering in cold pentane.[191] The growth of the particles was terminated by addition of oleic acid, yielding air-stable surfactant-coated Fe clusters ranging in size from 2 to 12 nm. TEM analysis showed that the larger particles were prolate spheroids and the smaller ones spherical. Heating of these oleic acid ligated particles caused oxidation of Fe by the carboxyl

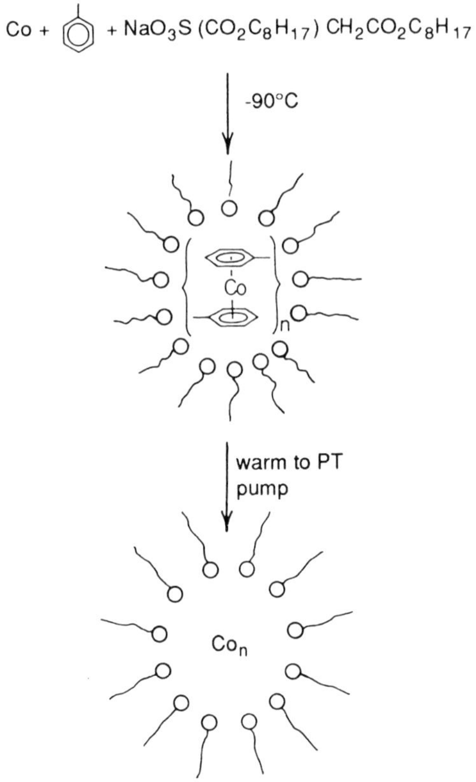

$$Co + \bigcirc + NaO_3S\,(CO_2C_8H_{17})\,CH_2CO_2C_8H_{17}$$

SCHEME 5-4 Toluene-solvated Co atoms in surfactant cavity followed by cluster growth and surfactant stabilization. Reprinted with permission from Andrews and Ozin, *Chem. Mater.* **1**, 186 (1989). Copyright 1989 American Chemical Society.

groups to yield Fe_3O_4 at 360°C followed by reduction of α-Fe by carbon/ hydrogen present (processes due to the presence of the oleic acid). Magnetic properties ranged from saturation magnetization of 12.3 emu/g for the fresh particles to 200 emu/g for heat treated, while coercivities ranged from $H_c = 60\ O_e$ for the fresh to 20 O_e for heat treated. A filtration process that allowed fractionation of the particles into 8–12 and 2–8 nm size regimes was developed.

If oleic acid was replaced with a long chained thiol, thiol-stabilized Fe clusters could be prepared.[192]

$$Fe + pentane \xrightarrow[\text{77 K}]{\text{codep.}} \xrightarrow{R \sim SH} Fe_n(S \sim R)_x + H_2$$

Heating of this material causes a reaction with the thiol to form an FeS-coated Fe cluster,

$$(Fe)_n(S{\sim}R)_x \xrightarrow[400°C]{heat} \quad \text{Fe} \quad \text{FeS}$$

which served to protect the Fe core from environmental degradation. A novel feature of this core-shell system is that Fe is ferromagnetic while FeS is antiferromagnetic. Magnetic exchange interactions between the two phases of these nanoscale particles had the effect of increasing coercivity compared with surfactant-coated material.

Similar experiments have allowed the synthesis of a series of core-shell nanoscale particle structures, Fe/Li, Fe/Mg, Fe/FeF$_2$, Fe/Fe$_3$O$_4$, and Fe/surfactant, and further discussion can be found later in this chapter in Section IIIA.

A variation on this solvent codeposition (SMAD) method has allowed stable nonaqueous colloidal Pd solutions to be prepared in acetone, isopropanol and ethanol. Simply by cocondensing Pd vapor with acetone followed by slow warming, colloidal particles of about 8 nm were formed and remained suspended. The particles were stabilized toward agglomeration by solvation (steric effects) and by scavenging of electrons to form negatively charged particles (electronic stabilization). The colloidal particles are calling "living colloidals" because solvent removal induced growth to form films. These films still contained some of the organic solvent, part of which could be removed by heat treatment to yield conducting metallic films.[193]

Palladium and Pt colloids are also formed by the condensation of Pd or Pt vapor into cold solutions of iso-butylaluminoxane oligomers in methylcyclohexane.[194] The particles were <10 Å for Pt and 18 Å for Pd, and adsorbed CO efficiently. Studies of the solutions by IR and ^{13}C NMR indicated the presence of both linear and bridged carbonyls. The carbonylated Pt colloid was easily transformed into the molecular cluster $[Pt_{12}(CO)_{24}]^{2-}$ by reaction with water. Similarly, ethene adsorption was studied, and NMR revealed exchange between adsorbed and free C$_2$H$_4$.

3. Catalysts

Perhaps the most important application of solvated metal atoms as cluster precursors is in the synthesis of heterogeneous catalysts.[195] In this procedure, cluster growth is arrested by trapping the growing clusters on solid supports such as SiO_2, Al_2O_3, C, or MgO. Such catalysts are obtained in a form where the metal clusters are extremely small and in a reduced state and therefore immediately ready for use. Thus, no high-temperature hydrogen reduction step is necessary, as in conventional catalyst preparations.

Catalysts prepared by this metal vapor method have been called "solvated metal atom dispersed" or SMAD catalysts, and a review of this work has appeared elsewhere.[195] A brief description of monometallic SMAD catalysts appears below, while the important subject of bimetallic SMAD catalysts is considered in this chapter in Section IIIA.

Monometallic Ni/Al_2O_3, Ni/MgO, Ni/C, and Ni/SiO_2 SMAD catalysts were under investigation beginning in 1976,[87] and two significant papers appeared in 1978.[112,113] The small particle size (15 Å up to 30 Å), with tenacious thermal stability, and high inherent reactivity made further investigations desirable.

The preparation of SMAD catalysts has been described in detail in Chapter 2. Codeposition of metal and solvent followed by warmup, melting and solvation, contact with catalyst support, followed by continual warming and stirring of the solvent-support slurry leads to impregnation of the support with the metal.[196] An alternative method is to evaporate the metal directly into a cold liquid solvent in a rotary reactor.[197,198] In this way the solvated metal atoms contact the solid support as the metal evaporation is ongoing. Metal loading, time and temperature of impregnation, surface area of support, and acidity of support can have efforts on ultimate particle size and catalytic behavior.[199,200]

The solvent employed is perhaps the most critical feature of all. During the cluster growth stage the forming metal particles exhibit high reactivities with the host solvent. As discussed earlier, this inherent reactivity may be due to defect sites and unusual morphology. Therefore, carbonaceous fragments of the solvent become incorporated into the trapped metal particles.

In one series of experiments six solvents were compared in the preparation of SMAD Ni/Al_2O_3 catalysts,[197] THF, toluene, trifluoromethylbenzene, pentane, perfluorocyclobutane, and xenon. General trends were: the more polar and reactive solvent THF yielded SMAD particles of smaller crystallite size, more incorporation of carbonaceous material, less catalytic activity, but more catalytic selectivity. The least reactive solvent, xenon, yielded large Ni crystallites with low surface areas and thereby low catalytic activities. With fluorinated solvents, invariably some metal fluorides formed and catalytic activities were diminished due to MF_2 surface coatings.[201] The best results were found for solvents of intermediate or low polarity and reactivity, especially pentane and toluene. In these cases the desired balance between cluster growth and cluster interaction/stabilization with solvent can be achieved.

Characterization and catalytic properties of the SMAD catalysts have involved a wide variety of spectroscopic methods (XPS, XRD, Mössbauer, EXAFS), gas adsorption methods, and catalytic probe reactions shown below.[195,202]

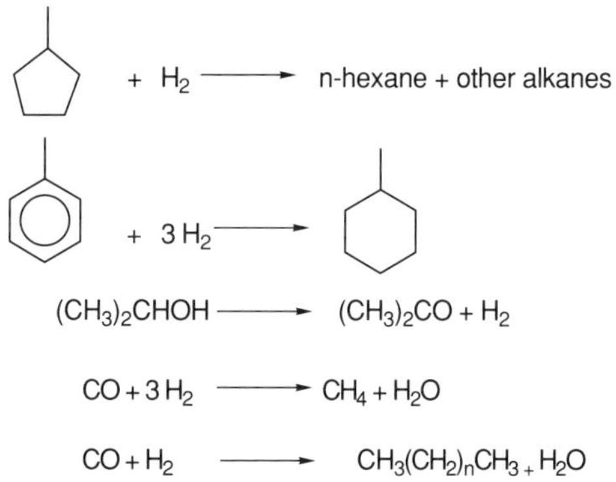

$$(CH_3)_2CHOH \longrightarrow (CH_3)_2CO + H_2$$

$$CO + 3H_2 \longrightarrow CH_4 + H_2O$$

$$CO + H_2 \longrightarrow CH_3(CH_2)_nCH_3 + H_2O$$

On supports such as Al_2O_3, SiO_2, and MgO, the Ni clusters/particles possessed metallic surface areas of usually over 50 m^2/g and often over 100 m^2/g. However, no crystalline nature could be detected by XRD. And it seemed that cluster/particle size could be controlled by adjusting the metal concentration in the SMAD solvent; lower concentrations yielded smaller particles. Likewise, higher-surface-area supports yielded smaller particles. Catalytic activities in general were incredibly high.

All of these studies allowed a picture to develop with regard to cluster growth from these organic media, especially from toluene, the most studied system: (a) The solvated atoms nucleate at surface O–H sites or Lewis acid sites on the catalyst support (Al_2O_3, SiO_2, MgO, or C). More surface sites lead to smaller metal clusters, since the growth of the clusters proceeds until metal in the solution is depleted (see Fig. 5-2). (b) The clusters/particles deposited are amorphous and particles sizes are generally <25 Å.

An additional novel feature of the SMAD method is the possibility of forming metal clusters inside the pores of zeolites.[203] Also, it may be possible by careful control of the decomposition of certain organometallic precursors to prepare monoatomic, diatomic, or other very small clusters on the support surfaces.[204]

III. Bimetallic and Binuclear Systems

A. Bimetallics

Morse and co-workers have reported on two-photon ionization of supersonic jet-cooled gas-phase Ni–Pt, Ni–Pd, and Pd–Pt dimers.[205,206a] The D_o(NiPt) was determined to be 2.798 ± 0.003 eV while only approximate values for NiPd (1.46 eV), Pd_2 (<1.41 eV), and PdPt (1.98 eV) were obtainable. A comparison of the platinum group dimers and the coinage metal (Cu, Ag, Au) dimers is given and demonstrates the increased importance of d-orbital contributions to the bonding of the platinum group dimers as one moves down the periodic table. The anomolous weak bonding of the palladium-containing diatomics is believed to be due to the very stable $4d^{10}5s^o$ ^1S_o ground state of atomic Pd. Rough estimates of other bond strengths led to an order of Pt_2 > NiPt > Ni_2 > PdPt > NiPd > Pd_2, which are in agreement with the pattern of diatomic abundances observed when the equimolar NiPdPt alloy was subjected to laser vaporization. Similar studies have been reported for NiAu, PtCu, and NiCu.[206b]

Returning again to matrix conditions, a major advantage of the atom-clustering process at low temperatures is that all metal combinations can be expected to bind together, even those that are normally immiscible in the bulk. Thus, the preparation of clusters of immiscible metals has become a possibility, and one of these first applications has been the preparation of novel nanoscale magnetic materials.[207,208] Iron and lithium atoms were trapped in cold, frozen pentane, and upon warming, atom aggregation took place. In this way, Fe–Li clusters formed upon warming. Kinetic control of this growth in cold, liquid pentane led to the formation of α-Fe crystallites of about 3 nm diameter, and these were surrounded by a matrix of nanocrystalline Li metal. Overall particle size was about 21 nm. This ultrafine powder was pyrophoric and with a surface area of 140 m^2/g. Controlled oxidation and heat treatments caused the conversion of the "plum-pudding" clusters to core-shell structures with the α-Fe crystallites on the inside and Li metal or Li oxides on the outside (Fig. 5-6).

Choice of proper heating and oxidizing conditions allowed the control of α-Fe crystallite size ranging from 3 to 32 nm. And, most importantly, the α-Fe crystallite could be protected by the Li_2O/Li_2CO_3 coating and therefore were stable in the atmosphere for many months.

Similar behavior has been found for metastable Fe–Mg clusters.[209] In this case a beautiful protective coating of MgO–Mg could be engineered, and the internal α-Fe crystallites were completely protected from the atmosphere.

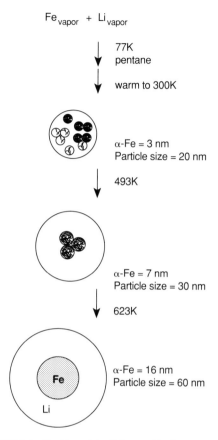

Fe_{vapor} + Li_{vapor}

77K
pentane

warm to 300K

α-Fe = 3 nm
Particle size = 20 nm

493K

α-Fe = 7 nm
Particle size = 30 nm

623K

Fe

Li

α-Fe = 16 nm
Particle size = 60 nm

FIGURE 5-6   Fe–Li particles "plum pudding" structure.[208]

The ultimate goal was to produce protected ultrafine metallic particles of iron metal. The heat processing of metastable clusters of immiscible metals worked very well in this regard.

B. Bimetallic Solvated Metal Atom Dispersed Catalysts

A significant innovation in recent years dealing with metal atom chemistry is the finding that at low temperature, unusual *bimetallic* clusters can be built up on catalyst supports. The catalytic properties of these bimetallic SMAD catalysts are quite remarkable.[195]

The morphology of the bimetallic particles is dependent on the reaction rates of solvated metal atoms with the surface of the catalyst support employed. The idea is to solvate two metals simultaneously. Upon exposure of this solution to a catalyst support, nucleation of one metal or both metals could take place, and layered structures or alloy-like particles could form. And since the nucleation processes take place at low temperatures (−20 to −70°C), bimetallic metastable particles can form, even for those metals that do not form thermodynamically stable alloys (immiscible metals).

Since the morphologies could be very different from one pair of metals to the next, catalytic properties are impossible to predict.

The cobalt–manganese combination was particularly interesting.[210–214] When a combination of 80% Co : 20% Mn was employed (metal loading of about 4%), the catalytic and chemisorption properties of the resultant Co–Mn/SiO$_2$ catalyst was greatly perturbed when compared with similarly prepared Co/SiO$_2$, Mn/SiO$_2$, or other ratios of Co to Mn/SiO$_2$ catalysts. Hydrogen chemisorption more than doubled. Catalytic

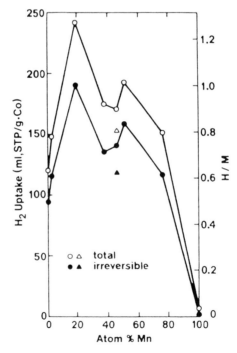

FIGURE 5-7   Total and irreversible H$_2$ adsorption at room temperature over a bimetallic SMAD Co–Mn/SiO$_2$ catalyst as a function of composition. Triangles, the catalyst was prepared by sequential codeposition instead of simultaneous codeposition.[195]

rates for hydrogenation of 1-butene (at $-60°C$) and for cyclopropane hydrogenolysis were 100-fold greater compared with Co/SiO$_2$ and 10,000-fold greater than Mn/SiO$_2$ (also prepared by the SMAD method) (see Fig. 5-2, 5-7, and 5-8).

Extensive studies of structure-sensitive catalytic probe reactions, EXAFS, XPS, and chemisorption allowed a morphological picture to be established. It was shown that toluene-solvated Mn reacted with surface O–H groups on SiO$_2$ first to form a manganese oxide layer. As the solution became exhausted of Mn, then toluene-solvated Co nucleated as particles on top of the MnO$_x$ sites. In this way a layered structure rich in surface cobalt was formed.

The Mn behaved as a sacrificial metal such that surface O–H groups were used up and were not available to oxidize (and deactivate) Co. In this way, highly active very small Co particles were available for catalytic processes. It also seems that the MnO$_x$ layer has some electronic effect on the Co enhancing the intrinsic catalytic properties of the cobalt.

Similar studies were carried out with Fe–Co, Fe–Mn, and with Co–Mn. Layered catalyst particle structures were formed here as well. A series of active Fischer–Tropsch catalysts that were very selective for terminal alkene formation (from CO + H$_2$) were obtained.[215]

With two metals possessing oxophilicities that are quite similar, such as Fe and Co, layered structures were not as cleanly formed. In fact alloy-like particles, perhaps with partial segregation, were obtained.[216] In this case the toluene-solvated iron atoms deposited before cobalt, iron being slightly more oxophilic, but the Fe and Co mixed more than compared with Co and Mn.

In a similar way, layered Fe–Mn/support particles could also be prepared from Fe and Mn–toluene solutions. The more oxophilic metal Mn deposited first. These results confirm that the nucleation step is the reaction of solvated metal atoms with surface OH groups.[216,217]

With this experience in preparing SMAD bimetallic catalysts, it was decided to use the method to make combinations that were already known to be important in the catalysis industry, in particular Pt–Re and Pt–Sn. Two approaches were used: first the "half-SMAD" method where a Pt/Al$_2$O$_3$ catalyst was prepared by conventional methods followed by

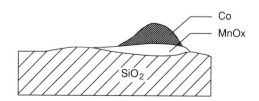

FIGURE 5-8 Proposed particle structure for Co–Mn/SiO$_2$.[195]

treatment with solvated Re atoms in toluene or solvated Sn atoms in THF. By the half-SMAD method the preexisting Pt particles could be partially covered with Re or Sn metals. Alternatively, the full-SMAD method refers to the technique of evaporating both metals simultaneously, solvating them, and allowing nucleation of the mixture. By these methods the supported metal particles shown in Fig. 5-9 were prepared.

In conventional methods of preparation, the Re and Sn are usually present as oxides, not in the metallic state. Thus, the half-SMAD and full-SMAD procedures allowed the preparation of metallic Re or Sn on metallic Pt, or alloys with metallic Re and Sn on metallic Pt, and in the form of supported ultrafine metallic particles.[218–223]

Through extensive studies of all four of these catalyst structures by catalytic probe reactions, [119]Sn Mössbauer, XPS, EXAFS, and X-ray diffraction, it was learned that metallic Re and Sn did have significant effects on catalytic behavior and that their presence could control selectivity and activity in rather predictable ways. For example, half-SMAD Pt–Sn/Al_2O_3 catalysts were very active for heptane reforming to toluene and benzene, even more active than conventionally prepared for full-SMAD Pt–Sn/Al_2O_3. This suggests that active catalytic sites may be located at the interface of the Pt and Sn metals. In contrast, the full-SMAD Pt–Sn/Al_2O_3 catalysts were less catalytically active but were very selective for formation of toluene from n-heptane with very little unwanted hydrogenolysis (splitting of C–C bonds). This suggests an ensemble effect where Sn atoms alloyed with Pt break up Pt atom ensembles needed to carry out hydrogenolysis reactions.

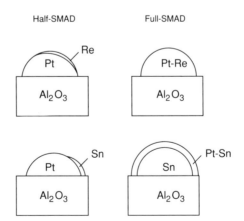

FIGURE 5-9   Structures of Pt–Re/Al_2O_3 and Pt–Sn/Al_2O_3 half SMAD and full SMAD bimetallic catalyst particles.

Thus, the SMAD method of preparing small bimetallic particles continues to be a promising technique for the production/engineering of a wide variety of unusual new materials, including ultrafine alloy particles or layered structures.[195]

C. Metal Carbides and Oxides, and Hydrides

Metal carbide molecules are also attracting some interest, which is discussed in some detail in Chapter 4. For the late transition metals only three diatomic carbide molecules are presently known, CoC, RhC, and PtC.[224–229] In the case of CoC, this molecule was prepared in frozen argon by codeposition of Co vapor and C vapor (from tantalum cells).[224] Utilizing ESR, the ground state was determined to be $^2\Sigma$. This odd electron species with 13 valence electrons localizes only a small portion of unpaired electron density on the Co and so the major orbital contributions to the open-shell electron wavefunction are cobalt $3d\sigma$ and/or $4p\sigma$ and carbon $2p\sigma$ orbitals. For comparison, the ground-state electronic configurations for RhC is also $^2\Sigma$, whereas for PtC it is $^1\Sigma$.

Nickel clusters produced from Ni atom aggregation in cold pentane are exceedingly reactive with organic materials, such that controlled exposure of these Ni clusters to alkanes, alkenes, or esters leads to the formation of Ni_3C nanoscale particles.[230] The complete conversion to Ni_3C occurred about 200°C. Similar results were obtained with palladium. Comparisons with other forms of reactive nickel or palladium particles revealed that the metal vapor-produced clusters were most reactive. It was concluded that in order for clean conversion to metal carbide to occur, activated forms of metal and carbon are necessary. In fact, as Ni atoms form clusters in the presence of cold pentane, defective very reactive Ni clusters do form, and their partial reaction with pentane at low temperature initially satisfies the need for active carbonaceous groups.

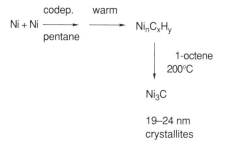

Matrix-isolated FeO, NiO, and CoO were produced by sputtering of the metals during the operation of a hollow cathode discharge on O_2.[231a]

This technique ensures that metal atoms and oxygen atoms are able to react and be cooled upon condensation in freezing argon. The ground states of these matrix-isolated molecules and ν_{M-O} frequencies are shown below:

FeO	$^5\Sigma$	880 cm^{-1}
CoO	$^4\Sigma^-$	854 cm^{-1}
NiO	$^3\Sigma^-$	837 cm^{-1}

The dioxides CoO_2, RhO_2, and IrO_2 have also been investigated in matrices.[231b]

Metal-deficient iron oxide clusters have been formed in the gas phase by reactions of O_2 with Fe_n clusters.[232] Interestingly, FeO seems to be a favored product, for example, FeO, Fe_2O_2, Fe_3O_3, Fe_4O_4, Fe_9O_{10}, and $Fe_{22}O_{24}$. This particular stoichiometry is very similar to that of bulk FeO, but what is surprising is that the more stable bulk oxides Fe_2O_3 and Fe_3O_4 do not seem to be favored as gas-phase molecules or clusters.

One report on the cocondensation chemistry of FeO vapor has appeared.[233] When FeO gas was cocondensed with cyclopentadiene, ferrocene was efficiently formed, with water as probable by-product.

$$FeO + CpH \rightarrow Cp_2Fe + H_2O$$

In cocondensations with ethene, propene, or styrene, small yields of coupling products were obtained; for example, styrene yielded *cis, trans*-stilbene (ratio of 1 : 2).

Ultrafine antiferromagnetic FeF_2 particles (< 10 nm) were prepared by IR multiphoton decomposition of $Fe(CO)_5$ using SF_6 as a sensitizer and reactant. The magnetic properties were examined by Mössbauer spectroscopy.[234] At low temperatures the magnetic hyperfine field decreased faster with increasing temperature than the hyperfine field of bulk FeF_2. This behavior of the fine particles appears to be consistent with collective magnetic excitations. Also, the transition from paramagnetic to antiferromagnetic states took place at a higher temperature and over a broader range as compared to the bulk. It was found that ultrafine FeF_2 particles were sensitive to oxidation forming cubic-type iron oxide.[234]

And finally, CoH_2, RhH_2, and IrH_2 have been prepared in low temperature matrices and ESR spectra recorded.[235]

References

1. K. J. Klabunde, "Chemistry of Free Atoms and Particles," Academic Press, New York, 1980.

2. G. A. Ozin, M. P. Andrews, C. G. Francis, H. X. Huber, and K. Molnar, *Inorg. Chem.* **29,** 1068 (1990).

3. S. A. Mitchell and P. A. Hackett, *J. Chem. Phys.* **93,** 7813 (1990); see also p. 7822.

4. G. H. Jeong and K. J. Klabunde, *J. Chem. Phys.* **91,** 1958 (1989).

5. A. J. L. Hanlan and G. A. Ozin, *Inorg. Chem.* **16,** 2848 (1977).

6. G. A. Ozin and J. Garcia-Prieto, *J. Phys. Chem.* **92,** 318 (1988); see also p. 325.

7. J. Shaksemampour, R. Pyzalski, M. Vala, and J.-C. Rivoal, *J. Physique* **45,** 953 (1984).

8. R. G. Graham and R. Grinter, *J. Chem. Phys.* **91,** 6677 (1989).

9. M. Vala, J.-C. Rivoal, C. Grisolia, and J. Pyka, *J. Chem. Phys.* **82,** 4376 (1985).

10. M. Vala, M. Eyring, J. Pyka, J.-C. Rivoal, and C. Grisolia, *J. Chem. Phys.* **83,** 969 (1985).

11. B. Breithaupt, J. E. Hulse, D. M. Kolb, H. H. Rotermund, W. Schroeder, and W. Schrittenlacher, *Chem. Phys. Lett.* **95,** 513 (1983).

12. C. P. Barrett, R. G. Graham, and R. Grinter, *Chem. Phys.* **86,** 199 (1984).

13. R. Grinter and D. R. Stern, *J. Chem. Soc., Chem Commun.,* 41 (1982).

14. L. C. Allen, *J. Am. Chem. Soc.,* **111,** 9003 (1989).

15. L. B. Knight, Jr., S. T. Cobranchi, J. O. Herlong, T. Kirk, K. Balasubramanian, and K. K. Das, *J. Chem. Phys.* **92,** 2721 (1990).

16. L. B. Knight, Jr., J. O. Herlong, S. T. Cobranchi, and T. Kirk, *J. Chem. Phys.* **92,** 6463 (1990).

17. K. J. Klabunde, G. H. Jeong, and A. W. Olsen, "Selective Hydrocarbon Activation: Principles and Progress" (J. A. Davies, P. L. Watson, A. Greenberg, and J. F. Liebman, Eds.), Chap. 13, p. 433, VCH, New York, 1990.

18. M. R. A. Blomberg, P. E. M. Siegbahn, and M. Svensson, *J. Phys. Chem.* **95,** 4313 (1991).

19. W. E. Billups, M. M. Konarski, R. H. Hauge, and J. L. Margrave, *J. Am. Chem. Soc.* **102,** 7393 (1980).

20. G. A. Ozin and J. G. McCaffrey, *Inorg. Chem.* **22,** 1397 (1983).

21. G. A. Ozin, J. G. McCaffrey, and J. M. Parnis, *Angew. Chem., Int. Ed. Engl.* **25,** 1072 (1986).

22. G. A. Ozin, D. F. McIntosh, and S. A. Mitchell, *J. Am. Chem. Soc.* **103,** 1574 (1981).

23. G. A. Ozin and J. G. McCaffrey, *J. Am. Chem. Soc.* **104,** 7351 (1982).

24. (a) S.-C. Chang, R. H. Hauge, W. E. Billups, J. L. Margrave, and Z. H. Kafafi, *Inorg. Chem.* **27,** 205 (1988); (b) W. E. Billups, S. C. Chang, R. H. Hauge, and J. L. Margrave, *J. Am. Chem. Soc.* **115,** 2039 (1993); (c) J. J. Carroll and J. C. Weisshaar, *J. Am. Chem. Soc.* **115,** 800 (1993).

25. Z. H. Kafafi, R. H. Hauge, L. Fredin, W. E. Billups, and J. L. Margrave, *J. Chem. Soc., Chem. Commun.,* 1230 (1983).

26. (a) S. F. Parker, C. H. Peden, P. H. Barrett, and R. G. Pearson, *J. Am. Chem. Soc.* **106,** 1304 (1984); (b) M. R. A. Blomberg, P. E. M. Siegbahn, and M. Svensson, *J. Phys. Chem.* **96,** 9794 (1992).

27. Z. Kafafi, R. H. Hauge, W. E. Billups, and J. L. Margrave, *J. Am. Chem. Soc.* **109,** 4775 (1987).

28. Z. Kafafi, R. H. Hauge, and J. L. Margrave, *J. Am. Chem. Soc.* **107,** 7550 (1985).

29. D. Ball, Z. H. Kafafi, R. H. Hauge, and J. L. Margrave, *Inorg. Chem.* **24,** 3708 (1985).

30. D. W. Ball, Z. H. Kafafi, R. H. Hauge, and J. L. Margrave, *J. Am. Chem. Soc.* **108,** 6621 (1986).

31. J. A. Bandy, M. L. H. Green, D. O'Hare, and K. Prout, *J. Chem. Soc., Chem. Commun.,* 1402 (1984).

32. J. A. Bandy, M. L. H. Green, and D. O'Hare, *J. Chem. Soc. Dalton Trans.*, 2477 (1986).
33. M. L. H. Green, D. S. Joyner, and J. M. Wallis, *J. Chem. Soc. Dalton Trans.*, 2823 (1987).
34. P. R. Brown, F. G. N. Cloke, and M. L. H. Green, *Polyhedron* **4**, 869 (1985).
35. R. G. Gastinger, B. B. Anderson, and K. J. Klabunde, *J. Am. Chem. Soc.* **102**, 4959 (1980).
36. K. J. Klabunde and R. Campostrini, *J. Fluorine Chem.* **42**, 93 (1989).
37. S. B. Choe, H. Kanai, and K. J. Klabunde, *J. Am. Chem. Soc.* **111**, 2875 (1989).
38. H. Kanai, S. B. Choe, and K. J. Klabunde, *J. Am. Chem. Soc.* **108**, 2019 (1986).
39. M. M. Brezinski, K. J. Klabunde, S. K. Janikowski, and L. J. Radonovich, *Inorg. Chem.* **24**, 3305 (1985).
40. M. M. Brezinski, J. Schneider, L. J. Radonovich, and K. J. Klabunde, *Inorg. Chem.* **28**, 2414 (1989).
41. M. M. Brezinski and K. J. Klabunde, *Organometallics* **2**, 1116 (1983).
42. K. J. Klabunde, B. B. Anderson, and K. Neuenschwander, *Inorg. Chem.* **19**, 3719 (1980).
43. S. T. Lin, R. N. Narske, and K. J. Klabunde, *Organometallics* **3**, 571 (1985).
44. S. L. Lin and K. J. Klabunde, *Inorg. Chem.* **24**, 1961 (1985).
45. Y. Tanaka, S. C. Davis, and K. J. Klabunde, *J. Am. Chem. Soc.* **104**, 1013 (1982).
46. M. W. Eyring and L. J. Radonovich, *Organometallics* **4**, 1841 (1985).
47. T. J. Grosens and K. J. Klabunde, *J. Organomet. Chem.* **1**, 564 (1982).
48. S. T. Lin, R. N. Narske, and K. J. Klabunde, *Organometallics* **3**, 571 (1985).
49. S. T. Lin, T. J. Groshens, and K. J. Klabunde, *Inorg. Chem.* **23**, 1 (1984).
50. (a) T. R. Bierschenk, M. A. Guerra, T. J. Juhlke, S. B. Larson, and R. J. Lagow, *J. Am. Chem. Soc.* **109**, 4855 (1987). (b) D. W. Firsich and R. J. Lagow *J. Chem. Soc. Chem. Commun.*, 1283 (1981). (c) M. A. Guerra, T. R. Bierschenk, and R. J. Lagow *Rev. Chim. Miner.* **23**, 701 (1986).
51. S.-C. Chang, R. H. Hauge, Z. H. Kafafi, J. L. Margrave, and W. E. Billups, *J. Am. Chem. Soc.* **110**, 7975 (1988).
52. S.-C. Chang, R. H. Hauge, Z. H. Kafafi, J. L. Margrave, and W. E. Billups, *Inorg. Chem.* **29**, 4373 (1990).
53. S.-C. Chang, R. H. Hauge, W. E. Billups, J. L. Margrave, and Z. H. Kafafi, *Inorg. Chem.* **27**, 205 (1988).
54. S.-C. Chang, Z. H. Kafafi, R. H. Hauge, W. E. Billups, and J. L. Margrave, *J. Am. Chem. Soc.* **107**, 1447 (1985).
55. G. J. Zimmerman, L. W. Hall, and L. G. Sneddon, *Inorg. Chem.* **19**, 3642 (1980).
56. G. J. Zimmerman and L. G. Sneddon, *J. Am Chem. Soc.* **103**, 1102 (1981).
57. R. P. Micciche and L. G. Sneddon, *Organometallics* **2**, 674 (1983).
58. R. P. Micciche, J. J. Briguglio, and L. G. Sneddon, *Organometallics* **3**, 1396 (1984).
59. R. P. Micciche, J. J. Briguglio, and L. G. Sneddon, *Inorg. Chem.* **23**, 3992 (1984).
60. (a) J. J. Bruguglio and L. G. Sneddon, *Organometallics* **4**, 721 (1985). (b) R. P. Micciche, P. J. Carroll, and L. G. Sneddon, *Organometallics* **4**, 1619 (1985).
61. J. J. Briguglio and L. G. Sneddon, *Organometallics* **5**, 327 (1986).
62. S. O. Kang and L. G. Sneddon, *Inorg. Chem.* **27**, 587 (1988).
63. S. O. Kang, P. J. Carroll, and L. G. Sneddon, *Organometallics* **7**, 772 (1988).
64. S. O. Kang, P. J. Carroll, and L. G. Sneddon, *Inorg. Chem.* **28**, 961 (1989).
65. S. A. Locke and P. B. Shevlin, *Organometallics* **3**, 217 (1984).
66. A. D. Berry, *Organometallics* **2**, 895 (1983).
67. G. Cardenas and P. B. Shevlin, *J. Org. Chem.* **49**, 4726 (1984).

68. S. F. Parker, C. H. Peden, P. H. Barrett, and R. G. Pearson, *Inorg. Chem.* **22,** 2813 (1983).
69. D. Ritter and J. C. Weisshaar, *J. Am. Chem. Soc.* **112,** 6425 (1990).
70. S. A. Mitchell and P. A. Hackett, *J. Chem. Phys.* **93,** 7822 (1990).
71. G. A. Ozin and W. J. Power, *Inorg. Chem.* **19,** 3860 (1980).
72. D. N. Cox, R. Roulet, and G. Chapuis, *Organometallics* **4,** 2001 (1985).
73. S. D. Ittel, *J. Organomet. Chem.* **195,** 331 (1980).
74. S. D. Ittel, F. A. Van Catledge, and J. P. Jesson, *J. Am. Chem. Soc.* **101,** 3874 (1979); and references therein.
75. J. R. Blackborow, U. Feldhoff, F. W. Grevels, R. H. Grubbs, and A. Miyashita, *J. Organomet. Chem.* **173,** 253 (1979).
76. N. Hao, J. F. Sawyer, B. G. Sayer, and M. J. McGlinehey, *J. Am. Chem. Soc.* **101,** 2203 (1979).
77. L. K. Beard, M. P. Silvon, and P. S. Skell, *J. Organomet. Chem.* **209,** 245 (1981).
78. G. Vitulli, A. Raffaelli, P. A. Costantino, C. Barberini, F. Marchetti, S. Merlino, and P. S. Skell, *J. Chem. Soc., Chem. Commun.,* 232 (1983).
79. E. S. Kline, Z. H. Kafafi, R. H. Hauge, and J. L. Margrave, *J. Am. Chem. Soc.* **107,** 7559 (1985).
80. E. S. Kline, Z. H. Kafafi, R. H. Hauge, and J. L. Margrave, *J. Am. Chem. Soc.* **109,** 2402 (1987).
81. L. H. Simons and J. J. Lagowski, *J. Organomet. Chem.* **249,** 195 (1983).
82. U. Zenneck, *Angew. Chem. Int. Ed. Engl.* **29,** 126 (1990).
83. H. Schäufele, H. Pritzkow, and U. Zenneck, *Angew. Chem. Int. Ed. Eng.* **27,** 1519 (1988).
84. M. B. Freeman, L. W. Hall, and L. G. Sneddon, *Inorg. Chem.* **19,** 1132 (1980).
85. (a) J. J. Schneider, R. Goddard, and C. Kruger, *Organometallics* **10,** 665 (1991); (b) J. J. Schneider, R. Goddard, S. Werner, and C. Kruger, *Angew. Chem. Int. Ed. Eng.* **30,** 1124 (1991).
86. (a) J. J. Schneider *Angew. Chem. Int. Ed. Eng.* **31,** 1392 (1992); (b) K. H. Theopold, J. J. Kersten, A. L. Rheingold, C. P. Casey, R. A. Wiedenhoefer, and C. E. C. A. Hop, *Angew. Chem. Int. Ed. Eng.* **31,** 1341 (1992); see also p. 471 (1992).
87. A discussion can be found in Ref.[1] and the original paper on this subject is K. J. Klabunde, H. F. Efner, T. O. Murdock, and R. Ropple, *J. Am. Chem. Soc.* **98,** 1021 (1976).
88. U. Zenneck and W. Frank, *Angew. Chem. Int. Ed. Eng.* **25,** 831 (1986).
89. S. F. Parker and C. H. F. Peden, *J. Organomet. Chem.* **272,** 411 (1984).
90. P. D. Morand and C. G. Francis, *Organometallics* **4,** 1653 (1985).
91. G. A. Ozin, K. Coleson, and H. Huber, *Organometallics* **2,** 415 (1983).
92. T. J. Groshens, B. Henne, D. E. Bartak, and K. J. Klabunde, *Inorg. Chem.* **20,** 3629 (1981).
93. B. J. Henne and D. E. Bartak, *Inorg. Chem.* **23,** 369 (1984).
94. S. D. Ittel and C. A. Tolman, *J. Organomet. Chem.* **172,** C47 (1979).
95. G. Vitulli, G. Uccello-Barnetta, P. Pannocchia, and A. Raffaelli, *J. Organomet. Chem.* **302,** C21 (1986).
96. D. C. Staplin and R. W. Parry, *Inorg. Chem.* **18,** 1473 (1979).
97. L. J. Radonovich, M. W. Eyring, T. J. Groshens, and K. J. Klabunde, *J. Am. Chem. Soc.* **104,** 2816 (1982).
98. D. W. Ball, R. H. Hauge, and J. L. Margrave, *High Temp. Sci.* **25,** 95 (1988).
99. J. W. Kauffman, R. H. Hauge, and J. L. Margrave, *High Temp. Sci.* **17,** 237 (1984).
100. G. W. Smith and E. A. Carter, *J. Phys. Chem.* **95,** 10828 (1991).

101. J. B. Chenier, M. Histed, J. A. Howard, H. A. Holy, H. Morris, and B. Mile, *Inorg. Chem.* **28**, 4114 (1989).

102. J. D. Woollins, *Polyhedron, 939 (1987).*

103. A. J. Cornish, M. F. Lappert, J. J. MacQuitty, and R. K. Maskell, *J. Organomet. Chem.* **177**, 153 (1979).

104. S. C. Wang and G. E. Ehrlich, *J. Chem. Phys.* **94**, 4071 (1991).

105. J. J. Schneider, R. Goddard, and C. Kruger, *Organometallics* **10**, 665 (1991).

106. J. W. Grover, W. J. Herron, M. T. Coolbaugh, W. R. Peifer, and J. F. Garvey, J. Phys. Chem. **95**, 6473 (1991).

107. K. Kimoto, Y. Kamiya, M. Nonoyama, and R. Uyeda, *Jpn. J. Appl. Phys.* **2**, 702 (1963).

108. S. Leutwyler and U. Even, *Chem. Phys. Lett.* **84**, 188 (1981).

109. B. A. Scott, R. M. Plecenik, G. S. Cargill, III, T. R. McGuire, and S. R. Herd, *Inorg. Chem.* **19**, 1252 (1980).

110. (a) Y. Wang, Y. X. Li, and K. J. Klabunde, *in* "Selectivity in Catalysis" (M. E. Davis and S. L. Suib, Eds.), ACS Sym. Series, No. 517, Chap. 10, p. 136, 1993; (b) K. J. Klabunde and Y. X. Li, *in* "Selectivity in Catalysis" (M. E. Davis and S. L. Suib, Eds.), ACS Sym. Series, No. 517, Chap. 7, p. 88, 1993.

111. (a) K. J. Klabunde, Y. X. Li, and B. J. Tan, *Chem. Mater.* **3**, 30 (1991); (b) B. J. Tan, K. J. Klabunde, and P. M. A. Sherwood *J. Am. Chem. Soc.* **113**, 855 (1991).

112. K. J. Klabunde, S. C. Davis, H. Hattori, and Y. Tanaka, *J. Catal.* **54**, 254 (1978).

113. K. J. Klabunde, D. Ralston, R. Zoellner, H. Hattori, and Y. Tanaka, *J. Catal.* **55**, 213 (1978).

114. E. A. Rohlfing, D. M. Cox, and A. Kaldor, *Bull Am. Phys. Soc.* **28** (1983).

115. E. A. Rohlfing, D. M. Cox, and A. Kaldor, *J. Phys. Chem.* **88**, 4497 (1984).

116. M. Knickebein, S. Yang, and S. J. Riley, *J. Chem. Phys.* **93**, 94 (1990).

117. E. K. Parks, T. D. Klots, and S. J. Riley, *J. Chem. Phys.* **92**, 3813 (1990).

118. M. B. Kniebkebein and W. J. C. Menezes, *J. Chem. Phys.* **94**, 4111 (1991).

119. E. K. Parks, B. J. Winter, T. D. Klots, and S. J. Riley, *J. Chem. Phys.* **94**, 1882 (1991).

120. (a) T. D. Klots, B. J. Winter, E. K. Parks, and S. J. Riley, *J. Chem. Phys.* **95**, 8919 (1991); (b) E. K. Parks, B. J. Winter, T. D. Klots, and S. J. Riley, *J. Chem. Phys.* **96**, 8267 (1992); (c) M. S. Stave and A. E. DePristo, *J. Chem. Phys.* **97**, 3386 (1992).

121. F. Vergand, D. Fargues, D. Oliver, L. Bonneviot, and M. Che, *J. Phys. Chem.* **87**, 2373 (1983).

122. S. Kajiwara, S. Ono, K. Honma, and M. Uda, *Philos. Mag., Lett.* **55**, 215 (1987).

123. R. E. Benfield, P. P. Edwards, and A. M. Stacey, *J. Chem. Soc. Chem. Commun.*, 525 (1982).

124. D. C. Johnson, R. E. Benfield, P. P. Edwards, W. J. H. Nelson, and M. D. Vargas, *Nature (London)* **314**, 231 (1985).

125. S. R. Drake, P. P. Edwards, B. F. G. Johnson, J. Lewis, E. A. Marseglia, S. D. Obertelli, and N. C. Pyper, *Chem. Phys. Lett.* **139**, 336 (1987).

126. S. R. Drake, P.P Edwards. B. F. G. Johnson, J. Lewis, D. Obertelli, and N. C. Pyper, *J. Chem. Soc., Chem. Commun.*, **1191**, (1987).

127. J. C. Rivoal, C. Grisolia, J. Lignieres, D. Kreisle, P. Fayet, and L. Wöste, *Z. Phys. D: At. Mol. Clusters* **12**, 481 (1989).

128. H. Tatewaki, M. Tomonari, and T. Nakamura, *J. Chem. Phys.* **88**, 6419 (1988).

129. M. Tomonari and H. Tatewaki, *J. Chem. Phys.* **88**, 1828 (1988).

130. K. K. Das and K. Balusubranian, *J. Chem. Phys.* **93**, 625 (1990).

131. S. K. Gupta, B. M. Nappi, and K. A. Gingerich, *Inorg. Chem.* **20**, 966 (1981).

132. J. P. Bucher, D. C. Douglass, and L. A. Bloomfield, *Phys. Rev. Lett.* **66**, 3052 (1991).

133. W. A. DeHeer, P. Milani, and A. Chatelain, *Z. Phys. D: At. Mol. Clusters* **19**, 241 (1991).
134. W. A. DeHeer, P. Milani, and A. Chatelain, *Phys. Rev. Lett.* **65**, 488 (1990).
135. V. Papefthymion, A. Kostikas, A. Simopoulos, D. Niarchos, S. Gangupadyay, G. C. Hadjipanayis, C. M. Sorensen, and K. J. Klabunde, *J. Appl. Phys.* **67**, 4487 (1990).
136. S. Gangopadhyay, G. C. Hadjipanayis, B. Dale, C. M. Sorensen, K. J. Klabunde, V. Papaefthymion, and A. Kostikas, *Phys. Rev. B*. **45**, 9778 (1992).
137. S. Ohnuma, A. Kunimoto, and T. Masumoto, *J. Appl. Phys.* **63**, 4243 (1988).
138. V. D. Parkhomenko, A. T. Kolodyazhnyi, Y. D. Galivets, and K. B. Pokholok, *Poroshk. Metall.* (*Kiev*) **86** (1990); CA 113(4): 27247.
139. J. Wu, H. Lu, Y. Du, X. Gav, and T. Wang, *Wuli Xuebao* **37**, 2044 (1988). [In Chinese]
140. S. Ohnuma, Y. Nakanouchi, C. D. Graham, Jr., and T. Masumoto, *IEEE Trans. Mag.* **22**, 1098 (1986) and **21**, 2038 (1985).
141. C. Kaito and K. Fujita, *J. Cryst. Growth* **79**, 132 (1986).
142. H. Dya, *J. Appl. Phys.* **30**, 1775 (1991).
143. K. J. Lawson and D. J. Stephenson, *J. Mater. Sci. Lett.* **10**, 699 (1991).
144. K. Hayakawa and S. Iwama, *J. Cryst. Growth* **99**, 188 (1990).
145. S. Kasukabe, *J. Cryst. Growth* **99**, 196 (1990).
146. T. Majima, T. Miyahara, and M. Takami, *Reza Kagaku Kenkyu* **11**, 13 (1989) CA 112 (22): 207577y. [In Japanese]
147. H. Nagai, K. Majima, M. Yokota, S. Masaru, T. Sawayama, T. Tsuchida, H. Umeda, T. Nagae, and H. Yamada, *Funtai Oyobi Funmatsu Yakin* **36**, 774 (1989) CA112(14):130928a. [In Japanese]
148. T. Majima, *Reza Kenkyu* **17**, 358 (1989). [In Japanese]
149. S. Ohno and M. Uda, *Nippon Kinzoku Gakkaishi* **53**, 946 (1989) CA 111(26):237208x. [In Japanese]
150. B. T. Chang and S. J. Kim, *Chem. Lett.* **8**, 1353 (1989).
151. I. Ibaraki, T. Araya, S. Hioki, R. Okada, and M. Kanamaru, *Inst. Phys. Conf. Ser.* **89**, 199 (1988).
152. T. Hayashi, *Hyomen Kagaku* **8**, 358 (1987), CA109(13):109851y. [In Japanese]
153. M. Oda, *Hyomen Kagaku*, **8**, 335 (1987), CA108(18):153014f. [In Japanese]
154. T. Takewaki, S. Kkato, K. Takeuchi, Y. Makide, and T. Toninaga, *Reza Kagaku Kankyu* **8**, 71 (1986), CA106(18):146957w. [In Japanese]
155. A. R. Thoelen, *Philos. Mag. A* **53**, 259 (1986).
156. V. I. Novikov, I. A. Repin, A. N. Semenikhin, S. V. Svirida, and L. I. Trusov, *Fiz. Tverd. Tela* (*Leningrad*) **27**, 2179 (1985), CA103(26):224685 K. [In Russian]
157. J. T. Yardley and A. Gupta, *Reza Kenkyu* **12**, 394 (1984), CA 102(6):53527h. [In Japanese]
158. T. Hayashi and T. Nagayma, *Nippon Kagaku Kaishi* (6), 1050 (1984). [In Japanese]
159. V. I. Novikov, L. I. Trusov, V. N. Lapovok, and T. P. Geleishvili, *Poroshk. Metall. Kiev* **29** (1984), CA101(6):42050a. [In Russian]
160. T. Yoshida and K. Akashi, *Trans. Jpn. Inst. Met.* **22**, 371 (1981).
161. K. Ohshima, S. Yatsuya, and J. Hanada, *J. Phys. Soc. Jpn.* **50**, 3071 (1981).
162. S. C. Richtsmeier, E. K. Parks, K. Lin, L. G. Pobo, and S. J. Riley, *J. Chem. Phys.* **82**, 3659 (1985).
163. E. K. Parks, G. C. Nieman, L. G. Pobo, and S. J. Riley, *J. Chem. Phys.* **88**, 6260 (1988).
164. E. K. Parks, B. H. Weiller, P. S. Bechthold, W. F. Hoffman, G. C. Nieman, L. P. Pobo, and S. J. Riley, *J. Chem. Phys.* **88**, 1622 (1988).
165. S. J. Riley, *Z. Phys. D: At. Mol. Clusters* **12**, 537 (1989).

166. R. L. Whetten, D. M. Cox, D. J. Trevor, and A. Kaldor, *J. Phys. Chem.* **89**, 566 (1985).
167. M. R. Zakin, R. O. Briekman, D. M. Cox, and A. Kaldor, *J. Chem. Phys.* **88**, 6605 (1988).
168. D. P. Onwood and A. L. Campanion, *J. Phys. Chem.* **89**, 3777 (1975).
169. (a) M. E. Geusic, M. D. Morse, and R. E. Smalley, *J. Chem. Phys.* **82**, 590 (1985); (b) M. R. A. Blomberg, P. E. M. Siegbahn, and M. Svenson, *J. Phys. Chem.* **96**, 5783 (1992).
170. P. Fayet, M. J. McGlinchey, and L. H. Wöste, *J. Am. Chem. Soc.* **109**, 1733 (1987).
171. J. W. Lauher, *J. Am. Chem. Soc.* **100**, 5305 (1978).
172. D. M. P. Mingos and D. J. Wales, *J. Am. Chem. Soc.* **112**, 930 (1990).
173. (a) D. M. Cox, K. C. Reichmann, D. J. Trevor, and A. Kaldor, *J. Chem. Phys.* **88**, 111 (1988); (b) A. Goursot, I. Papai, and D. R. Salahub, *J. Am. Chem. Soc.* **114**, 7452 (1992).
174. (a) P. Fayet, A. Kaldor, and D. M. Cox, *J. Chem. Phys.* **92**, 254 (1990); (b) M. R. A. Blomberg, *J. Phys. Chem.* **96**, 5783 (1992).
175. D. J. Trevor, D. M. Cox, and A. Kaldor, *J. Am. Chem. Soc.* **112**, 3742 (1990).
176. R. L. Whetten, D. M. Cox, D. J. Trevor, and A. Kaldor, *J. Phys. Chem.* **89**, 566 (1985).
177. (a) D. J. Trevor, R. L. Whetten, D. M. Cox, and A. Kaldor, *J. Am. Chem. Soc.* **107**, 518 (1985); (b) P. Schnabel, M. P. Irion, and K. G. Weil, *J. Phys. Chem.* **95**, 9688 (1991); (c) E. N. Rodriguez-Arias, L. Rincon, and F. Ruette, *Organometallics*, **11**, 3677 (1992).
178. K. D. Bier, T. L. Haslett, A. D. Kirkwood, and M. Moskovits, *J. Chem. Phys.* **89**, 6 (1988).
179. M. Moskovits, Ed., "Metal Clusters," Wiley, New York, 1986.
180. W. E. Klotzbucher and G. A. Ozin, *Inorg. Chem.* **19**, 3767 (1980).
181. K. J. Klabunde and T. O. Murdock, *J. Org. Chem.* **44**, 3901 (1979).
182. S. C. Davis, S. Severson, and K. J. Klabunde, *J. Am. Chem. Soc.* **103**, 3024 (1981).
183. S. C. Davis and K. J. Klabunde, *Chem. Rev.* **82**, 153 (1982).
184. R. C. Baetzold, *Solid State Commun.* **44**, 781 (1982).
185. E. Shustoravich, R. C. Baetzold, and E. L. Muetterties, *J. Phys. Chem.* **87**, 1100 (1983).
186. A. W. Olsen, Ph.D. thesis, Kansas State University, 1989.
187. N. Kilmer, N. Mason, D. Lambrick, P. D. Hooker, and P. L. Timms, *J. Chem. Soc. Chem. Commun.*, 356 (1987).
188. M. P. Andrews and G. A. Ozin, *Chem. Mater.* **1**, 174 (1989).
189. M. A. Marcus and M. P. Andrews, *Appl. Phys. Lett.* **67**, 1076 (1990).
190. I. Nakatani, T. Furabayashi, T. Takahashi, and H. Hanaoka, *J. Magn. Magn. Mater.* **65**, 261 (1987).
191. C. F. Kernizan, K. J. Klabunde, C. M. Sorensen, and G. C. Hadjapanayis, *Chem. Mater.* **2**, 70 (1990).
192. K. Easom, Ph.D. thesis, Kansas State University, 1992.
193. G. Cardenas-Trivino, K. J. Klabunde, and E. B. Dale, *Langmuir*, **3**, 986 (1987).
194. J. S. Bradley, J. Miller, E. W. Hill, and M. Melchior, *ACS Symp. Ser.* **437**, 160 (1990).
195. K. J. Klabunde, Y. X. Li, and B. J. Tan, *Chem. Mater.* **3**, 30 (1991).
196. K. Matsuo and K. J. Klabunde, *J. Catal.* **73**, 216 (1982).
197. K. J. Klabunde and Y. Tanaka, *J. Mol. Catal.* **21**, 57 (1983).
198. L. F. Nazar, G. A. Ozin, F. Hugues, J. Godber, and P. Rancount, *J. Mol. Catal.* **21**, 313 (1983).
199. K. Matsuo and K. J. Klabunde, *J. Org. Chem.* **47**, 843 (1982).
200. S. J. Woo, J. Godber, and G. A. Ozin, *J. Mol. Catal.* **52**, 241 (1989).

201. D. H. Ralston and K. J. Klabunde, *J. Appl. Catal.* **3**, 13 (1982).
202. P. F. Meier, F. Pennella, K. J. Klabunde, and Y. Imizu, *J. Catal.* **101**, 545 (1986).
203. L. F. Nazar, G. A. Ozin, F, Hugues, J. Godber, and D. Rancount, *Angew. Chem. Int. Ed. Eng.* **22**, 624 (1983).
204. P. S. Skell and S. N. Ahmed, *J. Catal.* **125**, 525 (1990).
205. S. Taylor, E. M. Spain, and M. D. Morse, *J. Chem. Phys.* **92**, 2698 (1990).
206. (a) S. Taylor, E. M. Spain, and M. D. Morse, *J. Chem. Phys.* **92**, 2710 (1990); (b) E. M. Spain and M. D. Morse, *J. Chem. Phys.* **97**, 4605 (1991); (NiAu, PtCu), see also p. 4623 (NiCu) and p. 4641 (NiCu).
207. G. N. Glavee, C. F. Kernizan, K. J. Klabunde, C. M. Sorensen, and G. C. Hadjipanayis, *Chem. Mater.* **3**, 967 (1991).
208. G. N. Glavee, K. Easom, K. J. Klabunde, G. C. Hadjipanayis, and C. M. Sorensen, *Chem. Mater.* **4**, 1360 (1992).
209. D. Zhang and K. J. Klabunde, unpublished work.
210. K. J. Klabunde and Y. Imizu, *J. Am. Chem. Soc.* **106**, 2721 (1984).
211. Y. Imizu and K. J. Klabunde, in "Catalysis of Organic Reactions" (R. L. Augustine, Ed.), p. 225, Dekker, New York, 1985.
212. K. J. Klabunde and Y. Imizu, U.S. Patent, 4,588,708, 1986.
213. B. J. Tan, K. J. Klabunde, T. Tanaka, H. Kanai, and S. Yoshida, *J. Am. Chem. Soc.* **110**, 5951 (1988).
214. B. J. Tan, K. J. Klabunde, and P. M. A. Sherwood, *J. Am. Chem. Soc.* **113**, 855 (1991).
215. H. Kanai, B. J. Tan, and K. J. Klabunde, *Langmuir* **2**, 760 (1986).
216. B. J. Tan, K. J. Klabunde, and P. M. A. Sherwood, *Chem. Mater.* **2**, 186 (1990).
217. S. J. Woo, J. Godber, and G. A. Ozin, *J. Mol. Catal.* **52**, 241 (1989).
218. V. Akhmedov and K. J. Klabunde, *J. Mol. Catal.* **45**, 193 (1988).
219. Y. X. Li and K. J. Klabunde, *New J. Chem. (France)* **12**, 691 (1988).
220. Y. X. Li and K. J. Klabunde, *Hyperfine Interact.* **41**, 665 (1988).
221. Y. X. Li and K. J. Klabunde, *J. Catal.* **126**, 173 (1990).
222. Y. X. Li and K. J. Klabunde, *Langmuir* **3**, 558 (1987).
223. Y. X. Li, Y. F. Zhang, and K. J. Klabunde, *Langmuir* **4**, 385 (1988).
224. R. J. Van Zee, J. J. Bianchini, and W. Weltner, Jr., *Chem. Phys. Lett.* **127**, 314 (1986).
225. I. Shim and K. A. Gingerich, *Surf. Sci.* **156**, 623 (1985).
226. I. Shim and K. A. Gingerich, *J. Chem. Phys.* **81**, 5937 (1984).
227. A. Lagerquist, H. Neuhas, and R. Scullman, *Z. Naturforsch.* **20a**, 751 (1965).
228. B. Kaving and R. Scullman, *J. Mol. Spectrosc.* **32**, 479 (1969).
229. J. M. Brom, Jr., W. R. M. Graham, and W. Weltner, Jr., *J. Chem. Phys.* **57**, 4116 (1972).
230. P. Hooker, B. J. Tan, K. J. Klabunde, and S. Suib, *Chem. Mater.* **3**, 947 (1991).
231. (a) D. W. Green, G. T. Reedy, and J. G. Kay, *J. Mol. Spectrosc.* **78**, 257 (1979); (b) R. J. Van Zee, Y. M. Hamrick, S. Li, and W. Weltner, Jr., *J. Phys. Chem.* **96**, 7247 (1992).
232. S. J. Riley, E. K. Parks, G. C. Nieman, L. G. Pobo, and S. Wexlar, *J. Chem. Phys.* **80**, 1360 (1984).
233. G. Cardenas-Trivino, P. B. Shevlin, and K. J. Klabunde, *Inorg. Chim. Acta* **131**, 1 (1987).
234. K. Haneda, X. Z. Zhou, A. Morrish, R. J. Polland, and T. Majima, *Hyperfine Interact.* **54**, 551 (1990).
235. R. J. Van Zee, S. Li, Y. M. Hamrick, and W. Weltner, Jr., *J. Chem. Phys.* **97**, 8123 (1992).

Copper and Zinc Group Elements (Groups 11 and 12)

I. Copper and Zinc Group Metal Atoms (Cu, Ag, Au, Zn, Cd, Hg)

A. Occurrence and Techniques

All of these metals are relatively volatile. Mercury, of course, is so volatile that its vapors are ever present in laboratory vacuum systems and it is a problem in the lower atmosphere, due to its toxicity. The sources of Hg vapor in the atmosphere are industrial processes and volcanoes.[1]

The vapors of these metals and their ions, oxides, and sulfides have been detected in stars, as is true of most elements.[1] In the laboratory, the vapors are easily produced by a wide variety of methods, and these procedures and the properties of these elements have been reviewed earlier.[1] Likewise, the early chemistry described for these atoms has been reviewed, and in particular abstraction and oxidation addition reactions were considered in detail.[1]

B. Physical Properties and Theoretical Studies

It is rare that the Auger or XPS spectroscopic properties of gas-phase atoms can be determined. The first report of an Auger electron spectrum of Au atoms was made possible by the use of a new spectrometer with an inductively heated oven. The $4f$ binding energies for free Au atoms were determined to be 91.60 and 95.25 eV for the $4f\frac{7}{2}$ and $4f\frac{5}{2}$ levels, respectively. Furthermore, the values of 2.5 and 10.5 eV were obtained for the free atom to solid binding and Auger energy shifts, respectively.[2] In a similar way, Cu atoms have been studied. The free atom-binding energies for $2p\frac{3}{2}$ and $2p\frac{1}{2}$ were determined as 939.85 ± 0.15 and 959.45 ± 0.15 eV, respectively. This is a 2.5-eV shift from the solid state and 13.2 eV for the Auger shift.[3]

Zinc and Cd atoms have also been analyzed by XPS in the gas phase.[4] The binding energies measured were $Zn(2p\frac{1}{2}) = 1052.3$ eV, $Zn(2p\frac{3}{2}) = 1029.1$, $Cd(3d\frac{3}{2}) = 418.8$ eV, and $Cd(2d\frac{5}{2}) = 412.0$ eV. A decrease of 2.8 eV for $Zn(2p)$ and 3.3 eV for $Cd(3d)$ was found ongoing from the free atom to the bulk metal. This difference is interpreted with regard to extra atomic relaxation in solids.

The effects of noble-gas hosts on the properties of matrix-isolated Cu and Ag atoms have been studied by a variety of techniques, including magnetic circular dichroism (MCD) and optical absorption.[5] For the $2p \leftarrow {}^2S$ transition in Cu atoms in argon, MCD and optical spectra were recorded over the 13–25 K temperature range. An observed 25% reduction in the excited state spin–orbit coupling constant was attributed to the overlap of Cu_{4p} orbitals with the Ar_{3p} orbitals. Thus, a significant Cu–Ar interaction is evident. Furthermore, the matrix lattice modes seem to have an important influence on the spectroscopic appearance of matrix-isolated atomic species and on the fact that the Jahn–Teller effect may be a general phenomenon in matrices. Interestingly, it was found through calculations that the excited-state geometry of the Cu/rare gas cage is distributed via the Jahn–Teller effect in opposite ways in krypton and xenon. Again, spin–orbit coupling constant reduction was attributed to the overlap of the Cu_{4p} orbital with the np orbital of Ar, Kr, or Xe. Indeed, Cu was shown to be unique, exhibiting an excited-state spin–orbit coupling constant whose sign is dependent upon the matrix: positive in Ar and Kr and negative in Xe.[6]

Anisotropic hyperfine interactions of 12 Xe nuclei surrounding a matrix-isolated Cu atom (octahedral substitutional site) were computed using a multiple scattering local-density-functional method.[7] Agreement with experimental hyperfine splitting was within 25%. The Xe hyperfine tensor was found to be anisotropic and due to a weak "σ-type" complex between the Cu and Xe atoms.

Silver atoms have also come under considerable scrutiny in noble-gas matrices. A time-resolved fluorescence study of energy dissipation in matrix-isolated Ag atoms has been reported. A detailed model was established from a systematic analysis of rise and decay times of all the observed fluorescence bands. There is an overall dependence of decay time on wavelength, but this dependence was not observed within an individual emission band.[8]

Two photon excitation and fluorescence lifetimes of silver atoms in noble-gas matrices have been studied by Mehretcab et al.[9] This excitation method exposed bands different from one-photon absorption spectra, and these were interpreted to be due to $^2S \leftarrow {}^2D$ transitions. It was suggested that Ag(5P) atoms must significantly distort the surrounding lattice even along coordinates that do not correspond to exiplex formation.

When ESR and UV–VIS spectra of Ag atoms in N_2 matrices were recorded, two types of Ag atoms were observed under dilute conditions. The well-known triplet absorption $5s\ {}^2S \rightarrow 5p\ {}^2P$ was perturbed apparently due to the Ag atoms being in nanocrystalline sites as well as noncrystalline regions. Annealing to 20 K caused the absorption for Ag atoms in noncrystalline N_2 regions to disappear.[10]

X-ray irradiation of Ag atoms in Ar, Kr, and Xe matrices allowed emission spectra to be observed. The spectra consist of broad lines that were red shifted by about 1 eV relative to absorption lines. Annealing after X-ray irradiation caused thermoluminescence of all Ag emission lines. It was concluded that excitons were partly trapped at Ag sites and gave rise to these emission spectra. However, part of the excitons led to production of trapped charge carriers that were released upon annealing. Different site-trapping configurations were implicated.[11]

Exciton trapping of Ag and Au atoms in rare-gas matrices has also been investigated.[12] A very interesting phenomenon was observed where fluorescence of the Ag and Au atoms took place during excitation of matrix rare gas atoms, demonstrating that energy transfer from host to guest was occurring. It was concluded that exciton trapping led to an excitation of the same inner shell states that are populated by *direct* inner shell excitation of the metal atoms.

Related studies on the mechanism of relaxation of excited-state Hg atoms in N_2 matrices have suggested that the interaction between one atom of mercury and one molecule of nitrogen is sufficient to explain the experimental data. The temperature dependence of the spectrum, quantum yield, and lifetime of the UV emission from the exciplex $(Hg \cdot N_2)^*$ was explained by two independent temperature effects on mixing and branching ratios between the electronic and vibrational relaxation of vibrationally hot $(Hg \cdot N_2)$.* Theoretical ab initio calculations on the HgN_2 system suggest the existence of a weakly bound complex in the excited state correlating to $Hg(^3P) + N_2(^1\Sigma_g^+)$ in the linear Hg–N–N configuration, and this complex probably plays an important role in the different relaxation paths of $Hg(^3P)$.[13,14]

Laursen and Carland have reported on the frequency shifts that Hg, Cd, and Zn undergo when trapped in rare gas matrices at 12 K.[15] As expected, based on earlier work with transition-metal atoms,[16] the shifts to higher energy plotted approximately linearly with inert gas polarizability, but these authors showed that the *slope* of the line depended on the transition, $^1P_1 \leftarrow {}^1S_o$ or $^3P_1 \leftarrow {}^1S_o$. In other words, the matrix-induced frequency shift is also dependent on the excited state in question. First singlet excited states of the gas-phase M–rare gas (M = Zn,Cd,Hg) complexes, $C^1\Pi$, which correlate with M $^1P_1$ + rare gas, are strongly bound, whereas triplet states should be more weakly bound. In a matrix the analogous effect was observed. The excited singlet interacts more strongly than the triplet and therefore shows greater variation as the rare gas changes from Ar to Kr to Xe. A discussion of matrix shifts vs gas-phase shifts (in the presence of rare gases) indicated that blue shifts are best explained as a repulsive effect for Ar, rather than the M–Xe interaction being especially attractive. Thus, matrix shifts usually are to the blue and depend on the multiplicity of the excited state and the polarizability of the rare gas. On the other hand, gas-phase shifts are usually to the red compared with the free, gas-phase metal atom, and this is attributed to greater interaction of the upper electronic state with the rare gas in question.[17–19]

Multiphoton ionization of gas-phase Hg atoms and Hg_2 complexed to NH_3 has been reported.[20] As discussed earlier in Chapters 4 and 5, it has been shown that with some irregularities, generally ionization energies (IE) fall as cluster size increases, eventually to the work function of the bulk metal. However, an opposite trend has been reported for Ag_n in liquid solvents.[21] This was explained in terms of a better solvation stabilization of smaller (ionized) Ag_n^+ species. Thus, the comparison of *solvated* gas-phase metal atoms/clusters is of interest. In the case of Hg $(NH_3)_m$ and $Hg_2(NH_3)_m$, it was found that the ammonia solvation caused a significant lowering of the IE^s, as m goes from 1 to 6, IE fell more and more, although the first NH_3 molecule had the greatest effect, and IE decreased about 3 eV overall. For $Hg_2(NH_3)_m$, the effect was similar but less pronounced. And for the same m, the ionization onsets of $Hg(NH_3)_m$ are *lower* than those of $Hg_2(NH_3)_m$. Thus, the Hg^+ species is stabilized most strongly by NH_3 solvation.

C. Chemistry

1. Abstraction

Earlier work[1] showed that Cu and Ag atoms were effective halide abstraction reagents when codeposited with boron halides or organic halides. Recent work has taken a different direction, and perhaps the most significant results deal with the formation of M—CH_2 complexes:

$$M + CH_2N_2 \rightarrow M\text{-}CH_2 + N_2.$$

In the case of Cu, both Cu—CH_2 and N_2Cu—CH_2 form spontaneously upon codesposition onto a window at 12 K. Interestingly, the N_2Cu—CH_2 complex can be dissociated upon photolysis with visible light and photoassociated with UV light in Ar matrices.[22] However, in N_2 matrices only N_2Cu—CH_2 was observed, and it could not be photodissociated. These species were spectroscopically characterized, and for Cu=CH_2 the $\nu_{Cu\text{-}C}$ was found at 614.0 cm^{-1} (Table 6-1). This can be compared with the Zn=CH_2 species obtained when the Zn—N_2CH_2 adduct was photolyzed with visible light in an Ar matrix.[23] A slightly lower $\nu_{M\text{-}CH_2}$ value was found for the Zn analog, suggesting a weaker bonding mode. Photolysis of the Zn=CH_2 species with UV light caused a hydrogen shift to form H—Zn≡CH, and for this system the progressively higher $\nu_{Zn\text{-}C}$ could be compared with Zn—CH_3, Zn=CH_2 and Zn≡CH (Table 6-1).

Knight and co-workers[24] prepared Cu=CH_2 by the laser evaporation of copper in the presence of cyclopropane followed by matrix trapping, as well as by coevaporation of copper and carbon in the presence of H_2. They were able to obtain an ESR spectrum that indicated that the unpaired electron occupied the carbon p orbital primarily, which was perpendicular to the C_{2v} molecular plane, confirming a ^2B ground state. Their results suggested that the Cu=CH_2 bond is very weak; the degree of bonding donation from C : $2p_x$ to Cu : $4p_x$ is almost neligible. The best picture for Cu=CH_2 is to consider the CH_3 radical as having a H replaced by a large copper ($4s^1$) atom.

Additional M atom abstraction chemistry has dealt with photoexcited metal atoms in matrices.[25,26] For example copper atoms react with H_2 upon UV photolysis.[25,26]

$$\begin{array}{c} \quad\quad\quad h\upsilon \\ \quad\quad\quad 310\ nm \\ Cu\ +\ H_2\ \xrightarrow{\quad\quad\quad}\ CuH\ +\ H \\ (^2p)\quad\quad\quad Kr, \\ \quad\quad\quad 10\text{-}12\ K \end{array}$$

$$\begin{array}{c} \quad\quad warm \\ CuH\ +\ H\ \xrightarrow{\quad\quad}\ CuH_2\ or\ Cu\ +\ H_2 \end{array}$$

Upon annealing, the product CuH was found to react with H atoms in the matrix with a very low E_a and negligible diffusion control. The reaction is the microscopic reverse of the photochemical reaction of excited Cu(^2P) atoms with H_2.

TABLE 6-1

IR Data for M—CH$_3$, M=CH$_2$, and HM≡CH Species

Species	ν_{M-C} (cm^{-1})	CH$_2$(s stretch) (cm^{-1})	CH$_2$(a stretch) (cm^{-1})	ν_{M-H} (cm^{-1})
Cu=CH$_2$	614.0	2960.7	3034.7	—
Zn=CH$_2$	513.7	2956.1	3047.2	—
Zn—CH$_3$	447.1	—	—	—
HZn≡CH	647.5	—	—	1924.4

Photosensitized reactions with Hg atoms in the ^{3}P state with ethyl halides in Kr matrices showed that some insertion of Hg into C—Cl and C—Br bonds occurred, a process absent in gas-phase photochemistry. However, the primary processes occurring were elimination reactions:[27]

$$Hg^* + EtCl \rightarrow HCl—C_2H_4 + Hg$$

$$Hg^* + EtF \rightarrow H_2 + CH_2=CHF + Hg.$$

It was also concluded that sensitizer and acceptor need not be close in proximity for an elimination process to occur, and energy transfer over distances as large as 27 Å was consistent with experimental results.

Similarly, Hg(^{3}P), Cd(^{3}P), and Zn(^{3}P) atoms react with vinyl halides in a krypton matrix.[28] The primary process in all cases was HX elimination and formation of a HX—C$_2$H$_2$ hydrogen-bonded complex. Insertion of metal atoms into C—Cl and C—Br bonds, but not C—H or C—F bonds, was also observed. This insertion photochemistry was explained by a mechanism that requires that the process occur on a triplet surface with the vinyl halide in a planar ground-state conformation.

On considering the chemistry of these excited-state metal atoms, the gas-phase photosensitization work over several decades should be recalled.[29] Recently, such studies have been taken up again, and mercury-sensitized dimerization of funtionalized organic compounds has been carried out in the presence of H$_2$. In this case H· was formed by the Hg(^{3}P) + H$_2$ reaction, and H·-abstracted hydrogen atoms from the organic compounds in question, or added to C=C bonds.[30] The resultant organic radicals dimerized. Some selectivity in organic synthesis was achieved by this method.

2. Oxidative Addition

A reaction of Au atoms with CF$_3$I resulted in a product apparently due to abstraction *and* oxidative addition.[31]

$$Au + CF_3I \longrightarrow (CF_3)_2 Au \underset{I}{\overset{I}{\diagup \diagdown}} Au(CF_3)_2$$

When a copper film coated with reactants was laser vaporized, Cu atoms reacted with CH_3F, HF, and DF. Matrix trapping of resultant fragments yielded CH_3CuF, $^{13}CH_3CuF$, HCuF, and DCuF, and these were studied by ESR.[32] Magnetic parameters obtained were used to analyze electronic structure trends in CuF_2, HCuF, CH_3CuF, $HCuCH_3$, and Cu—CO.

Copper atoms photoactivated in a CH_4 matrix reacted to form CH_3CuH.[33,34] The $^2P \leftarrow {}^2S$ excitation promotes this reaction. A competing photodimerization reaction due to the ^2D state of Cu in pure Ar or Kr was effectively quenched. Continued photolysis of Cu in pure CH_4 caused a photofragmentation of the CH_3CuH and CH_3Cu and H were formed.

When laser-evaporated Cu or Ag atoms were codeposited with CH_4/Ar on a window at 12 K, both metals were consumed with a reasonably high efficiency.[35a] Products produced were a mixture of CH_3MH, CH_3M, and possibly M—H. The interesting feature about this work is that reaction occurred at all since thermally evaporated Cu/Ag atoms have not shown any evidence of reaction. There are two possible explanations for the higher reactivity of laser-evaporated atoms. First, it is possible that the laser flash causes photoexcitation of previously deposited Cu or Ag atoms. The Ar—Cl excimer laser employed has a wavelength of 301 nm, and both matrix-isolated Cu and Ag atoms do have absorption bands near 301 nm. And it may be significant that Au atoms, which were consumed less efficiently in a CH_4/Ar matrix, do not have absorption bands in that region. Thus, photoactivation of matrix-isolated atoms by stray laser light is a viable explanation. However, a second possibility is that the higher kinetic energy of laser-evaporated atoms provides a way to surmount the activation barriers.

Laser pulses often can operate with a very high power density, perhaps 10^7–10^9 W/cm².[35b] Much of this energy can be converted to kinetic energy of the ablated atoms or ions. For example, using the Stefan–Boltzmann Law:[35c]

$$\text{Power} = P = \sigma T^4, \qquad \text{where } \sigma = 5.67 \times 10^{-12} \text{ W/cm}^2/\text{K}^4$$

$$T = \sqrt[4]{P/\sigma}, \qquad \text{where } P = 10^7 \text{ to } 10^9 \text{ W/cm}^2$$

$$\text{Average kinetic energy} = \langle E \rangle = \tfrac{3}{2} kT, \qquad \text{where } k = 1.38 \times 10^{-23} \text{ J/K}$$

$$\text{so } \langle E \rangle = \tfrac{3}{2} k \sqrt[4]{P/\sigma}$$

$\langle E \rangle$ can range from 5 to 15 eV when $P = 10^7$–10^9 W/cm² (or 36,000–115,000 K).

In contrast, thermal evaporation employs 700–2000 K, so $\langle E \rangle = 0.09$ to 0.26 eV. It is clear from this calculation that laser vaporization/ablation can yield atoms and ions with plenty of kinetic energy to cause chemical reactions upon striking a cold matrix molecule. Thus, reactions that are not possible with thermal atoms can occur.

That Cu and Ag atoms are in d^9s^1 electronic states, believed to be the most reactive configuration, lends another possible explanation.[35] Further work will be necessary to clarify these interesting anomalies.

Copper, Ag, and Au atoms exhibited a fascinating chemistry with allene. Codeposition at 77 K allowed an addition to the internal carbon:[36a]

$$
\begin{array}{c}
\text{M} + \text{CH}_2 = \text{C} = \text{CH}_2 \xrightarrow[\text{codep}]{77\text{K}} \text{CH}_2 - \text{C} - \text{CH}_2 \\
\backslash \\
\text{M}
\end{array}
$$

$$
\text{M} = \text{Cu, Ag, Au} \quad \xrightarrow{\ X\ } \quad \text{M} - \text{CH}_2 - \text{C} = \text{CH}_2
$$

These results are in contrast to addition of other radical species that often add to the terminal carbon to form vinyl radicals.

Lagow and co-workers have extended their chemistry of $\cdot\text{CF}_3$ and $\cdot\text{SiF}_3$ radicals to Cd and Zn atoms.[36b] Ultilizing a radiofrequency discharge on Si_2F_6 (or C_2F_6), SiF_3 radicals were produced and cocondensed with Cd or Zn vapors to form $\text{Cd(SiF}_3)_2$ and $\text{Zn(SiF}_3)$. When diglyme was added to the cadmium compound, a stable, useful reagent $\text{Cd(SiF}_3)_2 \cdot$ diglyme that could be used to transfer the SiF_3 moiety to other metals, such as nickel or palladium, was obtained.[36c,d]

3. Simple Orbital Mixing

a. O_2 and CO   Codeposition of Cu, Ag, and Au at 77 K with O_2/ adamantane yields paramagnetic $1:1$ $\text{M}:\text{O}_2$ adducts that have been characterized by ESR.[37] Most of the unpaired spin density resides in a π^*-orbital on the two oxygen atoms, but with a low but measurable unpaired s-spin density on the metal atom. It was determined that $\text{Ag}—\text{O}_2$ and $\text{Au}—\text{O}_2$ have side-on bonding to O_2, while $\text{Cu}—\text{O}_2$ has end-on bonding. This difference is probably attributable to the closer proximity of the $4d$ and $5d$ orbital energy levels to the π_{xx} level. Kasai and Jones[38] have reported a similar matrix isolation ESR study, but in rare gas matrices at 4 K. Both $\text{M}—\text{O}_2$ and $\text{M(O}_2)_2$ adducts were detected. The complexes were formulated as $\text{M}^+(\text{O}_2)^-$ and $\text{M}^+(\text{O}_2)_2^-$ ion pairs, and $\text{Cu}—\text{O}_2$ and $\text{Ag}—\text{O}_2$ were presented as end-on bonded while $\text{Au}—\text{O}_2$ was side-on bonded. The differences in structure (Howard et al. vs Kasai et al.) were attributed to the different matrix host and temperature. This sensitivity to different matrix conditions is definitely a drawback to matrix isolation work when it comes to structural analogies.

Silver atoms and CO form adducts at 77 K in hydrocarbon matrices.[39] Several complexes including Ag—CO, $Ag(CO)_2$ (bent), $Ag(CO)_3$ (two conformers), and $Ag_n(CO)_x$, were formed. This system is more complex than the analogous Cu/CO system and this may be attributed to the greater mobility of Ag in the forming matrix.[40,41]

b. Alkenes/Alkynes As a follow-up to their many papers on M atom–ethylene complexes published earlier,[1] Ozin and co-workers have reported optical spectra and SCF–X_α–SW calculations.[42] The paramagnetic $Cu–C_2H_4$, $Ag–C_2H_4$, and $Au–C_2H_4$ complexes exhibit UV–VIS spectra that could be compared, experiment vs theory. Generally good agreement was found and bonding arguments were presented. A summary of the total percentage charge found in the intersphere and outersphere regions for the $5a$, $6a$, and $3b_2$ orbitals for each of the three $M–C_2H_4$ complexes was given.

More recently, Howard *et al.* have reported ESR studies of $Cu–C_2H_4$, $Cu(C_2H_4)_2$, and $Cu(C_2H_4)_3$ in hydrocarbon matrices.[43] ESR spectra for these paramagnetic species at 77 K suggest a side-on C_{2v} structure for $Cu–C_2H_4$ and a D_{2h} structure for $Cu(C_2H_4)_2$. This work and earlier work[44] suggest that the bonding involves two dative interactions, electron donation from the ethylene π-orbital into an empty sp_z hybridized orbital on Cu and back donation of d-electron density from the metal $3d_{yz}$ orbital into the ethylene π^*-orbital. The unpaired electron is located in the sp_z orbital directed away from the ligand.

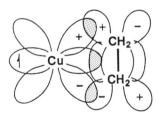

A direct comparison of $Ag(C_2H_4)_n$ complexes shows that $Ag–C_2H_4$ and $Ag(C_2H_4)_2$ form at 77 K in hydrocarbon matrices.[45] However, for $Ag–C_2H_4$ the unpaired spin density has more s character compared with Cu. For $Ag(C_2H_4)_2$ the structure appeared to be a side-on bonded species with D_{2h} geometry with the ligands adapting an eclipsed conformation on either side of the Ag atom. Unpaired spin density resides largely in a $4p$ metal orbital.

Propene has also been matrix deposited with Cu, Ag, and Au atoms.[46] ESR absorption of M–C_3H_6 complexes at 4 K in Ar indicated that $Cu(C_3H_6)$, $Cu(C_3H_6)_2$, $Au(C_3H_6)$, and $Au(C_3H_6)_2$ were present, but for Ag no complexes were formed.

It was concluded that the propene complexes bond with Cu and Au in exactly the same way as ethylene. It is apparent, however, that the formation of such complexes through dative bonding is extremely sensitive to the energy levels of the interacting orbitals, and this is most evident when comparing Ag–C_2H_4 complexes (that do form) with Ag–C_3H_6 complexes (that do not form).

Kasai has also compared bonding of ethylene, propene, and acetylene with Au atoms, by matrix isolation ESR.[47,48] Interestingly, the acetylene reacted spontaneously to yield a vinyl radical species rather than π-complexes with dative bonding.[47–49]

$$Au \;+\; HC \equiv CH \;\longrightarrow\; \begin{array}{c} H \\ \diagdown \\ C = C \\ \diagup \quad \diagdown \\ Au \qquad H \end{array}$$

c. Ammonia and NO_2 Copper atoms codeposited with NH_3/Ar form a Cu–NH_3 complex. Only the 1:1 stoichiometry was observed, but even a small amount of NH_3 was sufficient to retard photoaggregation of Cu to Cu_2, indicating that NH_3 efficiently scavenged Cu atoms. Photolysis at >400 nm caused no changes. However, UV photolysis caused an insertion (probably due to the ^2P state of Cu) to form $HCuNH_2$, and continued photolysis caused further fragmentation to $CuNH_2$, CuH, and H atoms. However, the $HCuNH_2$ species could be driven back to the original adduct by 400-nm light.[50a] In the case of NO_2, a charge transfer complex $Cu^+NO_2^-$ was formed along with $Cu(NO_2)_2$, which is probably $Cu^+(NO_2^-)(NO_2)$:[50b]

$$Cu + NH_3 \;\longrightarrow\; Cu\text{–}NH_2 \underset{\geq 400 \text{ nm}}{\overset{UV}{\rightleftarrows}} HCuNH_2$$

$$\geq 400 \text{ nm.}$$

d. RCN, PN, SiS, GeO, and CH_2 Both HCN and RCN have been codeposited with Cu and Ag atoms in a rotating cryostat (77 K, adamantane matrix).[51,52] Both end-on and side-on bonding was observed, with most of the unpaired spin density residing on the metal.

$$M - N \equiv CR$$

$$M - \overset{\displaystyle N}{\underset{\displaystyle C}{|||}}$$

However, some metal atoms reacted to form organometallic iminyls such as CuCH=N· and AgCH=N·. These species could be intermediates in the reaction of Cu/Ag with HCN to produce H_2 and metal cyanide.

More unusual or exotic ligands have also come under study. For example, SiO–metal atom complexes are of interest due to the analogy between CO and SiO and its relevance to silicon–metal film technology. Schnöckel and co-workers have reported such a study.[53] Silver atoms were codeposited with monomeric SiO in excess Ar, and Ag(SiO) and Ag_2(SiO) were detected by IR. Isotopic shifts implied a strongly bent structure.[53] Similarly, Howard and co-workers obtained ESR spectra of Ag–PN, Ag–SiS, and Ag–GeO complexes in frozen argon.[54] In the Ag–PN complex, about 60% of the unpaired $5s$ spin density is on the Ag nucleus. Calculations suggest that a Ag–P–N bond angle of 97.4° would be preferred. However, the AgGeO and AgSiS should have triangular structures, although experimental evidence for these species was not clear.

II. Copper and Zinc Group Metal Clusters

A. Occurrence and Techniques

Free clusters of these metals are quite fragile, and so they are rarely observed in the environment of cool stars or in the upper atmosphere. However, the stability of these metals in their metallic zero-valent state, especially gold, allows for their observation in the environment in their bulk metallic state. This simple fact has had a remarkable effect on the history of chemistry since Cu, Ag, and particularly Au were established as special. Thus, the ease with which zero-valent metal clusters form has allowed the synthesis of a vast array of cluster sizes in various media. Recent developments will now be reviewed; first small cluster synthesis and then large cluster (colloids) synthesis based on metal atom aggregation (classical methods for forming metal colloids by metal ion reduction are not covered in much depth).

Cryophotochemistry, a term coined by Ozin and co-workers, is based on the idea that metal atoms isolated in cold rare gas matrices can be selectively photodimerized/trimerized. Photolytic excitation of the isolated metal atom is followed by conversion of this energy to matrix warming when the excited metal atom relaxes to its ground state. Matrix softening in the vicinity of the metal atom allows slight movement of the atom in the matrix. Continued photolysis into an atomic absorption band causes further movement until a second metal atom collides with the first, and a dimer is formed. Overtime, since the dimers are generally not absorbing the light and are less mobile than atoms, the atoms can be converted to dimers and trimers.[55,56] However, it has been demonstrated that Ag_3 can be photolytically converted to Ag_2 and Ag as well.[55] Photochemical vs thermal aggregation processes have been compared.[56] Both processes are complex with regard to kinetics, exhibiting pronounced deviations from simple second-order kinetics. The inhomogeneous nature of the matrix is apparently the reason for these complications.

As described in Chapter 2, supersonic He–metal vapor beams can generate metal clusters in the gas phase. One of the first applications of this laser evaporation method was for the preparation and study of Cu_1–Cu_{29} clusters and subsequent determination of ionization energies.[57] In this study it was shown that IE Cu = 7.72 ev, Cu_2 = 7.89, Cu_9 = 5.58, and irregular falling off of IE was observed as cluster size went up. The alternation in ionization threshold behavior was attributed to an even/odd alternation in the electronic structure of the Cu clusters with the highest occupied molecular orbital of the even clusters being considerably more strongly bonding.

Cold Au_3 clusters were prepared in a similar way, and two-photon ionization spectroscopy was carried out.[58] Various possible assignments for the upper and lower states of Au_3 were considered, and the best possibility is the $A^4E' \leftarrow X^2E'$ transition. Apparently both the upper and lower states undergo a Jahn–Teller distortion. Presumably a equilateral D_{3h} structure is involved.

Lindsay and co-workers have described an intricate and powerful new method for matrix isolating mass selected silver and gold clusters in rare gas matrices.[59] Silver cluster cations were produced by sputtering, mass selected in a quadruple mass filter, and then codeposited with Kr in the presence of low-energy electrons, which neutralized the cations. Silver dimers and trimers were studied by excitation and fluorescence spectroscopy. A "soft landing" technique that allowed less than 25% fragmentation of the dimer to monomers was developed. This technique enabled a 10^6 : 1 enrichment of the desired cluster over impurity clusters, and therefore it was possible to assign unambiguously the absorption and emission bands for Ag_3 in a rare gas matrix. This new "soft landing" technique is a significant advance. However, as the authors point out, the translational energies of the impinging clusters are 10 times that of cluster bond energies, so the term soft landing may better be understood in terms of a series of soft collisions that efficiently transfer the excess energy to the matrix.

Larger clusters (fine particles) of Cu, Ag, Au, Zn, and Cd have been prepared by the well-known gas-evaporation method (see Chapter 2). A summary of recent developments is given later in this chapter. At this time it is appropriate to mention the early reports of Wada[60,61] in which many metals were evaporated in He or Xe pressures of 2–16 Torr, and ultrafine particles of Cu, Ag, Au, Zn, and Cd were prepared. An important addition/modification on the gas-evaporation technique was reported by Kimura and Bando in 1983.[62] Here the metal clusters were swept into organic solvents and nonaqueous colloidal solutions were obtained. A series of metals, including Cu, Ag, and Au, were studied. The solvent employed was usually ethanol, although methanol and some alkanes were also tested. Further work was carried out by Teo and co-workers[63] where Au and Ag particles were trapped in EtOH. Analysis by TEM of the particles before and after entering the EtOH showed: 10- to 60-Å clusters before entering the solution, and 30–100 Å (Au) and 50–200 Å (Ag) after. The size increase occurred by aggregation in the EtOH medium, and this took place upon warming from $-78°C$ to room temperature.

Still another method that has proven to be particularly useful is a variation on the solvated metal atom dispersion (SMAD) technique.[64] Herein metal atoms are codeposited with organic solvents at 77 K followed by warming and subsequent cluster formation by atom aggregation. This approach has been particularly useful with Au, and colloidal solutions in acetone, ethanol, isopropanol, DMSO, and DMF were obtained. A novel feature of these solutions is the absence of impurities; only gold particles and solvent are present. For further discussion see Sections III and IV, particularly for work with Au.

An additional procedure for cluster formation in solution involves the photolytic decomposition of metal azides, for example $Hg(N_3)_3^-$.

$$Hg(N_3)_3^- \xrightarrow{h\nu} Hg_n + 3N_2 + N_3^-$$

It was found that Hg atoms were stabilized/solvated for a short period of time in ethanol, but that colloidal mercury also formed.[65]

B. Physical Properties and Theoretical Studies

A number of studies of Cu_2 and Cu_3 where production has been accomplished by cluster beam and matrix aggregation techniques have been reported.[66–78] Smalley and co-workers produced ultracold Cu_2–Cu_{13} in a pulsed cluster beam.[67] Ionization using an Ar–F laser showed that the ionization energies of the smaller clusters alternated above and below the 6.4-EV laser photon energy depending on whether the cluster contained an even (high IE) or odd (low IE) number of atoms.

Extended Hückel calculations, theoretical study of electron affinities, optical spectra, laser spectroscopy, ESR, and other studies have probed the interplay between electronic and geometrical structure especially for Cu_3, believed to be a fluxional molecule.

Lindsay et al.[79] have summarized much of the theoretical and experimental work and carried out an ESR study of Cu_3 in a N_2 matrix. The spectra indicate that for Cu_3, about 36% of the isotropic spin density is on one nucleus and 13% on each of the remaining two nuclei. These data indicate that Cu_3 exists in N_2 as an acute angled isomer with ground-state $^2A_1$. For both Cu_3 and Ag_3, the ground state is $^2B_2$ in organic matrices but $^2A_1$ in N_2, again pointing out the subtle but important effects of different matrix environments.

The Cu_7 cluster has also been ESR analyzed in low-temperature matrices, particularly in Ne at 4 K.[80] The complex spectrum was best fit to a structure possessing two equivalent ^{63}Cu nuclei. Further but weaker splitting is due to five equivalent ring nuclei, and thus a pentagonal bipyramid was proposed as the correct structure, which gains support from other work on Cu_7 as well as alkali metal Mg clusters (see Chapter 3).

Small clusters of silver and gold have received less attention than copper. However, the Ag_3 species is presumed to be distorted as an acute triangular structure in a solid N_2 matrix.[81,82] Also, Ag_3,[83] Ag_5,[84] and Ag_7[85] have been studied. ESR spectra of the trimer, as with Cu_3, were similar to those obtained for the alkali metal trimers (Chapter 3). In Ag_5 and Ag_7 essentially the same spectra were obtained, and it is possible that Ag_7 was the actual species in both studies. The data supported a pentagonal bipyramid in a $^2A_2''$ ground state with D_{5h} symmetry.[85]

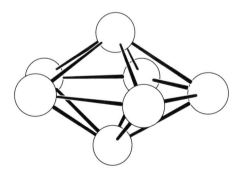

Comparison of spin densities for Na_7, K_7, and Ag_7 indicates that these clusters are closely related in structure, and it was concluded that both series (alkali and copper group) can be treated as having one s electron interacting by a suitable pseudopotential with an inner closed-shell core. Gold clusters Au_3[86] and Au_4[87] have also been studied. Howard et al. trapped Au_3 in benzene at 77 K and the ESR observed consisted of 16 sets of quartets, which suggested a slightly bent structure with a $^2B_2$ (C_{2v}) ground state.[86] Multiconfigurational SCF calculations on Au_4 suggested that the ground state should be a closed-shell $^1A_1$ state. According to these calculations, Au_4 should be more stable than Au_3 or Au_2.[87]

Optical spectroscopy of Au_3 was possible by depositing size-selected clusters in Kr.[88] As discussed earlier, sputtering, mass selection, and "soft-landing" accompanied by neutralization allowed about 5×10^{13} Au_3 species to be deposited in 34 min. Good quality fluorescence spectra for both Au_2 and Au_3 were obtained, and this allowed unambiguous identification of bands (Table 6-2):

TABLE 6-2
Summary of Absorption and Emission Bands for
Au, Au_2, and Au_3[a]

Species	Absorption	Emission	
Au	233[b]	453	816
	261	453	816
Au_2	215[b]	280	750
	251[b]	**665**	810
	299	**665**	**810**
	347	**665**	**810**
Au_3	233	**510**	
	258	**510**	
	281	**510**	
	308[b]	**510**	
	388	580	**660**
	458	580	820

[a]Boldface type denotes dominant or characteristic bands.[88]
[b]Due to doublets (averaged).

Ultraviolet photoelectron spectra of gas-phase copper, silver, and gold clusters anions have been reported.[89] A direct estimate of the vertical electron affinities (EA) as a function of cluster size was possible. The large even/odd alteration of EA in small clusters was found to largely disappear as the cluster size exceeded 40 atoms. And the predicted spherical shell closing at cluster 58 was evident for silver and gold.

Wöste and co-workers have examined gas-phase mercury clusters in an attempt to probe the transition from metallic bonding to van der Waals interactions, with change in cluster size.[90a] Similar considerations for Cd_2

and Zn_2 weakly bound dimers prompted Rodriquez and Eden to study bound free emission spectra and photodissociation.[90b]

Ozin, Kevan, and their coworkers have devised the means of producing and trapping Ag atoms and Ag clusters in zeolite frameworks and have made comparisons with rare gas low-temperature matrices.[91–94] Silver atoms appear to be weakly bound to the zeolite lattice, according to ESR studies compared with water/ethanol-solvated Ag atoms. By increasing silver loading and applying appropriate conditions, Ag_2^+ and Ag_3^+ have been detected by optical spectroscopy.

C. Chemistry

In the studies of bare metal clusters one of the most remarkable findings has been the appearance of electronic shell closings. These shell closings appear for spherical clusters of M_n, where $n = 2, 8, 18, 20, 34, 40$, etc.[95a] These closed-shell clusters have higher thermodynamic stabilities, and generally this is reflected in higher abundances, high ionization energies, low electron affinities, and low reactivities compared with open-shell clusters.

Could open-shell clusters be closed by addition of a chemical reagent? This idea was probed by the study of Cu_6, Cu_7^+, Cu_{17}^+, and other clusters using experimental and theoretical approaches.[95a] The metal atoms were modeled as one-electron systems using effective core potentials and core polarization potentials and chemisorption of CO at various cluster sites modeled. Experiments were carried out on size-selected clusters using FT–ICR. Thermalized neutral or charged clusters were allowed to chemisorb one CO molecule and the relative stability of the resultant adduct was probed by collision-induced dissociation using Ar gas. Table 6-3 gives the energies of chemisorption of CO in electron volts. Considering CO as a two-electron ligand, and each Cu atom as donating one electron, shell closing should occur for Cu_6CO and Cu_7^+CO. The theoretical modeling shows this to be true, and experiments tend to support these findings. Larger clusters, such as Cu_{15}^+CO, $Cu_{17}CO$, and Cu_{19}^+CO all appear to be relatively stable species with Cu_{17}^+CO the maximum, again confirming that the chemisorption of CO could cause shell closing. However, the effect is less noticeable in small clusters, apparently because the lone pair electrons can be added to the unfilled $1p$ shell of the cluster.

Theoretical considerations of CO chemisorption on Cu clusters was reported by Nygren and Siegbahn,[95b] and theoretical treatment of the electronic structure of Cu_6, Ag_6, and Au_6 and their positive ions has been reported by Liao and Balasubramanian.[95c]

TABLE 6-3
Energies of Chemisorption of CO on Cu_n
and Cu_n^{+} [a]

Cluster	Neutral	Cation
Cu_1	0.10	1.18
Cu_2	0.60	1.06
Cu_3	0.93	1.00
Cu_4	1.03	1.04
Cu_5	0.46	1.19
Cu_6	1.06	1.23
Cu_7	0.59	1.42
Cu_8	0.44	0.93
Cu_9	0.72	0.88
Cu_{10}	0.31	0.66

[a]After Smalley and co-workers.[95]

The gas-phase oxidation of silver clusters with ozone has been studied by Gole et al.[96] Chemiluminescent emission from AgO, as well as Ag_2O and Ag_xO were monitored. These data combined with thermodynamic considerations suggest that Ag_3 or larger clusters must be responsible. A large energy release is possible because of the low trimer, $Ag–Ag_2$ bond strength, and possibly some vibrational excitation in the trimer.

$$Ag_3 + O_3 \rightarrow AgO^* + Ag_2O_2.$$

Also, it is more likely that clusters of odd atom combination would react to yield the observed fluorescence since the process $Ag_n \rightarrow Ag_{n-1} + Ag$ requires less energy if n is odd, because the potentially closed-shell even atom combination should show more stability.

Baetzold has studied the chemisorption of halogens on supported Cu and Ag clusters by photoelectron emission.[97a] Silver was evaporated onto a carbon substrate and average cluster size determined by TEM. It was found that different sized clusters gave different rates of chemisorption of halogens from $CHCl_3$. Smaller Ag clusters (<40 atoms) were more reactive than larger clusters, bulk films, or single crystals, supporting the idea that small clusters (in this case Baetzold is referring to Ag_n when n = 3–200 atoms) have more reactive sites. However, little size dependence on reactivity was observed for analogous Cu_n clusters. This effect was explained in terms of the orbital interactions between p orbitals of the halogen and d orbitals of the metal cluster. Thus, halogen-p metal d orbitals are more separated in energy for copper than for silver. Similar results were found when small clusters of silver and nickel were deposited on graphite or alumina surfaces and were allowed to react with O_2, H_2S, and CO.[97b] All clusters were found to be more reactive than bulk Ag or Ni, and it was found that O_2 dissociation on smaller clusters was more favored and less temperature dependent than for large clusters or bulk metal. The change in reactivity was greatest at the metal-insulator transition. These results again reinforce the idea that reactivity is dependent on cluster size.

The growth of Cu and Zn by atom aggregation in alkanes matrices leads to reaction of the growing clusters with the host (see Chapters 4 and 5 for discussion of this with regard to early and late transition-metal atoms/clusters).[35] After codeposition at 77 K followed by slow warming, incorporation of carbonaceous material is one measure of the efficiency of this reaction vs particle growth. Under the same experimental conditions it was found that Cu and Zn particles were much less reactive than Cr, Mn, Co, Ni, or Pd. This trend is basically in agreement with the predictions of Baetzold.[98,99] Thus, as the number of d-electrons increases in passing to the later transition elements, activation energy for dissociative adsorption of alkane C–H bonds should increase. This is apparently related to the need for vacant as well as filled/partially filled metal orbitals for a better synergistic interaction with the C–H or C–C bond.

III. Bimetallic and Binuclear Systems

A. Molecules

Resonant two-photon ionization spectroscopy has been used to study the jet-cooled CuAg dimer. Four vibrational band systems were observed, three of which were rotationally resolved. The ground state is $X^1\Sigma^+$ derived from the $3d_{Cu}^{10}4d_{Ag}^{10}\sigma^2$ configuration. A bond length of 2.3735 $\pm$ 0.0006 Å was estimated. Similarly, CuAu was analyzed and its ground state determined to be $X^1\Sigma^+$ with an estimated bond length of 2.3302 $\pm$ 0.0006 Å. It is interesting that the Cu–Au band length is estimated to be *longer* than the Cu–Ag bond length. Next, the AgAu dimer was analyzed, but in less detail.[100–104]

Morse and co-workers have moved further with two-photon ionization spectroscopy into the study of the coinage metal mixed trimers.[100] For Cu_2Ag, 47 vibrational bands were assigned, whereas for Cu_2Au only seven bands were observed, and this low number may be explained by the occurrence of predissociation. In the case of CuAgAu, 92 vibrational bands corresponding to excitations of three totally symmetric vibrational modes in the C_s point group were assigned. For all three molecules it was argued that the resonant two-photon ionization caused the removal of an electron from a filled $3d^{10}$ subshell of Cu (or $5d^{10}$ subshell of Au) and placed it in a weakly antibonding s-type orbital.

The CuAgCu trimer has also been matrix isolated and investigated by ESR.[105] The silver atom was at the center of this molecule. It was found that the CuAgCu species was slightly bent with a $^2B_2$ ground state (C_{2v} symmetry).

Copper–silver pentamers as detected by ESR trapped at 77 K in a cyclohexane matrix have also been reported.[106] The CuAg$_4$ and Cu$_2$Ag$_3$ species exist as trigonal bipyramidal arrays of metal atoms; CuAg$_4$ has one Cu atom and one Ag atom that bear most of the unpaired spin density. For Cu$_2$Ag$_3$ the two Cu atoms bear most of this spin density.

Of considerable interest is the work of Weltner and co-workers on the matrix isolations of dimers of Mo or W with Cu, Ag, or Au (also see Chapter 4 for discussion of CrCu, CrAg, and CrAu).[107] In argon at 4 K, MoCu, MoAg, WCu, WAg, and WAu were all found to have $^6\Sigma$ ground states. It might be expected that Cr and Mo would behave similarly since their atomic ground states are d^5s^1 electronic configurations. However, W has a d^4s^2 configuration although the d^5s^1 is only about 3000 cm^{-1} (0.4 eV) higher in energy. Thus, it would be expected that Cr, Mo, and W would also form s–s bonds to Cu, Ag, and Au atoms, all of which have $d^{10}s^1$ ground states. The simplest picture would be an s–s bond with all five unpaired electrons remaining mainly on the Cr, Mo, or W atom. Indeed this type of bonding was verified.

Although not related to "free clusters," molecular cluster compounds of gold and silver have recently been synthesized with comparatively high nuclearities, for example $[p\text{-tol}_3P]_{12}Au_{18}Ag_{20}Cl_{14}$. Similarly a Au$_{13}Ag_{12}Cl_7$ core has been reported.[108,109] High nuclearity Au–Sn cluster compounds have also been reported, and this could have considerable significance for understanding of Pt–Sn catalytic materials.[110]

Copper chloride molecules (vapor) have been codeposited with H$_2$/Ar yielding new IR bands that were assigned to a CuCl complex with hydrogen coordinated in a side-on fashion: Cu(n^2-H$_2$)Cl. This serves as a simple model compound for other n^2-H$_2$ complexes, and all IR bands were assigned for H$_2$, HD, and D$_2$ complexes.[111a] In related work, Knight et al.[111b] have laser-vaporized copper in the presence of HF or CH$_3$F and trapped CH$_3$CuF and HCuF in cold matrices (reactive laser vaporization method). ESR analysis of the various isotopes of these species

$$Cu + CH_3F \xrightarrow[\text{vaporize}]{\text{laser}} CH_3CuF$$

were reported and compared with CuF$_2$, HCuF, CH$_3$CuH, and Cu-CO.

B. Clusters/Particles

The solvated metal atom dispersion method has been used to prepare colloidal Au–Sn particles.[112] Both Au and Sn were simultaneously evaporated and the atoms cocondensed at 77 K with a variety of solvents including acetone, ethanol, toluene, and pentane. Upon warming atom agglomeration occurred, and colloidal particles precipitated. Particle sizes and crystallite sizes were measured with variation in solvent/M ratio, matrix warming rate (after codeposition), solvent polarity, and Au/Sn ratio. The primary conclusion from this work was that some selectivity can be realized when bimetallic phases (AuSn, Au_5Sn, Sn) are formed under these low-temperature growth conditions. This requires that in the early stages of cluster/particle growth thermodynamic control is also important; the process is not completely kinetically controlled and therefore is not simply based on statistics of atom–atom collisions. There must be some reversibility in Au–Sn bond formation somewhere along the cluster growth path (see Fig. 6-1).

$$Au(solv) + Sn(solv) \longrightarrow AuSn(solv) \tag{1}$$

$$2\ Au(solv) + AuSn(solv) \longrightarrow Au_2Sn_2(solv) \tag{2}$$

$$Au_2Sn_2(solv) + AuSn(solv) \longrightarrow\!\!\!\!\rightarrow Au_nSn_n(solv) \tag{3}$$

$$Au_nSn_n(solv) + solv \longrightarrow \underset{\substack{\text{stabilized toward}\\ \text{further growth}}}{Au_nSn_n(R)_x(solv)} \tag{4}$$

- Mobility decreases with size. Therefore k_3 decreases.

- Mobility is low at high viscosities. Therefore, k_4 competes with k_3 more effectively.

- Interdiffusion of Au and Sn does not occur from particle to particle, because each particle is a separate entity due to the solvation shell.

FIGURE 6-1   Gold–tin cluster growth in low-temperature solvents.[112]

IV. Particles and Films

A. Occurrence and Techniques

Andrews and coworkers codeposited Cu and Cu/Ag with SF_6.[113] Upon warming, Cu and Cu–Ag particles formed and were isolated in relatively large amounts. Using EXAFS as a probe, it was concluded that these SMAD particles were disordered, and addition of 12% Ag caused a stiffening of the sample against internal vibrations. There also appeared to be some segregation of Cu and Ag in the bimetallic particles.

Colloidal Au has been prepared by the SMAD technique. Gold atoms codeposited with acetone, ethanol, THF, trimethylamine, DMF, and DMSO upon warmup and cluster growth formed stable purple solutions of colloidal gold.[114] The novel feature of these materials is that only solvent and gold particles are present, no by-products of reduction processes or other impurities. Stabilization of the colloidal particles, which were usually about 60 Å in size, occurred by solvation by the polar organic solvent (a form of ligation or steric stabilization) and by electronic effects wherein the particles scavenged electrons from solvent to form anionic particles and cationic solvent molecules both of which were effectively solvated.[115]

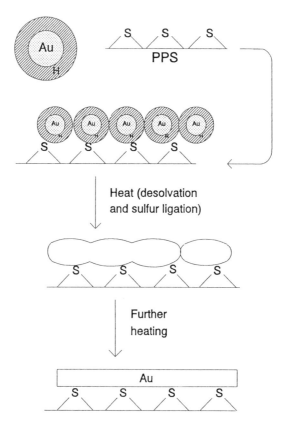

FIGURE 6-2 Adherence of gold films to polyphenylenesulfide. (Gold–acetone colloid where H stands for acetone-derived organic coating.)[119]

The "purity" of these colloidal solutions allows them to have a unique property. The particles are "living colloids" in the sense that upon solvent removal gold films are produced.[114-121] The films are best prepared by "spray painting." Some organic solvent is incorporated into the films, which can be partially removed by heat treating. When the underlying substrate can provide stabilization by ligation to Au atoms, the best adherence was observed, for example, with polyphenylenesulfide (see Fig. 6-2).[119]

XPS studies of gold films prepared from Au–acetone SMAD solutions have provided some understanding of the steps involved in film growth.[118] The film material appeared metallic but contained carbonaceous residue. This "Au–carbon" species exhibited a positive shift in the core level binding energy. By applying a biasing potential, separation of Au_{4f} peaks of the surface species from bulk gold was possible (Fig. 6-3). The results

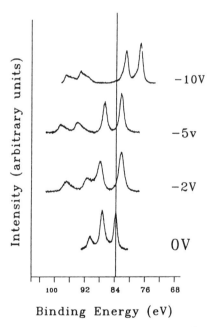

Binding Energy (eV)

FIGURE 6-3 Au_{4f} spectra of a SMAD Au–acetone film under the influence of an applied negative bias potential. The bias potential is indicated on the right-hand side of the diagram. Applying a positive bias is equivalent to drawing electrons away from the surface clusters while applying a negative bias is equivalent to pumping electrons into the surface clusters. The carbonaceous layer around the surface gold clusters acts as a selective barrier to the direction of electron flow, providing a greater resistance to electron flow from bulk gold to the surface gold clusters as compared to the flow of electrons in the opposite direction.[118]

of the biasing experiments are consistent with a model of the surface metal cluster being negatively charged and coated with solvent/solvent fragments (which agree with electrophoresis experiments on the starting colloidal solutions). These clusters slowly amalgamate to bulk-like metal particles and then films. The nature of the substrate and temperature both play a role in determining the rate of film formation and adherence qualities.

A fascinating extension of the SMAD preparation of nonaqueous gold colloids was found in the use of perfluorocarbon solvents, in particular with perfluoro-n-tributylamine $(CF_3CF_2CF_2CF_2)_3N$, PFTA, trademark fluoroinert).[119,121] Upon warmup of a Au–PFTA matrix a dark brown colloidal solution was formed. An interesting feature of the Au–PFTA colloidal solutions was found upon extraction of the solution with acetone. The heavier layer of PFTA–Au was quickly converted to a colorless solution while the upper layer of acetone received the Au particles and formed a purple solution. Thus the Au–PFTA colloidal particles were "soluble" in acetone. However, PFTA/acetone fragments remained bound to the particles, as was demonstrated by IR, NMR, and chemical studies. Models of the acetone- and PFTA-derived particles are shown in Fig. 6-4.

The Au–PFTA/acetone solutions could be used to prepare metallic gold-like films. However, these films remained soluble in acetone or other polar organic solvents. The particles are both lyophilic (solvent loving) and lyoselective.[121] They also possess some of the properties of very large ligand-stabilized molecular clusters. Both EXAFS and XRD measurements indicate that they possess a metallic gold core. Thus, the fluorocarbon moieties coordinate only to surface atoms.

Finally, metal vapor methods have also been employed for the preparation of metal colloids in polymer matrices. For example, deposition of Ag vapor into liquid polybutadiene at room temperature yielded stable colloidal silver.[122] Other liquid polymers behaved similarly, for example, polyolefins, oligoolefins, vinyl and siloxane polymers, crown ethers, and low-molecular-weight ether solvents. Employing optical absorption spectroscopy as a probe, it was determined that the Ag particle sizes did not vary much with silver loading. Apparently, the rate of Ag atom clustering in these media is commensurate with the rate with which polymer molecules can form conformations that form a protective layer around the particle. The coating then prevents further particle growth. Thus, the optical absorption increased linearly with loading (within limits) but λ_{max} did not change. For example, Ag atoms deposited into liquid squalene at 260 K formed colloids about 70 Å in diameter. These particles were stable at 260 K but grew larger at 300 K.

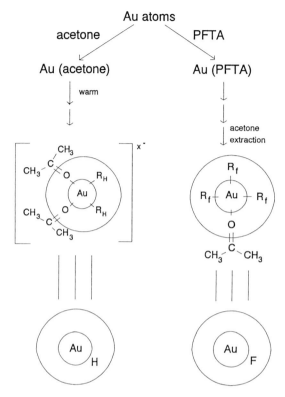

FIGURE 6-4   Gold atom agglomeration in acetone and PFTA (R_H, fragments of acetone; R_f, fragments of PFTA).[119]

One application of these colloidal polymer solutions was deposition of the ultrafine particles on carbon supports followed by removal of the polymer by solvent wash (Scheme 6-1).

Andrews and Kuzyk showed that Ag particles smaller than 100 Å could be grown in solutions of poly(methylmethacrylate) (PMMA) in THF at 145 K. Solid films of the Ag–PMMA composite were formed by spinning and ranged from 1 to 1000 μm thickness.[123] This approach to preparing nonlinear optical films has an advantage over conventional silver salt reduction methods since it avoids ionic by-products of the reduction procedure.

A slightly different approach to producing colloidal metals in polymers, where metal atoms have been codeposited with monomers, has been reported.[124] Thus, colloidal solutions of Cu, Ag, Au, and Pd in

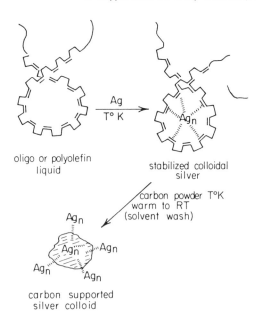

oligo or polyolefin
liquid

stabilized colloidal
silver

carbon powder T°K
warm to RT
(solvent wash)

Ag_n

Ag_n Ag_n

Ag_n Ag_n

carbon supported
silver colloid

SCHEME 6-1   Reprinted with permission from Andrews and Ozin, *Chem. Mater.* **1**, 186 (1989). Copyright 1989 American Chemical Society.

styrene and methylmethacrylate have been prepared by codeposition of metal vapor and monomer vapor at 77 K, followed by warming to room temperature.

These colored solutions were then treated with a conventional free radical polymerization catalyst such as AIBN. Using gold as an example, particle sizes ranged from 70 to 150 Å and did not change upon polymerization of the monomer at 60°C. The metal-doped polymers appeared very homogeneous. With gold, light-purple to deep-purple transparent solids were obtained. With palladium, gray solids were obtained.

Microtoming of the solids allowed good TEM analysis, and this showed that the metal particles accumulated in some regions while other regions were much less densely populated. So apparently during the polymerization process some weak forces that encourage the metal particles to move into the same regions are present, and these particles are trapped there still in segregated form before agglomeration or amalgamation can occur.[124]

Olsen and Kafafi have successfully incorporated gold clusters into polyacetylene film derivatives by similar techniques and were the first to produce interesting nonlinear optical effects in this way.[125]

In related work, ultrafine Ag particles have been prepared by vacuum evaporation on to running-oil substrates.[126] The oils were siloxanes or hydrocarbons. Particles of 30 Å average were obtained and tended to coagulate together. Vacuum removal of the oils left thin metallic-like films as low as 80°C, but which contained significant amounts of organic material.

The copper and zinc metal groups have attracted a great deal of attention by people working in gas-aggregation and ion-cluster beam technology. A summary of the recent literature is given, much of it in tabular form (Table 6-4).

TABLE 6-4

Fine Particle and Thin-Film Formation by Gas-Aggregation and Ion-Cluster Beam (ICB) Technology

Material	Method	Comments	References
Au film	ICB	Au_3^+ was a dominant cluster.	128
Au film	ICB	Atomically flat Au films prepared on polyimide and SiO_2, with high reflectivity for x rays.	129
Au film	ICB	Energy-analyzed Au_2^+ and Au_3^+ were deposited on SiO_2, MgF_2, and Si. Film-substrate interface was markedly influenced by accelerating voltage.	130
Au film	ICB	Combination of high-resolution electron beam lithography and ICB has enabled 25-nm Au lines to be drawn on Si.	131
Ag film	ICB	Ag_n ($n = 100–500$) bombarded an amorphous C Surface; migration of Ag atoms calc.; some clarification of deposition mechanism of three stage film growth.	132
Au particles	Vac. deposition	High-resolution TEM showed 150-Å sphere with some icasahedrons. XPS detected surface Au_2O, Au_2O_3.	133
Ag cluster/films	ICB	Ag cluster ions of over 100 μm diameter were generated. The "droplets" did not freeze before impact on a room temperature substrate.	134
Au films	ICB	120-Å films were produced at 2–3 Å/sec.	135
Ag films	ICB	Ag was evaporated by heating with vibrating and convection to increase evaporation rate. The vapor was then nozzle ejected, ionized, and accelerated.	136
Au films	ICB	XPS studies indicate that Au is more bulk-like when high acceleration voltages are used.	137
Au films	ICB	Au deposited on Al substrate. A high cohesive strength was achieved even if a surface oxide layer was present on the substrate.	138

(continued)

TABLE 6-4 (*continued*)

Material	Method	Comments	References
Au films	ICB	Liquid Au ion source evaluated. Clusters of about 1000 atoms with a charge of approx 20$^+$ were found.	139 140
Zn–Fe–O Pb–O Bi–O Ba–Bi–O	ICB	Reactive ionized-cluster beam deposition. Film orientation could be controlled by adjusting electron current and acceleration voltage. Charge and kinetic energy are controlling factors.	141
Cu, Au, Ni films	Electron beam deposition	Au–Ni bilayers deposited on Cu. Diffusion of Ni through Au studied. Small deposition rate favors formation of crystalline films and makes a more effective diffusion barrier.	142
ZnS on Ge PbF$_2$ on ZnS	ICB	High transmission films with high adhesive strengths. Acceleration voltage play an important role.	143
CdTe films CdT/PbTe	ICB	ICB epitaxy on Si(111) and InSb(111). The CdTe films were high quality of good crystallinity and effective as a buffer for CdTe/PbTe structures.	144
CdTe	Electron beam evaporation	Films has a cubic zinc blend structure with strong (111) orientation.	145
CdTe	ICB	Advantage of ICB preparation method is low temperature. Zincblende or Wurzite structure could be controlled depending on if Cd or Te clusters were ionized.	146 147
ZnS	ICB	Crystallinity is dependent on accelerating, voltage of the ionized clusters. Highly crystalline films can be grown.	148
ZnO	Reactive ICB	Zn clusters were oxidized. Epitaxial growth of ZnO was possible on sapphire substrates.	149
Cu, Ag, Pb	ICB	Two of the original papers on ICB; describes the cluster source developed for deposition and eptiaxy. The new techniques are named ICBD (ion-cluster beam deposition) and ICBE (ion-cluster beam epitaxy).	150 151

The most recent studies of the ICB method have been concerned with the effect of the substrate on which the film grows. For example, Cu-ionized clusters were deposited on Si(111) and Si(100) at room temperature in ultrahigh vacuum.[127] The growth of the Cu films was initially two dimensional but after greater than 10 monolayers a three-dimensional structure appeared. And 100-nm-thick films had quite a uniform crystal structure with epitaxial relation to the substrate. No evidence for intermixing or silicide formation was found. Ion acceleration improved the smoothness of the film. Generally the surface structure of the underlying substrate can have a strong effect on film structure.[152,153]

Cluster size measurements for Ag and Ge ICB samples have been analyzed by a time-of-flight mass spectrometer. Most of the clusters were made of fewer than 35 atoms.[154] Earlier work on ICB of Ag and Ge showed that large clusters could only be formed in nozzle beams generated under extreme conditions.[155] Clusters of about 100 atoms were produced by evaporation into an inert gas. With proper choice of source pressure and orifice diameter, intense cluster beams could be generated, and from 5 to 25% of the clusters could be ionized during the process. The best crystalline films were produced when the average cluster size was relatively small. For Ag deposited on Si, oriented growth was possible with the Si substrate at room temperature. However, for Ge on Si, 230°C was required.

The growth of particles in inert gas is affected by gravity according to experiments of Webb.[156] Gravity-driven convection increases the coalescence rate, leading to larger sizes and broader distribution. In microgravity, smaller particles and narrower size distribution should be achievable.

Another notable feature of the ICB method was reported in a patent by Friedman *et al.*[157] A narrowly mass-selected beam of metal cluster ions (25–106 atoms/cluster) was accelerated to a critical velocity and used to bombard a selected area of a surface (work piece) at a preselected rate. Since cluster ion beams do not penetrate as deeply as do beams of atoms or small molecules, modifications to the surface can be selectively directed. This method can be used to compress an area of the workpiece, change the grain size in the surface, or to coat selected areas.

B. Physical Properties

Gas aggregation has been used to prepare Ag particles of 1–4 nm. The forming particles were examined *in situ* in a molecular beam by electron diffraction. In this way it was shown that multiply twinned particles formed for sizes below 3 nm. Larger sizes were of *fcc* structure. Icosahedral and decahedral structures coexisted in the beam in about equal proportions. These results clearly show a crystal morphology change with particle growth.[158]

Heat capacities of small Ag particles have been examined at 1.6 and 13 K. Particle size was 21 nm and heat capacity was lower than that of bulk silver at 1.6 K, but larger than bulk silver at 13 K. The plot of heat capacity vs temperature showed maxima and minima. These phenomena were interpreted as due to lattice vibrations within a particle.[159]

The surface chemistry of ultrafine silver particles differs from that of bulk silver. Particles 7 nm in diameter were suspended in He gas and O_2 adsorption carried out at room temperature. The sticking coefficient was determined as a function of coverage.[160]

An interpretation for the growth of icosahedral ultrafine particles of Ag has been given.[161] These particles consist of multiply twinned pairs consisting of 20 tetrahedra having cubic structure accompanied by un-avoidable strains and dislocations. This embryo could have icosahedral symmetry and the 20 triangular facets of the icosahedron have the same intraplanar symmetry with the (111) plane of the cubic close-packed struc-ture. Thus, they provide a seed crystal for (111) epitaxial growth of the cubic crystal.

The dynamic behavior of ultrafine particles of Au at the level of atomic resolution was studied by TEM.[162] A real-time video-recording system revealed that the 20-Å gold particles changed shape constantly through an internal transformation from a single crystal to a twinned crystal and back again. The transitions were induced to some extent by irradiation by the electron beam. These morphological changes were abrupt and took place in less than 0.1 sec.

Plasmon absorption bands of small metal particles are frequently ana-lyzed and interpreted in terms of particle size; as particle size decreases the λ_{max} shifts to higher energy. However, such data must be interpreted with caution, since surface effects can cause shifts in λ_{max}. In fact, Ag colloids have been studied under a variety of conditions where different "supports" were compared. It was shown that surface interactions of the Ag particles with the embedding matrix could have significant effects on the position of the plasmon absorption band.[163–165]

References

1. K. J. Klabunde, "Chemistry of Free Atoms and Particles," Academic Press, New York, 1980; and references therein.
2. S. Aksela, M. Harkoma, M. Pohjola, and H. Aksela, *J. Phys. B.* **17**, 2227 (1984).
3. S. Aksela and J. Sivonen, *Phys. Rev. A.* **25**, 1243 (1982).
4. M. S. Banna, D. C. Frost, C. A. McDowell, and B. Wallbank, *J. Chem. Phys.* **68**, 696 (1978).
5. M. Vala, K. Zeringue, J. ShakhsEmampour, J.-C. Rivoal, and R. Pyzalski, *J. Chem. Phys.* **80**, 2401 (1984).

6. M. Vala, K. Zeringue, J. ShekhsEmampour, J.-C. Rivoal, and R. Pyzalski, *J. Chem. Phys.* **80**, 2401 (1984).
7. S. M. Mattar, *J. Phys. Chem.* **92**, 3360 (1988).
8. H. Wiggenhauser, W. Schroeder, and D. M. Kolb, *J. Chem. Phys.* **88**, 3434 (1988).
9. A. Mehretcab, J. R. Andrews, A. B. Smith, III, and R. M. Hochstrasser, *J. Phys. Chem.* **86**, 888 (1982).
10. E. Görlach, R. Rosendahl, U. Klein, H. J. Stöckmann, and A. Schrimpf, *Ann. Phys. Leipzig* **47**(7), 568 (1990).
11. R. Dersch, B. Herkert, M. Witt, H. J. Stöckmann, and H. Ackermann, *Z. Phys. B: Condens. Matter* **80**, 39 (1990).
12. A. Schrimpf, B. Herkert, L. Manceron, U. Schriever, and H. J. Stöckmann, *Phys. Status Solidi B* **165**, 469 (1991).
13. C. Crépine and P. Millié, *Chem. Phys.* **133**, 377 (1989).
14. (a) C. Crépin, F. Legay, N. Legay-Sommaire, and A. Tramer, *Chem. Phys.* **136**, 1 (1989); (b) C. Crépin and A. Tramer, *J. Chem. Phys.* **97**, 4772 (1992).
15. S. Laursen and H. E. Cartland, *J. Chem. Phys.* **95**, 4751 (1991).
16. G. H. Jeong and K. J. Klabunde, *J. Chem. Phys.* **91**, 1958 (1989).
17. D. J. Funk, A. Kvaran, and W. H. Breckenridge, *J. Chem. Phys.* **90**, 2915 (1989); and references therein.
18. D. J. Funk and W. H. Breckenridge, *J. Chem. Phys.* **90**, 2927 (1989).
19. T. Tsuchizawa, K. Yamanouchi, and S. Tsuchiya, *J. Chem. Phys.* **89**, 4646 (1988); and references therein.
20. C. Dadonder-Lardeux, C. Jouvet, M. Richard-Viard, and D. Solgadi, *Chem. Phys. Lett.* **170**, 153 (1990).
21. M. Mostafavi, J. Amblard, J. L. Marignier, and J. Belloni, *Z. Phys. D* **12**, 31 (1989).
22. S.-C. Chang, Z. H. Kafafi, R. H. Hauge, W. E. Billups, and J. L. Margrave, *J. Am. Chem. Soc.* **109**, 4508 (1987).
23. S.-C. Chang, R. H. Hauge, Z. H. Kafafi, J. L. Margrave, and W. E. Billups, *J. Chem. Soc. Chem. Commun.*, 1682 (1987).
24. L. B. Knight, Jr., S. T. Cobranchi, J. Petty, and D. P. Cobranchi, *J. Chem. Phys.* **91**, 4587 (1989).
25. G. A. Ozin and C. Gracie, *J. Phys. Chem.* **88**, 643 (1984).
26. G. A. Ozin, J. Garcia-Prieto, and S. A. Mitchell, *Angew. Chem. Int. Ed. Engl.* **21**, 380 (1982).
27. H. E. Cartland and G. C. Pimentel, *J. Phys. Chem.* **93**, 8021 (1989).
28. H. E. Cartland and G. C. Pimentel, *J. Phys. Chem.* **94**, 536 (1990).
29. E. C. R. Steacie, "Atomic and Free Radical Reactions" 2nd ed., Van Nostrand–Reinhold, Princeton, NJ, 1954.
30. C. A. Muedes, R. R. Ferguson, S. H. Brown, and R. H. Crabtree, *J. Am. Chem. Soc.* **113**, 2233 (1991).
31. J. L. Margrave, K. H. Whitmore, R. H. Hauge, and N. J. Norem, *Inorg. Chem.* **29**, 3252 (1990).
32. L. B. Knight, Jr. S. T. Cobranchi, B. W. Gregory, and G. C. Jones, Jr., *J. Chem. Phys.* **88**, 524 (1988); and references therein.
33. G. A. Ozin, J. G. McCaffrey, and J. M. Parnis, *Angew. Chem. Int. Ed. Engl.* **25**, 1072 (1986); and references therein.
34. G. A. Ozin, J. M. Parnis, S. A. Mitchell, and J. Garica-Prieto, *J. Am. Chem. Soc.* **107**, 8169 (1985); and references therein.
35. (a) K. J. Klabunde, G. H. Jeong, and A. W. Olsen, *in* "Selective Hydrocarbon Activation: Principles and Progress" (J. A. Davies, P. L. Watson, A. Greenberg, and J. F.

Liebman, Eds.), Chap. 13, p. 433, VCH, New York; (b) W. I. Linlor, *in* "Laser Interaction and Related Plasma Phenomena" (H. J. Schwarz and H. Hora, Eds.), pp. 173–185, Plenum, New York, 1971; (c) A. T. Goble and D. K. Baker, "Elements of Modern Physics," pp. 107, 169, Ronald, New York, 1962.

36. (a) J. H. B. Chenier, J. A. Howard, and B. Mile, *J. Am. Chem. Soc.* **107**, 4190 (1985); (b) M. Guerra, T. R. Bierschenk, and R. J. Lagow, *J. Am. Chem. Soc.* **108**, 4103 (1986); (c) M. A. Guerra and R. J. Lagow, *J. Chem. Soc. Chem. Commun.*, 65 (1990); (d) M. Guerra, T. R. Biershenk, and R. J. Lagow, *J. Chem. Soc. Chem. Commun.*, 1550 (1985).

37. J. A. Howard, R. Sutcliffe, and B. Mile, *J. Phys. Chem.* **88**, 4351 (1984).

38. P. H. Kasai and P. M. Jones, *J. Phys. Chem.* **90**, 4239 (1986).

39. J. H. B. Chenier, C. A. Hampson, J. A. Howard, and B. Mile, *J. Phys. Chem.* **92**, 2745 (1988).

40. H. Huber, E. P. Kundig, M. Moskovits, and G. A. Ozin, *J. Am. Chem. Soc.* **97**, 2097 (1975).

41. P. Kasai and P. M. Jones, *J. Am. Chem. Soc.* **107**, 813 (1985).

42. D. F. McIntosh, G. A. Ozin, and R. P. Messmer, *Inorg. Chem.* **19**, 3321 (1980).

43. J. A. Howard, H. A. Joly, and B. Mile, *J. Phys. Chem.* **94**, 1275 (1990).

44. P. H. Kasai, D. MeLeod, Jr., and T. Watanabe, *J. Am. Chem. Soc.* **102**, 179 (1980).

45. J. A. Howard, H. A. Joly, and B. Mile, *J. Phys. Chem.* **94**, 6627 (1990).

46. P. H. Kasai, *J. Am. Chem. Soc.* **106**, 3069 (1984).

47. P. H. Kasai, *J. Am. Chem. Soc.* **105**, 6704 (1983).

48. P. H. Kasai, *J. Phys. Chem.* **92**, 2161 (1988).

49. J. A. Howard, R. Sutcliffe, and J. S. Tse, *Organometallics*, **3**, 859 (1984).

50. (a) D. W. Ball, R. H. Hauge, and J. L. Margrave, *Inorg. Chem.* **28**, 1599 (1989); (b) D. Worden and D. W. Ball, *J. Phys. Chem.* **96**, 7167 (1992).

51. J. A. Howard, R. Suticliffe, and B. Mile, *J. Phys. Chem.* **88**, 5155 (1984).

52. J. A. Howard, R. Sutcliffe, H. Dahmare, and B. Mile, *Organometallics* **4**, 697 (1985).

53. T. Mehner, H. Schnöckel, M. J. Almond, and A. J. Downs, *J. Chem. Soc. Chem. Commun.*, 117 (1988).

54. J. A. Howard, R. Jones, J. S. Tse, M. Tomietto, P. L. Timms, and A. J. Seeley, *J. Phys. Chem.* **96**, 9144 (1992).

55. G. A. Ozin, H. Huber, and S. A. Mitchell, *Inorg. Chem.* **18**, 2932 (1979).

56. S. A. Mitchell and G. A. Ozin, *J. Phys. Chem.* **88**, 1425 (1984).

57. D. Powers, S. G. Hansen, M. E. Geusic, D. L. Michalopoulos, and R. E. Smalley, *J. Chem. Phys.* **78**, 2866 (1983).

58. G. A. Bishea and M. D. Morse, *J. Chem. Phys.* **95**, 8779 (1991).

59. (a) W. Harbich, S. Fedrigo, F. Meyer, D. M. Lindsay, J. Lignieres, J. C. Rivoal, and D. Kreisle, *J. Chem. Phys.* **93**, 8535 (1990); (b) W. Harbich, S. Fedrigo, J. Buttet, and D. M. Lindsay, *J. Chem. Phys.* **96**, 8104 (1992).

60. N. Wada, *Jpn. J. Appl. Phys.* **6**, 553 (1967).

61. N. Wada, *Jpn. J. Appl. Phys.* **7**, 1287 (1968).

62. K. Kimura and S. Bandow, *J. Chem. Soc. Jpn.* **56**, 3578 (1983).

63. Y. Saito, S. Nakahara, and B. K. Teo, *Inorg. Chem. Acta* **148**, 21 (1988).

64. K. J. Klabunde, Y. X. Li, and B. J. Tan, *Chem. Mater.* **3**, 30 (1991).

65. H. Kunkely and A. Vogler, *Polyhedron* **8**, 2731 (1989).

66. Cations of the dimers have also been matric isolated (Cu_2^+, Ag_2^+, Au_2^+). R. J. VanZee and W. Weltner, Jr., *Chem. Phys. Lett.* **162**, 437 (1989).

67. D. E. Powers, S. G. Hansen, M. E. Geusic, A. C. Puiu, J. B. Hopkins, T. G. Dietz, M. A. Duncan, P. R. R. Langridge-Smith, and R. E. Smalley, *J. Phys. Chem.* **86**, 2556 (1982).

68. V. E. Bondybey and J. H. English, *J. Phys. Chem.* **87**, 4647 (1983).
69. G. A. Ozin, S. A. Mitchell, D. F. McIntosh, S. M. Mattar, and J. Garcia-Prieto, *J. Phys. Chem.* **87**, 4651, 4666 (1983).
70. J. L. Gole, J. H. English, and V. E. Bondybey, *J. Phys. Chem.* **86**, 2560 (1982).
71. S. C. Richtsmeler, D. A. Dixon, and J. L. Gole, *J. Phys. Chem.* **86**, 3937 (1982).
72. B. J. Winter, E. K. Parks, and S. J. Riley, *J. Chem. Phys.* **94**, 8618 (1991).
73. D. P. Dihella, K. V. Taylor, and M. Moskovits, *J. Chem. Phys.* **87**, 524 (1983).
74. J. A. Howard, K. F. Preston, R. Sutcliffe, and B. Mile, *J. Phys. Chem.* **87**, 536 (1983).
75. Z. Zwanziger, R. L. Whetten, and E. R. Grant, *J. Phys. Chem.* **90**, 3298 (1986).
76. C. W. Bauschlicher, Jr., S. R. Langhoff, and P. R. Taylor, *J. Chem. Phys.* **88**, 1041 (1988).
77. D. M. Kolb, H. H. Rotermund, W. Schrittenlacher, and W. Schroeder, *J. Chem. Phys.* **80**, 695 (1984).
78. A. T. Amos, P. A. Brook, and S. A. Moir, *J. Phys. Chem.* **92**, 733 (1988).
79. D. M. Lindsay, G. A. Thompson, and Y. Wang, *J. Phys. Chem.* **91**, 2630 (1987); and references therein.
80. R. J. VanZee and W. Weltner, Jr., *J. Chem. Phys.* **92**, 6976 (1990).
81. J. A. Howard and B. Mile, "Electron Spin Resonance," Specialist Periodical Reports, Vol. 11B, London, 1988).
82. C. Kernizan, G. A. Thompson, and D. M. Lindsay, *J. Chem. Phys.* **82**, 4739 (1985).
83. J. A. Howard, K. F. Preston, and B. Mile, *J. Am. Chem. Soc.* **103**, 6226 (1981).
84. J. A. Howard, R. Sutcliffe, and B. Mile, *J. Phys. Chem.* **87**, 2268 (1983).
85. S. B. H. Bach, D. A. Garland, R. J. VanZee, and W. Weltner, Jr., *J. Chem Phys.* **87**, 869 (1987).
86. J. A. Howard, R. Sutcliffe, and B. Mile, *J. Chem. Soc. Chem. Commun.*, 1449 (1983).
87. K. Balasabramanian, P. Y. Feng, and M. Z. Liao, *J. Chem. Phys.* **91**, 3561 (1989).
88. W. Harbich, S. Fedrigo, J. Buttet, and D. M. Lindsay, *Z. Phys. D: At. Mol. Clusters* **19**, 157 (1991).
89. K. J. Taylor, C. L. Pettiette-Hall, O. Cheshnovsky, and R. E. Smalley, *J. Chem. Phys.* **96**, 3319 (1992).
90. (a) C. Bréchignac, M. Broyer, Ph. Cahuzac, G. Delacretaz, P. Labastie, J. P. Wolf, and L. Wöste, *Phys. Rev. Lett.* **60**, 275 (1988); (b) G. Rodriquez and J. G. Eden, *J. Chem. Phys.* **95**, 5539 (1991).
91. G. A. Ozin, *J. Am. Chem. Soc.* **102**, 3301 (1980).
92. N. Narayama, A. S. W. Li, and L. Kevan, *J. Phys. Chem.* **85**, 132 (1981).
93. G. A. Ozin and R. Hugues, *J. Phys. Chem.* **87**, 94 (1983).
94. G. A. Ozin and D. F. McIntosh, *J. Phys. Chem.* **90**, 5756 (1986).
95. (a) M. A. Nygren, P. E. M. Siegbahn, C. Jin, T. Guo, and R. E. Smalley, *J. Chem. Phys.* **95**, 6181 (1991); (b) M. A. Nygren and P. E. M. Siegbahn, *J. Phys. Chem.* **96**, 7579 (1992); (c) D. W. Liao and K. Balasubramanian, *J. Chem. Phys.* **97**, 2548 (1992).
96. J. L. Gole, R. Woodward, J. S. Hayden, and D. A. Dixon, *J. Phys. Chem.* **89**, 4905 (1985).
97. (a) R. C. Baetzold, *J. Am. Chem. Soc.* **103**, 6116 (1989); (b) C. N. R. Rao, V. Vijaya Krishnan, A. K. Santra, and M. W. J. Prins, *Angew. Chem. Int. Ed. Engl.* **31**, 1062 (1992).
98. R. C. Baetzold, *Solid State Commun.* **44**, 781 (1982).
99. E. Shustorovich, R. C. Baetzold, and E. L. Muetterties, *J. Phys. Chem.* **87**, 1100 (1983).
100. G. A. Bishea, C. A. Arrington, J. M. Behm, and M. D. Morse, *J. Chem. Phys.* **95**, 8765 (1991).
101. G. A. Bishea, N. Marak, and M. D. Morse, *J.Chem. Phys.* **95**, 5618 (1991).

102. G. A. Bishea, J. C. Pinegan, and M. D. Morse, *J. Chem. Phys.* **95**, 5630 (1991).
103. G. A. Bishea and M. D. Morse, *J. Chem. Phys.* **95**, 5646 (1991).
104. G. A. Bishea and M. D. Morse, *Chem. Phys. Lett.* **171**, 430 (1990).
105. J. A. Howard, R. Sultcliffe, and B. Mile, *J. Am. Chem. Soc.* **105**, 1394 (1983).
106. J. A. Howard, R. Sutcliffe, and B. Mile, *J. Phys. Chem.* **88**, 2183 (1984).
107. Y. M. Hamrick, R. J. VanZee, and W. Weltner, Jr., *Chem. Phys. Lett.* **181**, 193 (1991).
108. B. K. Teo, H. Zhang, and X. Shi, *J. Am. Chem. Soc.* **112**, 8552 (1990).
109. B. Teo and H. Zhang, *Inorg. Chem.* **30**, 3115 (1991).
110. Z. Demidowicz, R. L. Johnston, J. C. Machell, D. M. P. Mingos, and I. D. Williams, *J. Chem Soc. Dalton Trans.,* 1751 (1988).
111. (a) H. S. Plitt, M. R. Bar, R. Ahlrichs, and H. Schnockel, *Angew. Chem. Int. Ed. Engl.* **30**, 832 (1991); (b) L. B. Knight, Jr., S. T. Cobranchi, B. W. Gregory, and G. C. Jones, Jr., *J. Chem. Phys.* **88**, 524 (1988).
112. Y. Wang, Y. X. Li, and K. J. Klabunde, *in* "ACS Symposium Series" (M. E. Davis and S. L. Suib, Eds.), p. 136, Vol. 517, Chap. 10, *Am. Chem. Soc.,* Washington, DC, 1993.
113. M. A. Marcus, V. E. Lamberti, and M. P. Andrews, *Physica B* **158**, 34 (1989).
114. S. T. Lin, M. T. Franklin, and K. J. Klabunde, *Langmuir* **2**, 259 (1986).
115. M. T. Franklin and K. J. Klabunde, *in* "High Energy Processes in Organometallic Chemistry" (K. S. Suslick, Ed.), ACS Symp. Ser. 333, Chap. 15, p. 246, *Am. Chem. Soc.,* Washington, DC, 1987.
116. G. Cardenas-Trivino and K. J. Klabunde, *Bull. Soc. Chem. Chile* **33**, 163 (1988).
117. G. Cardenas-Trivino, K. J. Klabunde, and B. Dale, "Modeling of Optical Thin Films," SPIE Vol. 821, p. 206, 1987.
118. B. J. Tan, P. M. A. Sherwood, and K. J. Klabunde, *Langmuir* **6**, 105 (1990).
119. K. J. Klabunde, G. Youngers, E. Zuckerman, B. J. Tan, S. Antrim, and P. M. Sherwood, *Eur. J. Inorg. Chem.* **29**, 227 (1992).
120. K. J. Klabunde, *Platinum Met. Rev.* **36**, 80 (1992).
121. E. B. Zuckerman, K. J. Klabunde, B. J. Olivier, and C. M. Sorensen, *Chem. Mater.* **1**, 12 (1989).
122. M. P. Andrews and G. A. Ozin, *Chem. Mater.* **1**, 174 (1989).
123. M. P. Andrews and K. Kuzyk, submitted for publication.
124. K. J. Klabunde, J. Habdas, and G. Cardenas-Trivino, *Chem. Mater.* **1**, 481 (1989).
125. A. W. Olsen and Z. H. Kafafi, *J. Am. Chem. Soc.* **113**, 7758 (1991).
126. Y. Matsuoka and H. Morinaga, *Oyo Butsuri* **49**, 250 (1980). [In Japanese]
127. M. Sosnowski, H. Usui, and I. Yamada, *J. Vac. Sci. Technol. A* **8**, 1470 (1990).
128. F. K. Urban, III and A. I. Bernstein, *Proc. SPIE-Int. Soc. Opt. Eng.* **1323**, 8 (1990). Also see *J. Vac. Sci.; Thin Solid Films* **193**, 92 (1990); *Technol. Al.* **9**, 537 (1991).
129. I. Yamada, G. H. Takaoka, H. Usui, F. Satoh, Y. Itoh, K. Yamashita, S. Kitamoto, Y. Namba, and Y. Hashimoto, *Nucl. Instrum. Methods Phys. Res. B* **B55**, 876 (1991).
130. F. K. Urban, III and M. I. Zahn, *J. Vac. Sci. Technol. A* **8**, 1453 (1990).
131. S. E. Huq, Z. W. Chen, R. A. McMahon, G. A. C. Jones, and H. Ahmed, *Microelec. Eng.* **11**, 343 (1990).
132. Y. Yamamura, *Nucl. Instrum. Methods Phys. Res. B* **45**, 707 (1990).
133. X. Chai, Y. Jiang, Z. Zhu, L. Xiao, T. Li, and R. Guan, *Huagong Yejin* **10**, 51 (1990). [In Chinese]
134. J. Gspann, *Nucl. Instrum. Methods Phys. Res. B* **37/38**, 775 (1989).
135. S. E. Huq, R. A. McMahon, and H. Ahmed, *Thin Solid Films* **163**, 337 (1988).
136. T. Tsukasaki, K. Yamanishi, and M. Yasunaga, *Jpn. Kokai Tokkyo Koho* (1988), CA 109 (22):202264'u). [In Japanese]

137. I. Yamada, H. Usui, H. Harumoto, and T. Takagi, *Mater. Res. Soc. Symp. Proc.* **101,** 195 (1988).
138. Y. Hashimoto, Y. Maeyama, and K. Machida, *Mater. Res. Soc. Symp. Proc.* **77,** 483 (1987).
139. A. Bahasadri, K. Pourrezaei, M. Francois, and D. Nayak, *Mater. Res. Soc. Symp. Proc.* **93,** 259 (1987).
140. M. Francois, K. Pourrezaei, A. Bahasadri, and D. Nayak, *J. Vac. Sci. Technol. B* **5,** 178 (1987).
141. A. Matsumoto, H. Sadamura, A. Inubushi, M. Okubo, S. Masuda, and K. Suzuki, *Mater. Sci. Monogr.* **38B,** 1421 (1987).
142. S. Benhenda, J. M. Guglielmacci, M. Gillet, and T. Pech, *Appl. Surf. Sci.* **28,** 215 (1987).
143. K. Yamanishi, H. Tsukazaki, and S. Yasunaga, *Vacuum* **36,** 157 (1986).
144. T. Takagi, H. Takaoka, Y. Kuriyama, and K. Matsubara, *Thin Solid Films* **126,** 149 (1985).
145. O. Caporaletti and M. R. Westcott, *Can. J. Phys.* **63,** 798 (1985).
146. H. Takaoka, Y. Kuriyama, K. Matsubara, and T. Takagi, *Ext. Abstr. Conf. Solid State Devices Mater.* **16,** 423 (1984).
147. H. Takaoka, K. Matsubara, and T. Takagi, *Proc. Int. Ion Eng. Congr.* **2,** 1241 (1983)
148. Y. Ohkawa, K. Oki, M. Wakitani, S. Miura, and S. Umeda, *Jpn. J. Appl. Phys.* **21,** 551 (1982).
149. K. Matsubara, I. Yamada, N. Nagao, K. Tominaga, and T. Takagi, *Surf. Sci.* **86,** 290 (1979).
150. T. Takagi, I. Yamada, and A. Sasaki, *Conf. Ser.—Inst. Phys.* **38,** 229 (1978).
151. T. Takagi, I. Yamada, and A. Sasaki, *J. Vac. Sci. Technol.* **12,** 1128 (1975).
152. G. H. Takaoka, J. Ishikawa, and T. Takagi, *J. Vac. Sci. Technol. A* **8,** 840 (1990).
153. I. Yamado, *Appl. Surf. Sci.* **43,** 23 (1989).
154. D. E. Turner, K. M. Lakin, and H. R. Shanks, *Mater. Res. Soc. Symp. Proc.* **128,** 125 (1989).
155. A. E. T. Kuiper, G. E. Thomas, and W. J. Schouten, *J. Cryst. Growth* **51,** 17 (1981).
156. G. W. Webb, *Mater. Res. Soc. Symp. Proc.* **87,** 197 (1987).
157. L. Friedman, R. J. Beuhler, M. W. Matthew, and M. Ledbetter, U.S. Pat. Appl. Aval. NTIS Order No. PAT-APPL-6-623-874, 1985 (CA104(10):80614g).
158. B. D. Hall, M. Flueli, R. Monot, and J. P. Burel, *Z. Phys. D: At. Mol. Clusters* **12,** 97 (1989).
159. H. Oya, T. Enoki, N. Wada, *J. Phys. Soc. Jpn.* **59,** 1695 (1990).
160. U. Mueller, A. Schmidt-Ott, and H. Burtscher, *Phys. Rev. Lett.* **58,** 1684 (1987).
161. S. Nara, *Phys. Status Solids B* **149,** 555 (1988).
162. S. Iijima and T. Ichihashi, *Phys. Rev. Lett.* **56,** 616 (1986).
163. K. P. Charle, F. Frank, and W. Schulze, *Ber. Bunsenges. Phys. Chem.* **88,** 350 (1984).
164. I. S. Radchenko, *Opt. Specktrosk.* **55,** 1036 (1983).
165. C. G. Granquist, N. Calander, and O. Hunderi, *Solid State Commun.* **31,** 249 (1979).

Boron Group (Group 13)

I. Boron Group Atoms (B, Al, Ga, In, Tl)

A. Occurrence and Techniques

The occurrence of these atoms in stars and the upper atmosphere (100–210 km) has been discussed, as have their vaporization characteristics.[1] Elemental boron is strikingly more difficult to evaporate than Al, Ga, In, or Tl, and a variety of e-beam and laser techniques have been described. Recent reports have described a Xe–Cl pulsed excimer laser technique that works very well (301 nm, 50–90 mJ/pulse) for the evaporation of small amounts of boron.[2–4] The importance of vapor-deposited Al films has been a driving force for the development of furnaces and crucibles that can contain corrosive molten aluminum.[1] Crucibles of BN–TiB_2, BN, dense graphite, B_4C, AlN, and other refractory materials have been developed for this purpose. The other elements of the group are relatively easy to vaporize from normal W–Al_2O_3 crucibles.

Techniques for study of the chemistry of these atoms at temperatures ranging from 4 to 700 K have also been developed. These include high-temperature fast flow reactors (HTFFR)[5] and the now familiar laser pulse/ flow system methods (see Chapter 2).

A study of the vapors coming off of electromagnetically levitated and laser-heated Mo, W, Zr, and Al samples has been reported.[6] Laser-induced fluorescence was used to probe atom concentrations and electronic states. Vaporization enthalpies were obtainable in some cases, and the authors discussed space-based applications of this method.

Intriguing results coming from laser ablation of Al_2O_3 have been reported. Laser-induced fluorescence was used to measure the energy distributions of Al atoms and Al/O molecules.[7] The excimer laser power was held to a minimum to lower the effects of gas-phase collisions. Kinetic energies for both species were quite high, about 4 eV for Al and 1 eV for Al/O. Surprisingly, the AlO rotational and vibrational energies were quite low, corresponding to about 600 K. These results imply that laser evaporation/ablation is not a boil-off process but rather an electronic ablation mechanism. Related studies have been reported by Huie and Yeung.[8] Particulate matter (atoms and clusters) ejected by laser ablation of metals was monitored from 90° scattering of probe laser light. Spacially and temporally resolved distribution could be recorded and distinct scattering maps obtained as a function of laser power, surface property, absorption characteristics, and metal volatility.

Atomic and cluster ions have been trapped and detected by Fourier transform mass spectrometry (FTMS). These species were produced by laser vaporization of metals and semiconductors. For most metals, only monoatomic ions were detected (most transition metals and Al). Cluster ions were detected for Si, Ge, Bi.[9]

Intense beams of metal ions and atoms of Al and Ni have been produced by a focused CO_2TEA laser.[10] Power densities of 107–109 W/cm^2 were realized. A novel source design was employed in which a parabolic reflector focused the laser beam on a small metal target and the resulting vapors sampled through a small aperture in the reflector. Thus, the vapors moved back along the laser beam's path but with a wider trajectory.

By variation in the laser power, pulsed metal atom beams were produced with kinetic energies variable between 0.1 and 10 eV.

B. Physical Properties and Theoretical Studies

Optical spectra of B and Al atom are interesting to compare.[3,11,12] A feature of these Group 13 atoms is the easy accessibility of a $p \rightarrow d$ transition in the near-UV range, and this is an especially helpful transition in determining cage symmetries in cold matrices (see Table 7-1).[11] Of course the disadvantage in this sense is the extremely high reactivity of atoms with a p^1 configuration; indeed B atoms have been described as the most reactive atom of all the elements.[2,3]

For B the matrix shift is significant for the $2p \rightarrow 3s$ transition, but much smaller for the $2p \rightarrow 2d$. Also for Al, the average matrix shift for the $3p \rightarrow 3d$ transition is much smaller than for the $3p \rightarrow 4s$ transition (for Xe there actually is a red shift), and this shows that the $3d$ level of Al is less affected by the surrounding matrix than the more extended $4s$ level.

TABLE 7-1
Matrix Shifts of UV Absorptions for B and Al Atoms (nm)

		Kr	Xe	Gas phase
	Ar			
B	214 ($2p \rightarrow 3s$)			250
	208 ($2p \rightarrow 2d$)			209
Al	338 nm ($3p \rightarrow 4s$)	362	370	394
	291 ($3p \rightarrow 3d$)	310	326	308
	287 ($3p \rightarrow 3d$)	303	319	
	285 ($3p \rightarrow 3d$)	298	316	

Aluminum atoms have been trapped within cold, gas-phase clusters of argon atoms.[13] Photoionization measurements on $(Al)(Ar)_n$ in a cold cluster beam were carried out when $n < 150$. The ionization energy of the free, gaseous atom (5.99 eV) decreased to 5.5 eV when $n \approx 55$. Simulations of the data by Monte Carlo calculations gave satisfactory agreement when a strong neutral Al–Ar interaction was assumed.

Electron impact ionization cross sections of Ga and In atoms have been measured over a range of 0–200 eV.[14] Vapors (atoms) of Ga and In were produced by neutralization of 3-keV ion beams with triethylamine or Xe. Absolute electron impact cross sections for single, double, and triple ionizations were determined and were found to be considerably larger than predicted. These results were explained by proposing that autoionization of Ga^+ and In^+ spontaneously yields Ga^{2+} and In^{2+}. Thus, the initial electron input ionization removed a d electron followed by autoionization.

Diatomic van der Waals molecules of Al atom–rare gas atoms have been prepared in the gas phase and spectroscopically characterized.[15,16a] Rotationally resolved spectra for Al–Ar were described for the $B^2\Sigma^+ \leftarrow X^2\Pi_{1/2}$ transition. Dissociation energies of approximately 440 cm^{-1} and 180 cm^{-1} were obtained for the B and X states, respectively. These results indicate that the electronically excited state is significantly more strongly bound than the ground state. Indeed, the internuclear separation decreased by 0.74 Å upon excitation, and such changes can be rationalized by an increase in polarization that accompanies electronic excitation of the Al atom. The ground-state bond strengths have been accurately calculated as $D_0[AlAr] = 122.4 \pm 4$ cm^{-1} and $D_0[AlKr] = 194.7 \pm 0.8$ cm^{-1}. These values combined with measured ionization energies allowed bond strengths of the ions to be determined: $D_0 [Al^+Ar] = 982.3 \pm 5$ cm^{-1} and $D_0[Al^+Kr] = 1528.5 \pm 2$ cm^{-1}.

C. Chemistry

1. Abstraction Processes

Gas-phase B($4p$, ^{2}P) reacts with H_2 and D_2 to form electronically excited BH and BD, passing through a highly excited BH_2 (BD_2) intermediate.[16b] Chemiluminescence rate constants for production of $A^1\pi$, $b^3\Sigma^-$, and $C'1\Delta$ electronic states were determined.

Gas-phase B atoms also react with epoxides by an abstraction process, forming an excited state $BO(A^2\Pi)$ molecule.[17] Several organic epoxides were studied under single-collision conditions in a beam gas apparatus, and the nascent vibrational distribution in the electronically excited BO species was determined. A statistical model that assumed that only those modes of the organic product that corresponded to conformation change in the pathway from epoxide to alkene were excited was presented.

Supersonic beams of Al atoms have been obtained by laser vaporization and reactive collisions with O_2, CO_2, and SO_2 probed.[18] The main product, AlO, was detected by laser-induced fluorescence. Collision energy ranges were 0.08–0.49 eV for Al plus O_2, 0.14–0.53 eV for Al plus CO_2, and 0.30–1.19 eV for Al plus SO_2. Oxygen abstraction occurred in all cases, and the spectral analysis of AlO allowed the dissociation energy of AlO to be determined as 5.26 ± 0.03 eV.

Gas-phase Ga atoms in the ground state have been produced by multiphoton excitation of $GaMe_3$.[19] Interaction of the Ga atoms with CF_3X, SF_6, C_2F_4, N_2O, C_2H_2, 1-butene, and $GaMe_3$ was monitored by laser-induced fluorescence (LIF).

Abstractions and association reactions took place. In several cases an equilibrium was observed between free Ga atoms and association complex. For example, the binding energy of Ga to C_2H_4 was determined as 9 ± 2 kcal/mol, and Ga to $GaMe_3 = 14 \pm 2$ kcal/mol. Rate constants for abstraction reactions were determined, and examples are shown:

$$Ga + C_2H_4 \rightleftarrows Ga-C_2H_4$$

$$Ga + CF_3X \rightarrow GaX + CF_3$$

$$X = I, \quad 2.8 \times 10^{-10} \text{ cm}^3/\text{molecule} \cdot \text{s}$$

$$Br, \quad 1.9 \times 10^{-10}$$

$$Cl, \quad 2.2 \times 10^{-12}$$

$$F, \quad 8 \times 10^{-15}.$$

These rate constants for CF_3I and CF_3Br are near the gas kinetic limit.

2. Oxidative Addition Processes

a. Hydrogen, Alkanes, and Silanes   Under matrix isolation conditions, B atoms have proved to be the most reactive species that has been investigated.[2,3] Upon codeposition of B atoms with CH_4/Ar, an efficient reaction yielding the $HBCH_3$ radical took place, the process apparently taking place by a very low Ea process.

$$B + CH_4 \xrightarrow[10\ K]{Ar} H\text{-}\dot{B}\text{-}CH_3$$

ground
state

Theoretical support for a low-activation barrier for the $M + CH_4 \rightarrow HMCH_4$ process has been predicted for B, Al, and B^-.[20] In rationalizing such a low barrier, a partially filled p-shell appears to be very important so that an empty p lobe can interact with the C–H σ^* orbital. Under identical matrix conditions, reactivity comparisons were made, which showed that ground-state B and B_2 reacted efficiently with CH_4, Al with less efficiency, while Ga, In, Si, C_2, and Si_2 did not react at all. Similarly, comparisons with ground-state Mg, Ca, Ti, Cr, Fe, Cu, Ni, Pd, Ag, Au, and Sn demonstrated that B and Al were by far the most reactive.[2,3,21] By comparing all these data it became clear that the important features that drive such low-temperature reactions center around the radical nature of the ^2P ground state (partially filled p shell) and the strengths of the B–H, Al–H, B–C, and Al–C bonds.

Parnis and Ozin have presented evidence that photoexcited Al atoms are necessary for matrix reaction with CH_4.[22] Indeed, excited state Al atoms do react more efficiently than ground state. However, numerous experiments using both thermally evaporated and laser-evaporated Al have shown that a certain fraction of ground-state Al atoms always react. It seems likely though, since some Al atoms do not react, that the approach of the Al atoms to the CH_4 molecule and temperature are critical features. And in such codeposition experiments, matrix temperature and concentration can be affected by deposition rates, efficiency of cooling, and other features.

The Al (^2P) reaction with SiH_4 has also been studied under matrix conditions.[23a] Evidence for the initial formation of a complex Al—SiH_4 was presented. Upon UV photolysis the association complex reacted by oxidation addition. Interestingly, the ESR/IR,

$$Al + SiH_4 \rightarrow Al\text{—}SiH_4 \xrightarrow{h\upsilon} H_3Si\text{—}Al\text{—}H$$

and optical properties of H_3SiAlH are very similar to those of H_3CAlH.

Gallium atoms codeposited with H_2 and CH_4 undergo reaction on photolysis to produce GaH and GaH_2, while CH_4 also reacts:[23b]

$$Ga + H_2 \xrightarrow[h\upsilon]{codeposit} GaH + HGaH$$
$$\text{bent, } 136 \pm 5°$$

$$Ga + CH_4 \xrightarrow[h\upsilon]{codeposit} CH_3GaH.$$

Interestingly, Ga_2 appeared to react spontaneously with H_2:

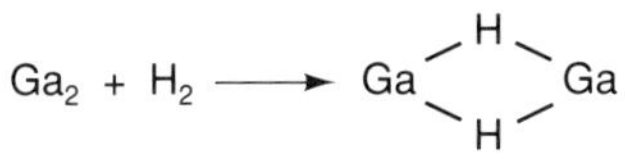

Knight et al.[23c] have also produced dihydride species in cold matrices and carried out ESR analyses on AlH_2, AlHD, AlD_2. In addition, $Al(OH)_2$ and $Al(OD)_2$ were matrix isolated.

b. Alkyl Halides, Silyl Halides, and Germyl Halides It would be expected that organohalides would be much more reactive with metal atoms than CH_4, and this is generally the case. However, the presence of nonbonding electron pairs on the halide ligand often allows a σ-complexation pathway to become dominant. Thus, when a series of metal atoms were compared in their reactivity toward CH_3Br at 12 and 77 K, it was found that Fe, Co, Ni, Pd, Cu, Ag, and Au did not react spontaneously or upon photolysis.[3,24] Comparisons of main group atoms showed that Mg, B, Al, Ga, and In did react,

$$In + CH_3Br \rightarrow CH_3InBr$$

whereas Zn, Tl, Ge, Sn, and Pb did not. Overall, these results yielded two important conclusions: (a) low first ionization energies allow higher reactivities, and therefore electron transfer is probably important:

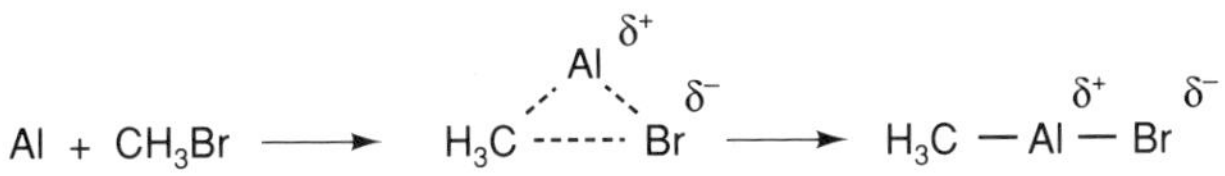

(b) sometimes clusters are necessary for reaction to occur, such as with Mg, Mg_2, Mg_3 (see Chapter 3).

In the case of Ga reacting with alkyl halides, a useful synthesis of R_2GaX compounds has been developed.[25] Thus, when Ga atoms were deposited with excess EtI or EtBr, complete reactions giving high yields of $(Et_2GaX)_2$ and $(EtGaX_2)_2$ took place. Gallium atoms were the reactive species in this preparation, probably as follows:

$$Ga + EtI \rightarrow EtGaI \rightarrow EtGaI_2 + Et\cdot$$

$$EtGaI + Et\cdot \rightarrow Et_2GaI.$$

Bierschenk and Lagow have also developed synthetic methods on the basis of metal atom reactions with alkyl radicals. A good example is the preparation of a series of CF_3M compounds.[26] In particular, $Tl(CF_3)_3$ was prepared by condensing metal atoms with CF_3 radicals on a cryogenic surface. The radicals were produced in a low-temperature glow discharge on CF_3CF_3. Using this technique, an "assist"

$$Tl + 3CF_3 \rightarrow Tl(CF_3)_3$$

is given so that radicals are the reacting species rather than alkyl halides or perfluoroalkane. Similar experiments have yielded crude product mixtures containing unstable $Ga(CF_3)_3$ and $In(CF_3)_3$. Many other metal atoms have also been studied, as described in other appropriate chapters.

Boron atoms also react vigorously with SiX_4 and GeX_4.[27] Rate constants for gas-phase reactions with SiF_4, $SiClF_3$, $SiCl_4$, $SiBr_4$, GaF_4, $GeCl_4$, and $GeBr_4$ were reported, employing a glow discharge flow tube apparatus at ambient temperature. Abstraction reactions are believed to dominate the chemistry:

$$B + SiX_4 \rightarrow BX + SiX_3$$

and

$$B + SiX_4 \rightarrow BX_2 + SiX_2.$$

Earlier studies[28] showed that CF_4 was essentially unreactive and that fluorine substitution reduced the reactivity of C–Cl bonds. In these reports a huge variation in reactivity was observed; for example,

$$SiF_4 \qquad k_2 = <1.2 \times 10^{-4} \ cm^3/molecule \cdot sec$$

$$GeBr_4 \qquad k_2 = 1.6 \times 10^{-10}.$$

These reactivity variations were rationalized in terms of the energy level of the LUMO of the silicon halide and strengths of the C–X, Si–X, or Ge–X bonds.

c. Water and Ammonia Under matrix conditions the interaction of B or Al atoms with H_2O could yield three types of products:

(a) $B + H_2O \rightarrow BO + H_2$, which has been observed in the gas phase by Gole and Pace.[29]

(b) Adduct formations such as $Al—OH_2$.

(c) Oxidative addition to yield $H—Al—OH$.

Indeed, upon deposition of B or Al with H_2O/Ar at 12 K there were spontaneous reactions that yielded the low-temperature stabilized HBOH and HAlOH.[30] Thus, B and Al react spontaneously at 12 K to form the insertion product. However, Ga, In, and Tl atoms formed $M—OH_2$ and $M_2—OH_2$ adducts. Upon photolysis the adducts of Ga and In converted to the HMOH species, but the Tl adduct did not. Extended photolysis caused decomposition of the HMOH species and $M—OH$ fragments were produced. Ground-state Al atoms also reacted with NH_3, but in this case both adduct formation and oxidative addition reactions were observed.[31]

$$Al + NH_3 \xrightarrow[\text{77 K}]{\text{adamantane}} Al(NH_3)_4 + Al(NH_3)_2 + HAlNH_2.$$

These products are probably intermediates in the formation of amide and hydrogen from the reaction of Al atoms and NH_3.

d. Oxygen and Ozone   Boron atoms react with O_2 under matrix conditions by rather complex pathways.[32a] Using a Nd–YAG laser for ablation of boron, Ar/O_2 codeposition at 11 K yielded linear BO_2 (oxidation insertion). The formation of smaller amounts of BO, BOB, B_2O_2, and BO_2^- was also evident. In addition, a strong band at 1512.3 cm^{-1} appeared on matrix annealing, and it was suggested that this was due to a B atom -O_2 complex.

Very similar results were reported in an analogous study where an Ar–Cl excimer laser was used or boron vaporization.[4] A major product indeed was linear BO_2, as confirmed by extensive isotope labeling experiments. The BO_2^- was also formed, but apparently by a different mechanism than BO_2 formation; the formation of BO_2^- was shown to involve oxygen isotope scrambling, whereas BO_2 was not. It was proposed that a B/O_3 reaction was responsible for BO_2^-. Also of interest was the reaction of BO_2 in the matrix with excess O_2. An adduct $O=B—O—O_2$, which was thoroughly characterized by isotope-labeling experiments, was a major product formed where higher O_2 concentrations were present in the matrix.

Similar results were observed for Al atom -O_2 reactions.[32b] The major product formed in solid argon is cyclic AlO_2. Minor products were also detected spectroscopically and assigned as linear OAlOAlO, AlOAlO, and OAlO. Thus, two products of 1:1 stoichiometry were formed:

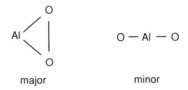

major   minor

Analogous experiments with Ga and In atoms indicated that GaO_2 and InO_2 were major products, but were probably of bent geometry:[32c]

Also observed were linear OGaO and OInO and linear GaOGaO and InOInO. It should be noted that all of the Group 13 atoms form linear insertion products OMO. However, boron did not form a side-on bonded complex as Al, Ga, and In did. In addition it was found that the insertion reaction took place readily with B, but with Al, Ga, and In, either excess thermal energy or photoexcitation was necessary to cause an efficient insertion. These studies also suggested that laser-vaporized atoms were more reactive, which probably is due to excess kinetic energy of the incoming atoms (see Chapter 6 for a more detailed discussion).

Continuing with matrix isolation conditions, it has been shown that Tl atoms react with O_2.[33] In N_2 and Ar matrices, the main initial product was $Tl^+O_2^-$, which is similar to results found earlier with Ga and In atoms. Various secondary products were also detected, such as the superoxide dimer TlO_2-O_2Tl (D_2d configuration) and TlO_2Tl (D_2h symmetry). Additional thallium atoms react with the TlO_2Tl species to yield the suboxide dimer $(Tl_2O)_2$.[34] Similarly, Ga, In, and Tl atoms codeposited with O_3 yielded primarily the charge transfer species $M^+O_3^-$, similar to the alkali- and alkaline-earth ozonides.[1] Vibrational bands for the O_3^- species in Ga and In were slightly higher in energy than for the $Tl^+O_3^-$ species, and this was attributed to more covalent character for the bonding in the Ga and In cases.

3. Simple Orbital Mixing

a. CO and CO_2 Weltner and co-workers have reported on the isolation of the B–CO molecule by codeposition of B atoms with CO/Ar.[35a] Spectral analysis by ESR in both Ne and Ar matrices showed that B–CO has a $^4\Sigma$ ground state, and the three unpaired electrons are predominantly on the boron atom. Ab initio calculations suggest a preferred ground state of $^4\Sigma^+$ for both BCO and the isocarbonyl BOC, but BCO is predicted to be about 70 kcal/mol more stable. The experimentally observed $\nu_{C\equiv O}$ value in argon was 2091 cm^{-1}. Overall the bonding picture for BCO is nearly that of a classic Lewis acid–base bonded complex with the three spins distributed in p_x^2 and $sp\sigma$ orbitals on boron. The formation of BCO from B(4s) and CO($^1\Sigma^+$) leads to donation of electron density from the lone pair on carbon to an empty sp orbital in an sp-hybridized boron atom.

No evidence for B(CO)$_2$ was found in this study, but Burkholder and Andrews[35b] did report evidence for the dicarbonyl as well as the linear dimer OCBBCO. The B(CO)$_2$ product was proposed to be a bent molecule.

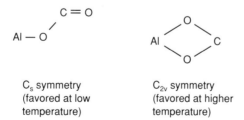

129°

For Al and Ga the dicarbonyls have also been detected, but no evidence was found for the monocarbonyl.[36-39] Kasai proposed that OC–Al–CO has a bent structure and a semifilled orbital where back donation from the Al$_p$ π-orbital into the π^*-orbitals of the CO molecules takes place. Chenier et al. also have detected Al(CO)$_2$ as well as Al$_2$CO and Al$_2$(CO)$_4$.[38] The Ga(CO)$_2$ species is also bent (C$_{2v}$ symmetry) and exhibits a $\nu_{C\equiv O}$ mode at 1930 cm^{-1}. Force constants for Al(CO)$_2$ compared with Ga(CO)$_2$ indicate that less charge (backbonding) occurs in the Ga species.

The codeposition of B or Al with CO$_2$ at 77 K has led to explosive (and spectacular) formation of CO and BO$_x$ or AlO$_x$. Light flashes are emitted and so one of these species must be formed in the excited state.[1] At lower temperatures, in frozen argon, an Al–CO$_2$ complex has been stabilized.[40] This species was found to reversibly interconvert between two geometrical isomers, possibly with these structures:

C $=$ O   O
／   ／   ＼
Al — O   Al   C
                            ＼   ／
                                O

C$_s$ symmetry   C$_{2v}$ symmetry
(favored at low   (favored at higher
temperature)   temperature)

Temperature studies indicate that the two isomers have an enthalpy difference of 1.55 ± 0.4 kJ/mol. And it was shown that for larger Al clusters, warming above 30 K caused formation of Al$_2$O and, presumably, CO.

b. Ethers Gas-phase studies of Al atom association with Me_2O, Et_2O, and THF have been reported and binding energies determined.[41] A pulsed visible laser was employed to dissociate $AlMe_3$ in a gas cell with Ar buffer gas. The resultant ground-state thermal Al atoms reacted with ethers with bimolecular rate constants approaching gas-kinetic values, which implies very small activation energy barriers. For each of the ethers an equilibrium was observed between free Al atoms and Al–ether complexes, and the temperature dependence of these equilibrium constants allowed estimation of binding energies: Al–OMe_2 (9.2 ± 0.6 kcal/mol); Al–OEt_2 (9.2 ± 1.2); Al–THF (10.8). Values for $\Delta S°$ ranged from 19.7 to 20.9 entropy units (cal/mol/K).

c. Alkenes, Alkynes, and Arenes The codeposition of Al atoms with ethylene near 77 K led to the formation of an aluminocycloalkane radical.[42] Since Al–alkene reactions did not produce detectable Al–CH–CHR radical species, it was suggested that this cyclodimerization reaction was concerted, probably going through an $Al(C_2H_4)_2$ complex. It was noted that two paramagnetic species were produced in this reaction, the more intense signal being assigned to the Al–C_2H_4 π-complex and the other to the cyclodimer product. Both species exhibited remarkable stability in an adamantane matrix (up to 343 K).

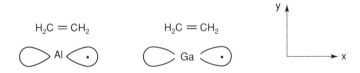

$$Al \; + \; 2C_2H_4 \; \longrightarrow$$

The results of Kasai and co-workers, working at 4 K in argon, were somewhat different. For Al and Ga with C_2H_4, only a monocomplex was formed with side-on bonding.[43,44] These complexes are held together primarily by the semifilled orbital, and unpaired spin density passed from the Al p_x (or Ga p_x) orbital into the π_y^*-orbital of ethylene.

$$H_2C = CH_2 \qquad\qquad H_2C = CH_2$$

With In atoms and C_2H_4 in argon, the main product at 12 K was an In–C_2H_4 complex with bonding very similar to that reported for the Al and Ga systems.[45]

In adamatane matrices as 77 K, propene with ground-state (^{2}P) Al atoms caused quite a different type of reaction, and ESR yielded evidence for a π-allyl/metal hydride:[46]

$$\text{Al} + \text{CH}_3\text{CH}=\text{CH}_2 \xrightarrow{\text{77K}} \begin{array}{c} \text{CH} \\ \diagup \cdots \diagdown \\ \text{CH}_2 \;|\; \text{CH}_2 \\ | \\ \text{Al} \\ | \\ \text{H} \end{array}$$

This represents the first direct evidence for such a species, which is believed to be an intermediate in the isomerization of alkenes. Also formed in these codeposition experiments were the allyl radical, and a dimethyaluminocyclopentane.

Under similar experimental conditions, 1, 3-butadiene with Al atoms yielded two products, a σ-bonded aluminum cyclopentene and an aluminum-substituted allyl:[47]

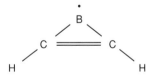

A theoretical study[48] of this reaction (Al + butadiene) suggested that the *cis*-form of 1,3-butadiene should yield the aluminocyclopentene, while the *trans*-form should yield in aluminomethylallyl, and these results are in accord with the earlier findings of Chenier and co-workers.

Acetylene codeposited with B, Al, and Ga atoms have been studied.[49] In the case of B atoms, the borirene radical was formed in frozen argon:[49b]

$$\begin{array}{c} \bullet \\ \text{B} \\ \diagup \quad \diagdown \\ \text{C} === \text{C} \\ \diagup \qquad\qquad \diagdown \\ \text{H} \qquad\qquad\quad \text{H} \end{array}$$

This was an interesting finding since both C—H and C≡C addition could be possible. Results were slightly different for Al and Ga. When C_2D_2 and Al were codeposited with neon, a purple matrix was formed, and ESR showed a clearly resolved sextet due to the hyperfine interaction of ^{27}Al(I = $\frac{5}{2}$). The adduct with C_2H_2 exhibited a spectrum where nonequivalent hydrogen atoms were present. It was concluded that *cis-* and *trans-*vinyl radicals that could be interconverted by UV inadiation were produced:

$$Al + C_2H_2 \longrightarrow \underset{Al}{\overset{H}{>}}C = C\overset{H}{\underset{\bullet}{<}} + \underset{Al}{\overset{H}{>}}C = \overset{\bullet}{C}\underset{H}{<}$$

Shaefer and co-workers[50] calculated the Al—C bond energies for the *cis-* and *trans-*vinyl radical adducts (8–9 kcal/mol), considered another isomer, the vinylidene adduct, and predicted this to be more stable with a Al—C bond energy of about 20 kcal.[51]

$$Al - \overset{\bullet}{C} = C\overset{H}{\underset{H}{<}}$$

Further theoretical treatment of B and Al atom interactions with C_2H_2 have shown that for B atoms, C–H insertion or addition to the C≡C bond have very low activation barriers. However, for Al atoms, C—H insertion does have a barrier, and HAlCCH would be less likely as a product at low temperature.[52] Benzene has also been treated with Al atoms by theoretical methods. The lowest energy structure was found to be the 1,4-addition complex, which agrees with experiment and should be bound by about 7.4 kcal/mol.[48] It was proposed that nearly a full negative charge would be transferred from Al to benzene. Experimental studies have shown that in the case of Al a strongly bonded complex was formed at 77 and 4 K. At the lower temperature, resolvable ESR hyperfine interactions with two equivalent protons were detected. It was concluded that 1,4-addition product was most likely and at 77 K this species appears to be fluxional.[53,54]

In the case of Ga, only a weakly interacting monocomplex was formed, more like a solvated Ga atom than a true bonding interaction.

II. Boron Group Clusters

A. Occurrence and Techniques

In 1982 Riley and co-workers described their apparatus for continuous generation of clusters of refractory metals (see Chapter 2). An oven was used to vaporize Al, Cr, Ni, Cu, and Ag. The atoms were entrained in a flow gas such as He or Ar and cluster growth took place. The clusters were quenched and trapped on a liquid N_2-cooled cell.[55] Smalley and co-workers also initially developed their supersonic metal cluster beam apparatus working with Al.[56] This powerful technique is described in Chapter 2 and has enabled the development of an entirely new field of chemistry.

The volatility of Ga and In metals has encouraged the development of liquid metal nozzle beam cluster sources (see Section IV of this chapter). Yang and Lu have made a careful study of homogeneous nucleation and growth of clusters resulting from an adiabatic cooling during the expansion of the metal vapor through a crucible nozzle.[57] Nucleation rates were evaluated for a variety of metals and semiconductors.

B. Physical Properties and Theoretical Studies

The B_2 and Al_2 dimers have received extensive analysis, both experimentally and theoretically.[58–60] For example, the ground-state bond energy for B_2 has been computed as 2.78 eV, and this is substantially larger than that calculated for Al_2 (1.40 eV). And in the case of B_2, five of the six excited electronic states that dissociate to ground-state atoms are bound, compared with only two for Al_2.[58] Also, ionization energies of various electronic states of B_2 have been computed and were in the range 9.0–9.2 eV.[59]

Several new electronic transitions of the B_2 molecule have been observed using a Corona excited supersonic expansion source.[60] Theory and experiment were compared regarding bond lengths in B_2 and were in rather good agreement: for example $b^1\Delta g$ 1.616 Å (exp) vs 1.627 (theory) and $3\pi g$ 1.496 (exp) vs 1.476 (theory).

Low-lying valence electronic states of Al_2 have been investigated by ab initio methods.[61] The two lowest states, of $^3\Sigma g^-$ and $^3\pi u$, were found to be almost degenerate. The lowest state $^3\Sigma g^-$ should exhibit a vibrational frequency at 354 cm^{-1} and should have a bond energy of 1.33 eV and a bond length of 2.51 Å.

Balasrubramanian[62a,b] has examined Ga_2 and In_2, by theoretical methods, and found, for example, the ground state of Ga_2 to be $^3\pi u$ with the $^3\Sigma g^-$ state only 410 cm^{-1} higher in energy. The calculated bond energy of the ground state was 1.0 eV and the bond length 2.762 Å. For In_2 the ground state was found to be $3\pi_u(O_u^-)$. Experimental work has detailed the electronic transitions in In_2.[62c]

Laser sputtering has produced B_2 for matrix isolation studies.[63] The $^{11}B_2$ molecule was trapped in neon and argon matrices at 5 K. ESR analysis indicated the ground state to be $^3\Sigma g^-$. Comparison of experimental hyperfine parameters with those calculated by ab initio and CI methods found good agreement for the dipolar component of the A tensor, but not for the small and difficult to calculate isotropic hfs.

Indeed, laser vaporization is a versatile method for production of many metal clusters.[64] A YAG laser was suitable for vaporizing both metals and semiconductors, and the technique is valuable for both gas-phase and matrix isolation studies. In particular, studies of LiBe and Al_2 were described and a general review of the method given.

Cluster polarizabilities have been compared with atoms by Milani and co-workers.[65] A laser evaporation source produced a collimated cluster beam that was deflected in an inhomogeneous electric field. The clusters were then photoionized and detected by a position-sensitive time-of-flight analyzer, and polarizabilities were calculated from the measured deflections. Aluminum atoms and Al_2 were compared: Al polarizability = 6.8 $\pm$ 0.3 $\times$ 10^{-24} cm^3 and Al_2 = 19 $\pm$ 0.2 $\times$ 10^{-24} cm^3.

Jet-cooled cluster anions of Al, Ni, Ag, and Sn have also been prepared by the laser evaporation–cluster beam method.[66] UV photodetachment measurements on mass-selected cluster anions showed the existence of a dramatic dependence on cluster size and differed substantially from respective spectra of bulk samples. In this way information about low-lying excited electronic states of the clusters, states that do not exist in the bulk were obtained. Photoelectron thresholds as estimates of electron affinities in part follow a simple electrostatic model.

Smalley and co-workers have also examined aluminum cluster anions Al_n^-, where n was 3–32. The clusters were prepared in a supersonic beam by laser vaporization and an Ar–F excimer laser (6.42 eV) used for photodetachment. The electron affinities were determined and these peaked at cluster sizes 6, 13, 19, and 23. These results are consistent with a shell model of electronic structure. A size-dependent increase in the photoelectron yield at about 5.0 eV binding energy could be due to the merging of the $3s$ and $3p$ valence bands.[67]

Recent ESR data on matrix-isolated B_3 and Al_3 have also been discussed[68,69] and compared with ab initio theoretical studies.[70–73] The ESR evidence confirmed a $^2A_1'$ ground state for triangular B_3 (D_{3h} symmetry), where the odd electron occupied a MO composed of in-plane $p\sigma$-orbitals. In all three matrices of Ne, Ar, and Kr, the B_3 molecule was rapidly tumbling. The interaction of B_3 with these matrices is small because it lacks a dipole moment.

In hydrocarbon matrices (77 K) the Al_3 species has been observed by ESR.[74] A quartet $^4A_2$ electronic state was observed. This molecule has three equivalent Al atoms, perhaps in an equilateral triangle geometry (D_{3h} symmetry) where three unpaired electrons occupy three almost degenerate orbitals constituted from the three $3p$ orbitals of the Al atom. However, an alternative is a fluxional isosceles triangular molecule of C_{2v} symmetry. The Al_3 species is reasonably stable and is persistent even at 273 K in adamantane.

Larger clusters such as Al_{13} have also been studied.[75] Equilibrium geometry, structural transformations, and trends in Na, Mg, Al, and Si were examined by theoretical methods. Elements were compared and conclusions drawn: (1) Na_n clusters ($n = 6-20$) have a preference for structures of fivefold local symmetry. The nondirectional character of the bonding makes compactness an important criterion. (2) The Mg_{13} species shows a tendency to form hexagonal rings. (3) The Al_{13} species is a special case. The distorted icosahedron seems to be highly favored energetically. Also, the Al_{13} cluster is the only one (compared to Na, Mg, and Si) whose electronic structure is compatible with structures derived from slight distortions of either the icosahedron or the anticubooctahedron. (4) As expected for more directional character of the bonds, the Si_{13} species does not prefer compact structures. The low-energy structures can be visualized as different cappings of trigonal prisms. The mobilities of Al cluster cations where Al_n^+ contains 5–73 atoms have been measured using injected ion drift tube techniques.[76] Clusters close to the electronic shell closing seem to have enhanced mobilities, and a structural transition appears to be associated with a shell closing with 138 valence electrons. Annealing temperatures are generally below room temperature. Shell structures of large Al_n clusters have also been probed using photoionization spectra.[77] Ionization energies were determined, and striking patterns were related to size-dependent ionization threshold effects and appeared to be a function of $N_e^{1/3}$ (where N_e is the number of valence electrons).

Cluster ions Al_n^- ($n = 3-50$) have been prepared by laser vaporization directly into a FT–MS (no He or Ar gas pulse).[78] Collision-induced dissociation was used to investigate fragmentation pathways and generally resulted in Al atom loss except for Al_7^-, Al_6^-, Al_3^-, which fragmented by electron detachment. These cluster anions were quite unreactive with CH_4 and N_2O, although some reacted with O_2 to form AlO_2^- and AlO^-. Even-numbered clusters ($n > 8$) reacted more rapidly then odd-numbered ones.

Photodissociation kinetics of such cluster ions has been studied by Jarrold and co-workers.[79] In this work Al_n^+ ($n = 7-17$) were studied over a broad energy range of 1.88–6.99 eV. Dissociation energies showed an overall increase with cluster size, but there are substantial oscillations around $n = 7-8$ and $n = 13-15$. The cohesive energies of the larger clusters ($n > 6$) were found to be in good agreement with the predictions of a simple model based on the bulk cohesive energy and cluster surface energy.

Magnetic properties of cold, gas-phase Al_n clusters ($n = 2-25$) have been investigated. Pulsed laser vaporization of aluminum inside the throat of a high-pressure pulsed nozzle produced thermal clusters of 2–25 atoms. The collimated cluster beam was passed through a Stern–Gerlach magnet and the deflected beam was quantitatively analyzed by specially resolved photoionization time-of-flight MS. Magnetic moments for Al_n, where $n < 9$, were generally found to be consistent with those predicted from spin and orbital moments of ground electronic states. As cluster sizes grew from Al_9 to Al_{25}, reduced magnetic moments/atom were observed.[80] The smaller clusters thus have magnetic moments that are consistent with the odd-atom clusters having doublet ground states, while even-atom clusters have either triplet or singlet ground states.

Small, spherical Sn and Ga clusters of 2–80 nm diameter have been prepared using a liquid metal source, trapped, and analyzed by scanning transmission microscopy.[81] Volume and surface plasmon excitations were determined vs cluster size. A blue shift in plasmon energy with decreasing cluster radius was clearly shown, and this was consistent with earlier experiments and predictions.

C. Chemistry

1. Electron Transfer/Abstraction/Oxidative Addition

Since it is next to impossible to separate out reaction types with these clusters, we combine several. First, theoretical treatments of the $B_3 + H_2$ reaction have suggested that hydrogen abstraction would be endothermic, while direct addition should have a slightly lower barrier, $B_3 + H_2 \rightarrow B_3H_2$.[82] The most stable isomer is predicted to be:

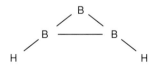

Jarrold and Bower have explored the chemistry of gas-phase Al_n^+ cluster ions with O_2[83] and D_2.[84] Their method employs collision-energy dependence of chemisorption using low-energy ion beam analysis. For D_2 with size-selected clusters, both chemical reactions and metastable adduct formations were observed, the latter species believed to be due to chemisorptions without the benefit of stabilizing subsequent collisions. Activation energies for adduct formation increased with cluster size increase and showed significant odd–even oscillations (larger Ea for odd-numbered clusters). The reduced Ea for even atom clusters was proposed to be due to the reduced repulsion interactions at the transition state due to the presence of an unpaired electron in the highest occupied MO. The main chemical reaction products of these endothermic reactions were $Al_{n-1}D^+$, Al_{n-2}^+, and Al^+ for smaller clusters and Al_nD^+ and $A_{n-1}D^+$ for larger clusters.

In the case of $Al_n + O_2$ a detailed analysis of the effects of cluster size and collision energy allowed several interesting conclusions:

(1) The total reaction cross sections increase with cluster size.
(2) The reaction mechanism involves chemisorption onto the cluster followed by rapid loss of two Al_2O molecules. After that, the remaining Al_{n-4}^+ species contains enough excess energy to eliminate one or more Al atoms.
(3) Energy disposal in the loss of Al_2O molecules is probably not statistical.
(4) Neutral Al_n and Al_n^+ react differently with O_2. The Al_nO_2 neutral adducts may simply be aluminum clusters with loosely bound O_2.

More detailed studies of these neutral clusters in a fast-flow reaction have been reported by Kaya and co-workers.[85,86] Thus reaction of Al_n ($n = 7$–24), formed by laser vaporization, with O_2 and NH_3 was studied. The O_2 *strongly* chemisorbed while NH_3 weakly chemisorbed on the clusters. Under multiple collision conditions, Al_9O_7 was a predominant species.

Boron cluster ions (B_n^+, $n = 1$–14) react with water, as determined in the gas phase by single collision dynamics.[87] Sequential etching took place. In most cases no activation barriers exist, and cross sections are generally large. For small cluster ions many reaction pathways were observed, but for $n > 6$ the product distributions were dominated by a single process:

$$B_n^+ + D_2O \rightarrow B_{n-1}D^+ + DBO.$$

When multiple collisions were possible a sequence of secondary etching took place where B atoms were continually replaced by hydrogen:

$$B_9D_3^+ + D_2O \rightarrow B_8D_4^+ + DBO$$

$$B_8D_4^+ + D_2O \rightarrow B_7D_5^+ + DBO$$

$$B_7D_5 + D_2O \rightarrow B_6D_6^+ + DBO.$$

Aluminum cluster cations were found to be much less reactive than boron analogs under similar conditions.

2. Simple Orbital Mixing

Cox *et al.* have examined the reactivity of gas-phase metal clusters toward CO, including aluminum.[88] Clusters up to 14 metal atoms were produced by pulsed laser vaporization. It was found that CO chemisorbed on most transition-metal clusters containing five or more metal atoms. Comparative studies showed that Fe, Mo, Cu, and Al were least reactive toward CO (no product peaks were observed for Cu and Al).

Further investigation of Al_n reactions with D_2, H_2O, O_2, CH_3OH, CH_4, and CO showed that for each reactant/Al_n pair, a unique dependence on cluster size and reaction rate was found.[89] Overall reactivity toward these molecules was found to be roughly in the order $O_2 > CH_3OH > CO > D_2O, > D_2 > > CH_4$.

When vapors of Ga and In (as well as Au, Pd, Cu, Ge, Sn, Sb, and Bi) were codeposited with styrene or methymethyacrylate, followed by warming from 77 to 300 K, colloidal metal particles that were solvated by these monomers formed. Particle sizes ranged from 40 to 90 Å generally.[90]

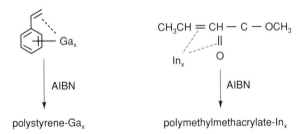

Upon addition of a polymerization catalyst such as azo-bis-isobutyrylnitrile (AIBN) and heating to 60°C, polymerization of the colloidal solutions took place. Metal loadings of 0.7 to 3% by weight could be achieved, and molecular weights of the polymethylmethacrylate ranged from 150,000 to 400,000. Indeed, the In, Pd, and Sb colloids allowed the formation of the highest MW. In some cases, the metal particles definitely affected the ultimate MW that could be achieved, and this apparently was due to metal–radical initiator interaction (trapping).

In the case of the styrene–metal colloids, MW of 100,000–160,000 were obtained for Ga, In, Sb, Sn, and Bi, while lower values of 22,000–72,000 were obtained with Au, Pd, Cu, and Ge.

III. Bimetallic and Binuclear/Trinuclear Systems

A. Bimetallic Clusters

Laser evaporation of Nb and Al (or Co and Al) yielded gas-phase bimetallic clusters, and these materials exhibited unusual selectivity in reactions with H_2. Aluminum clusters were not reactive with H_2, but when combined with Nb or Co, seemed to behave as an inhibitor or accelerator. Rigid geometry of mixed clusters may contribute to unreactivity.[91]

Similarly, laser-vaporized Al and C form mixed clusters, some of which were negatively charged. For Al itself Al_{13}^- was dominant, and a change in cluster ion distribution occurred at Al_{23}^- and Al_{24}^-. However, Al_7C was dominant in the mixed system, and the absence of $Al_{13}C$ was explained by the electronic structure in relation to the electronic shell model and particularly geometric structure.[92]

Alloy cluster beams of Ni_xAl_y and Ni_xCr_y were prepared by laser vaporization of bulk alloys.[93] Cluster distributions for clusters larger than about six atoms displayed abundances that could be reasonably well interpreted by using a simple statistical model for their formation. However, for smaller clusters, departure from statistical behavior was observed. These variations were attributed to ionization probabilities (for detection by MS), and to metal–metal bonding differences, especially for the smaller clusters.

Laser vaporization of Cu with Ga and In have yielded CuGa and CuIn diatomics.[94] A significant fraction of the atoms were formed in initially electronically excited or ionized states, and considerable concentrations of metastable excited atomic states were still observed after delays of 1 msec after the laser pulse. Eventually, Cu_2, CuGa, or CuIn were formed by collisional atomic recombination and were not products of the initial vaporization. Two new electronic transitions were observed and assigned to CuGa and CuIn. It was concluded that the CuGa and CuIn species are considerably more strongly bound than isoelectronic species such as Zn_2, Cd_2, or Hg_2 (which are very weakly bound).

B. Low-Valent Boron and Aluminum Halides and Oxides

Tacke and Schnöckel have reported on a fascinating new chemistry of AlCl.[95] A new, larger-scale synthesis was carried out by allowing HCl to react with molten Al:

$$Al_{(l)} + HCl_{(g)} \xrightarrow[<0.2 \text{ mbar}]{1200 \text{ K}} AlCl_{(g)} + \tfrac{1}{2}H_{(g)}$$
$$\text{yield} > 90\%.$$

The AlCl was condensed on the cold surface of a cryostat to give dark red AlCl. Upon warming to about 180 K a disproportionation reaction took place

$$3AlCl_{(s)} \xrightarrow{180 \text{ K}} AlCl_{3(s)} + 2Al_{(s)},$$

which is thermodynamically favored. Thus, solid $AlCl_{(s)}$ could only be stored for long periods below 77 K. The cocondensation of AlCl with 2-butyne/pentane led to dimeric 1,4-dialuminacyclohexadiene via carbene-like cycloadditions:[96]

The dialuminocyclohexadiene compound exists in a dimeric form as a staggered sandwich structure. Somewhat similar chemistry was observed when Al and dimethylbutadiene were cocondensed. A cyclic hexamer was isolated and structurally characterized.[97]

In the presence of polar solvents such as ether or toluene, a dark red solution of AlCl could be formed about 160 K. Standing at room temperature led to disproportionation and precipitation of Al metal over a few hours. However, addition of CH_3OH led to the formation of $(CH_3O)_2$ AlCl.

$$3AlCl \cdot (OEt_2)_x \text{ in toluene} \rightarrow AlCl_3 \cdot OEt_2 + 2Al$$

Analysis of ^{27}Al NMR spectra of this solvated species suggested that the AlCl $\cdot$ $(OEt_2)_x$ was not monomeric and may possess some Al–Al bonds.[98] Further chemistry of this interesting species has been elucidated. Especially interesting was treatment with $Mg(Cp^*)_2$, which yielded a tetrameric Al_4 cluster compound:

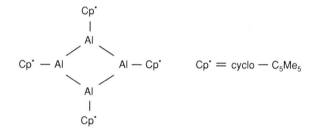

The Al_4 core formed a tetrahedron and each Al is capped by a Cp* ring. This is an extremely novel Al(I) organometallic and is reminicent of somewhat analogous Ga(I), In(I), and Th(I) arene complexes of Schmidbauer and co-workers, which have been prepared by more conventional organometallic methods:[99]

Analogous chemistry is being developed for gallium. Gaseous GaCl was prepared from Ga metal at 1200 K reacting with HCl.

$$Ga_{(l)} + HCl_{(g)} \rightarrow GaCl_{(g)} + \tfrac{1}{2}H_{(g)}$$

The GaCl was trapped as a red solid at 77 K. Warming to above 0°C yielded $GaCl_3$ and Ga metal.[100]

Schnockel and co-workers have also reported on the dimeric structures $(AlF)_2$, $(PN)_2$, and $(SiO)_2$.[101] The IR spectra of matrix-isolated species were reported and geometric and electronic structures compared with those calculated by ab initio SCF calculations. According to polarity, dimerization energy should increase in the order AlF > PN > SiO, if only dipole interactions are taken into account. However, experimentally the order is SiO > AlF > PN. These results were interpreted as evidence for a gain of additional covalent bonding during dimerization, particularly for SiO.

Further work on $(AlF)_2$, $OAlF$, and $(OAlF)_2$, as studied by matrix IR methods, indicated a D_{2h} structure for $(AlF)_2$, and for $(OAlF)_2$:[102]

$$
\begin{array}{cc}
\underset{Al}{\overset{F}{\diagup\diagdown}}Al & F-Al\underset{O}{\overset{O}{\diagup\diagdown}}Al-F \\
\end{array}
$$

Gas-phase AlO has also come under chemical study.[103] Multiphoton dissociation of $AlMe_3$ yielded Al atoms that then were allowed to react with N_2O, and Al and AlO formation were monitored by laser-induced fluorescence. The low reactivity of AlO with N_2O allowed N_2O to be used as buffer gas. Ground-state AlO ($^2\Sigma^+$) was found to exhibit a chemistry predominantly involving association complexes with CO_2, CO, C_2H_4, $Me_2C=CMe_2$, and ethylene oxide. Reactions were not observed with CH_4, H_2, isobutane, benzene, toluene, CF_3Cl, CF_3Br, or CCl_4 at room temperature. Large barriers to atom abstraction, even though thermodynamically very favorable, are responsible for this low reactivity.

C. Other Polynuclear Species

A series of reports on the production of unusual binuclear and trinuclear species by reactive laser sputtering have appeared. These species have been matrix isolated and usually analyzed by ESR. Table 7-2 summarizes some of the species produced, method, and pertinent information about them.

D. Gallium Arsenide Molecules and Clusters

Gallium arsenide clusters have been of particular interest in recent years, obviously due to the importance of GaAs semiconductor devices. Weltner and co-workers[114a] trapped Ga_2As_3 in frozen matrices by vaporizing a GaAs crystal in a relatively high pressure of Ar or Kr. ESR analysis allowed spectroscopic characterization of the Ga_2As_3 cluster, which contained two equivalent Ga and three equivalent As nuclei. Almost all of the unpaired spin density resides on the Ga atoms. The ground state of this Ga_2As_3 species is probably $^2A_2$ and its structure must be a trigonal bipyramid (Fig. 7-1).

TABLE 7-2

Matrix Isolated High-Temperature Species Produced by Reactive Laser Sputtering and Analyzed by ESR and Theoretical Treatments

Species	Method	Comments	Reference
Be^+	*Photoionization of BF gas*	$X^2\Sigma$ ground state	104
AlF^+	Photoionization of AlF gas	$^2\Sigma$ ground state	105
AlH^+	*Photoionization of AlH gas*	$^2\Sigma$ ground state	106
AlD^+	Photoionization of AlD gas	$^2\Sigma$ ground state	106
BC	Laser evaporation of B/C mixture	$^4\Sigma^-$ ground state	107
BH_2 (and CH_2^- and CH_2^+)	Electron bombardment of B_2H_6	$^2A_1$ electronic state	108
AlC	Laser evaporation of aluminum carbide	$^4\Sigma$ ground state, three unpaired electrons reside primarily on carbon	109
AlC_2	Laser evaporation of aluminum carbide	X^2A_1 ground state, unpaired spin density primarily on aluminum	109
GaP^+	GaP laser evaporation and photoionization of the resultant GaP gas	$X^4\Sigma$ ground state	110
$GaAs^+$	GaAs laser evaporation and photoionization of the resultant GaAs gas	$^4\Sigma^-$	111
BNB	Laser evaporation of boron nitride	$X^2\Sigma$ ground state, unpaired spin density mainly in sp orbitals on boron atoms.	112
BPd	Laser evaporation B and Pd mixtures	$X^2\Sigma$, charge transfer to Pd	113
AlPd	Laser evaporation of Al and Pd mixtures	$X^2\Sigma$, charge transfer to Pd	113

The electronic structure of molecular $GaAs_2$ has been probed by theoretical methods.[114b] Three electronic states, $^2B_2$, $^2A_1$, and $^2B_1$, were found with $^2B_2$ as lowest energy. The geometries of all three states are isosceles triangles. The $^2B_2$ state exhibited greater As–As bonding and ionic character, whereas the $^2A_1$ state exhibited greater Ga–As bonding.

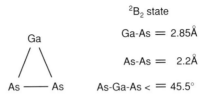

$^2B_2$ state

Ga-As = 2.85Å

As-As = 2.2Å

As-Ga-As < = 45.5°

Likewise, the electronic structures of small Ga_xAs_y clusters have been calculated using the local spin density method.[115] These calculations show that even-numbered clusters tend to be singlets, while odd-numbered systems are open shell, and such results are in agreement with electron affinity and ionization energy experimental determinations. In larger clusters ($x + y = 9$–11) the atoms prefer an alternating bond arrangement, and charge is transferred from Ga to As. Figure 7-2 shows the geometries of the most stable Ga_xAs_y clusters.

Additional calculations showed that all even-numbered clusters have closed-shell electronic structures.[116] Calculated electron affinities and ionization energies are distinctly different from neighboring odd-numbered clusters. Also, the even-numbered clusters ($n = 4$–10) are closed-shell singlet states with a substantial HOMO–LUMO gap. Sputtering, reactive evaporation, and direct laser vaporization of Al_2O_3, Ga_2O_3, and In_2O_3 have also yielded cluster ions.[117,118] The method used had little effect on the abundances and distributions observed. A series of ions corresponding to empirical formulas $MO(M_2O_3)_n^+$ and $M_2O_2(M_2O_3)_n^+$ for aluminum

$$Ga_2As_3$$

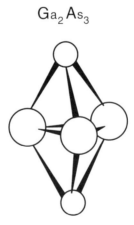

FIGURE 7-1   Geometry of the trigonal bipyramid structure of the Ga_2As_3 cluster (after Weltner and co-workers[114a]).

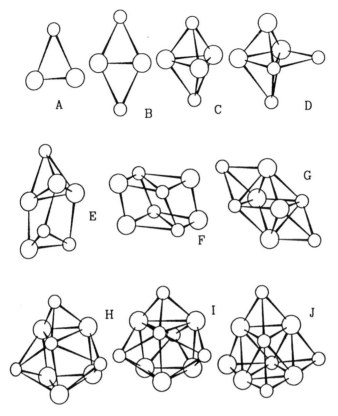

FIGURE 7-2   Geometry of the most stable Ga_xAs_y clusters (after Smalley and co-workers[115]). (A) $GaAs_2$, (B) Ga_2As_2, (C) Ga_2As_3, (D) Ga_3As_3, (E) Ga_3As_4, (F) Ga_4As_4, (G) Ga_4As_4, (H) Ga_4As_5, (I) Ga_5As_5, (J) Ga_5As_5.

oxide, and $M(MO)_n^+$ for gallium and indium oxides were observed, clearly showing the greater preference of Al for higher oxidation states. The Ga and In oxide ions were nonreactive with oxidizing gases while the Al oxide ions were highly reactive (with N_2O and NO_2).

Ultrafine powders of TiN and AlN have been produced by a reactive gas evaporation technique with electron beam heating. Low pressures of NH_3 or N_2 were allowed to react with the metal vapors, and TiN of cubic crystal structure and AlN of hexagonal crystal structure in particle sizes of less than 10 nm were obtained. The process is believed to take place first by surface nitridation of the molten metals, followed by vaporization of the nitride, and then condensation of the nitride particles.[119]

IV. Thin Films

There have been numerous reports of thin-film formation by deposition of atoms and clusters, especially aluminum. Table 7-3 summarizes these reports. The ionized cluster beam (ICB) method has been compared with several other deposition methods and often ICB was found to give more crystalline, smoother Al films. Yamada has reviewed these findings for Al, Cu, and Au.[120,121] More recently an apparatus has been developed for the production and study of ionized Ga cluster beams.[122] The graphite crucible containing the Ga was held at ground potential, and the vapors ejected through a nozzle, where clusters formed and were ionized by an electrical potential. Results showed that energetic cluster ions were only present when the crucible temperation was 1700°C or higher. The measured ion energies were consistent with the presence of clusters of hundreds of atoms ejected from the nozzle with thermal velocities.

A similar study of Ga_n species emitted from a liquid metal source was carried out by Barr.[123] Ionization of the species ejected from the nozzle source followed by MS analysis showed that monatomic Ga^+ was the dominant species. Small amounts of cluster were detected and their concentration did not seem to depend on source temperature. Clusters Ga_8 and Ga_{15} had particularly low abundances.

TABLE 7-3
Thin-Film Formations from Atoms and Clusters with Aluminum

Method	Comment	Reference
Ionized cluster beam (ICB)	ICB method minimizes radiation damage to film and allows high deposition rates. Low resistivity and highly oriented films were obtained.	124
Ion beam deposition (IBD) and ICB	Both methods yield films of comparable quality on Si substrates. IBD requires ultrahigh vacuum.	125
ICB	Al film on Ge(111) grows with its lattice parallel to the substrate lattice. On Ge(100) the Al lattice rotated 45°. Other heteroepitaxy also discussed.	126
ICB	Al deposited on Si(111) and showed Al(111) single-crystal structure. Film had high reflectivity and high resistance to oxidation.	127

(*continued*)

TABLE 7-3 (*continued*)

Method	Comment	Reference
Effusive and ICB	Surface roughness is a function of film thickness and usually not dependent on method of deposition.	128
Ion beam-assisted UHV evaporation	Effects of energies of Al atom and Kr/Xe bombardment on film quality were studied.	129
CVD of triisobutyl-aluminum/acetone clusters	Capillary nozzle cluster ion source. Acetone-triisobutylaluminum clusters accelerated and used to deposit Al films.	130
ICB and conventional vacuum deposition	Two deposition methods compared in electromigration failure; in some cases ICB was much better	131
ICB	Later annealing of ICB deposition film at 400°C caused small angle grain boundaries to disappear, and an almost perfect single crystal was formed of Al(111) on Si(111).	132
ICB compared with effusive beam deposition	The ICB method gives large grains which are single crystals.	133
ICB	Al deposited on thermal SiO_2 generated by steam oxidation and a p-doped SiO_2; masking results compared.	134
ICB	Epitaxial Al bicrystal films were formed on Si(100) at room temperature. Epitaxy occurs both from Si surface and at Al/Al grain boundaries	135
ICB	Al deposited on Si(100) has a unique grain structure consisting of only two crystal orientations. Both are (110) but are rotated 90° with respect to each other.	136
ICB	Al deposited on Si(111). Surface mobility of Al is anisotropic and is a function of acceleration voltage. This was explained with regard to probability of Al atoms to be absorbed or reflected at steps.	137
ICB	Al deposited on SiO_2. Al atom mobility studied by wire masking.	138
ICB	Al films formed epitaxially on Si substrates and had high thermal stability and long electromigration lifetimes. Annealing at 400°C caused the films to become large single crystals.	139

(*continued*)

TABLE 7-3 (*continued*)

Method	Comment	Reference
Computer simulations of ICB	Time-evolution stimulation for the study of Al_n ($n > 100$). Many Al atoms in the ionized cluster are reflected after a few collisions.	140
ICB	Al deposited on clean Si(111). The same high electromigration resistance and film structure was obtained with or without use of a nozzle crucible or application of ionization and acceleration voltages.	141
ICB	Large-scale applications possible and enabling Al epitaxial growth on Si despite the large lattice misfit.	142
ICB	Al was deposited on Si, CaF_2, GaAs, and Al_2O_3. Epitaxial and preferentially oriented films were formed.	143
ICB	Epitaxial Al(111)/Si(111) films were grown. Using a He^+ channeling technique, no observable strain was detected between Al and Si in spite of the 25% lattice mismatch.	144
ICB	A review of ICB and new results on epitaxial $Al/CaF_2/Si$ structures are discussed.	145
ICB	Epitaxial films on Si, GaAs, CaF_2, and Al_2O_3 exhibited remarkable stability.	146
ICB	Epitaxial films on Si(111) and Si(100). Kinetic energies and electronic charge effects studied.	147–150

References

1. K. J. Klabunde, "Chemistry of Free Atoms and Particles," p. 154, Academic Press, New York, 1980.
2. G. H. Jeong and K. J. Klabunde, *J. Am. Chem. Soc.* **108,** 7103 (1986).
3. G. H. Jeong, R. Boucher, and K. J. Klabunde, *J. Am. Chem. Soc.* **112,** 3332 (1990).
4. R. Boucher, Y. Wang, and K. J. Klabunde, *High Temp. Sci.* **31,** 87 (1991).
5. A. Fontijn and W. Felder, *J. Phys. Chem.* **83,** 24 (1979).
6. R. A. Shiffman and P. C. Nordine, *Mater. Res. Soc. Symp. Proc.* **87,** 339 (1987).
7. R. W. Dreyfus, R. Kelly, and R. E. Walkup, *Appl. Phys. Lett.* **49,** 1478 (1986).
8. C. W. Huie and E. S. Yeung, *Anal. Chem.* **58,** 1989 (1986).
9. W. D. Reents and V. E. Bondybey, *Chem. Phys. Lett.* **125,** 324 (1986).

10. H. Kang and J. L. Beachamp, *J. Phys. Chem.* **89**, 3364 (1985).
11. H. Abe and D. M. Kolb, *Ber. Bunseges. Phys. Chem.* **87**, 523 (1983).
12. W. R. M. Graham and W. Weltner, Jr. *J. Chem. Phys.* **65**, 1516 (1976).
13. R. L. Whetten, K. E. Schriver, J. L. Persson, and M. L. Hahn, *J. Chem. Soc. Faraday Trans.* **86**, 2375 (1990).
14. R. J. Shul, R. C. Wetzel, and R. S. Freund, *Phys. Rev. A: Gen. Phys.* **39**, 5588 (1989).
15. S. A. Heidecke, Z. Fu, J. R. Colt, and M. D. Morse, *J. Chem. Phys.* **97**, 1692 (1992).
16. (a) M. J. McQuaid, J. L. Gole, and M. C. Heaven, *J. Chem. Phys.* **92**, 2733 (1990); (b) X. Yang and P. J. Dagdigian, *J. Phys. Chem.* **97**, 4270 (1993).
17. M. K. Bullitt, R. R. Paladugu, J. DeHaven, and P. Davidovits, *J. Phys. Chem.* **88**, 4542 (1984).
18. M. Costes, C. Naulin, G. Dorthe, C. Vaucamps, and G. Nouchi, *Faraday Discuss. Chem. Soc.* **84**, 75 (1987).
19. S. A. Mitchell, P. A. Hackett, D. M. Rayner, and M. Cantin, *J. Phys. Chem.* **90**, 6148 (1986).
20. C. B. Lebrilla and W. F. Maier, *Chem. Phys. Lett.* **105**, 183 (1984).
21. K. J. Klabunde and Y. Tanaka, *J. Am. Chem. Soc.* **105**, 3544 (1983).
22. J. M. Parnis and G. A. Ozin, *J. Am. Chem. Soc.* **108**, 1699 (1986).
23. (a) M. A. Lefcourt and G. A. Ozin, *J. Am. Chem. Soc.* **110**, 6888 (1988); (b) Z. L. Xiao, R. H. Hauge, and J. L. Margrave, *Inorg. Chem.* **32**, 642 (1993); (c) L. B. Knight, Jr., J. R. Woodward, T. J. Kirk, and C. A. Arrington, *J. Phys. Chem.* **97**, 1304 (1993).
24. Y. Tanaka, S. C. Davis, and K. J. Klabunde, *J. Am. Chem. Soc.* **104**, 1013 (1982).
25. K. B. Starowieski and K. J. Klabunde, *Appl. Organomet. Chem.* **3**, 219 (1989).
26. T. R. Bierschenk and R. J. Lagow, *J. Organomet. Chem.* **277**, 1 (1984).
27. C. T. Stanton, S. M. McKenzie, D. J. Sardella, R. G. Levy, and P. Davidovits, *J. Phys. Chem.* **92**, 4658 (1988).
28. M. B. Tabacco, C. T. Stanton, D. J. Sardella, and P. Davidovits, *J. Chem. Phys.* **83**, 5595 (1985).
29. J. L. Gole and S. A. Pace, *J. Phys. Chem.* **85**, 2651 (1981).
30. R. H. Hauge, J. W. Kauffman, and J. L. Margrave, *J. Am. Chem. Soc.* **102**, 6005 (1980).
31. J. A. Howard, H. A. Joly, P. P. Edwards, R. J. Singer, and D. E. Logan, *J. Am. Chem. Soc.* **114**, 474 (1992).
32. (a) T. R. Burkholder and L. Andrews, *J. Chem. Phys.* **95**, 8697 (1991); (b) L. Andrews, T. R. Burkholder, and J. T. Yustein, *J. Phys. Chem.,* *96,* 10182 (1992); (c) T. R. Burkholder, J. T. Yustein, and L. Andrews, *J. Phys. Chem.* **96**, 10189 (1992).
33. B. J. Kelsall and K. D. Carlson, *J. Phys. Chem.* **84**, 951 (1980).
34. S. M. Sonchik, L. Andrews, and K. D. Carlson, *J. Phys. Chem.* **88**, 5269 (1984).
35. (a) Y. M. Hamrick, R. J. VanZee, J. T. Godbout, W. Weltner, Jr., W. J. Lauderdale, J. F. Stanton, and R. J. Bartlett, *J. Phys. Chem.* **95**, 2840 (1991); (b) T. R. Burkholder and L. Andrews, *J. Phys. Chem.* **96**, 10195 (1992).
36. P. H. Kasai and P. M. Jones, *J. Am. Chem. Soc.* **106**, 8018 (1984).
37. G. V. Chertikhin, I. L. Razhanskii, L. V. Serebrennikov, and V. F. Shevel'Kov, *Russ. J. Phys. Chem.* **62**, 1165 (1988).
38. J. H. B. Chenier, C. A. Hampson, J. A. Howard, and B. Mile, *J. Chem. Soc. Chem. Commun.,* 730 (1986).
39. J. A. Howard, R. Sutcliffe, C. A. Hampson, and B. Mile, *J. Phys. Chem.* **90**, 4268 (1986).
40. A. M. LeQuere, C. Xu, and L. Manceron, *J. Phys. Chem.* **95**, 3031 (1991).
41. J. M. Parnis, S. A. Mitchell, D. M. Rayner, and P. A. Hackett, *J. Phys. Chem.* **92**, 3869 (1988).

42. J. H. B. Chenier, J. A. Howard, and B. Mile, *J. Am. Chem. Soc.* **109**, 4109 (1987).
43. P. M. Jones and P. H. Kasai, *J. Phys. Chem.* **92**, 1060 (1988).
44. P. H. Kasai, *J. Am. Chem. Soc.* **104**, 1165 (1982).
45. L. Manceron and L. Andrews, *J. Phys. Chem.* **94**, 3513 (1990).
46. M. Histed, J. A. Howard, H. Morris, and B. Mile, *J. Am. Chem. Soc.* **110**, 5290 (1988).
47. J. H. B. Chenier, J. A. Howard, J. S. Tse, and B. Mile, *J. Am. Chem. Soc.* **107**, 7290 (1985).
48. M. McKee, *J. Phys. Chem.* **95**, 7247 (1991).
49. (a) P. H. Kasai, *J. Am. Chem. Soc.* **104**, 1165 (1982); (b) J. M. L. Martin, P. R. Taylor, P. Hassanzadeh, and L. Andrews, *J. Am. Chem. Soc.* **115**, 2510 (1993).
50. M. Trenary, M. E. Casida, B. R. Brooks, and H. F. Shaefer, III, *J. Am. Chem. Soc.* **101**, 1638 (1979).
51. R. W. Zoellner and K. J. Klabunde, *Chem. Rev.* **84**, 545 (1984).
52. J. R. Flores and A. Largo, *J. Phys. Chem.* **96**, 3015 (1992).
53. J. A. Howard, H. A. Joly, B. Mile, and R. Sutcliffe, *J. Phys. Chem.* **95**, 6819 (1991).
54. J. A. Howard, H. A. Joly, and B. Mile, *J. Am. Chem. Soc.* **111**, 8094 (1989).
55. S. J. Riley, E. K. Parks, C. R. Mao, L. G. Pobo, and S. Wexler, *J. Phys. Chem.* **86**, 3911 (1982).
56. T. G. Dietz, M. A. Duncan, D. E. Powers, and R. E. Smalley, *J. Chem. Phys.* **74**, 6511 (1981).
57. S. N. Yang and T. M. Lu, *J. Appl. Phys.* **58**, 541 (1985).
58. S. R. Langhoff and C. W. Bauschlicher, Jr., *J. Chem. Phys.* **95**, 5882 (1991).
59. P. J. Bruna and J. S. Wright, *J. Phys. Chem.* **94**, 1774 (1990).
60. C. R. Brazier and P. G. Carrick, *J. Chem. Phys.* **96**, 8684 (1992).
61. T. H. Upton, *J. Phys. Chem.* **90**, 754 (1986).
62. (a) K. Balasubramanian, *J. Phys. Chem.* **90**, 6786 (1988); [see also correction in **93**, 8388 (1989)]; (b) K. Balasubramanian and J. Li, *J. Chem. Phys.* **88**, 4979 (1988); (c) P. Bicchi, C. Marinelli, and R. A. Bernheim, *J. Chem. Phys.* **97**, 8809 (1993).
63. L. B. Knight, Jr., B. W. Gregory, S. T. Cobranchi, D. Feller, and E. R. Davidson, *J. Am. Chem. Soc.* **109**, 3521 (1987).
64. V. E. Bondybey, I. Fischer, and R. Schlachta, *Proc. Indian Acad. Sci. Chem. Sci.* **103**, 313 (1991).
65. P. Milani, A. Chatelain, and W. Dettear, *Helv. Phys. Acta.* **62**, 840 (1989).
66. G. Gantefoer, M. Gausa, K. H. Meiwes-Broer, and H. O. Lutz, *Faraday Discuss. Chem. Soc.* **86**, 197 (1988).
67. K. J. Taylor, C. L. Pettiette, M. J. Craycraft, O. Chesnovsky, and R. E. Smalley, *Chem. Phys. Lett.* **152**, 347 (1988).
68. Y. M. Hamrick, R. J. VanZee, and W. Weltner, Jr., *J. Chem. Phys.* **95**, 3009 (1991).
69. Y. M. Hamrick, R. J. VanZee, and W. Weltner, Jr., *J. Chem. Phys.* **96**, 1767 (1992).
70. A. Pellegatti, F. Marinelli, M. Roche, D. Maynau, and J. Malrieu, *J. Phys. Chem.* **48**, 29 (1987).
71. L. Hanley, J. Whitten, and S. Anderson, *J. Phys. Chem.* **92**, 5803 (1988).
72. F. Marinelli and A. Pellegatti, *Chem. Phys. Lett.* **158**, 545 (1989).
73. R. Hernandez and J. Simons, *J. Chem. Phys.* **94**, 2961 (1991).
74. J. A. Howard, R. Sutcliffe, J. S. Tse, H. Dahmane, and B. Mile, *J. Phys. Chem.* **89**, 3595 (1985).
75. U. Rothlisberger, W. Andreoni, and P. Giannozzi, *J. Chem. Phys.* **96**, 1248 (1992).
76. M. F. Jarrold and J. E. Bower, *J. Chem. Phys.* **98**, 2399 (1993).
77. M. Pellarin, B. Baguenard, M. Broyer, J. Lerme, J. L. Vialle, and A. Perez, *J. Chem. Phys.* **98**, 944 (1993).

78. R. L. Hettich, *J. Am. Chem. Soc.* **111**, 8582 (1989).
79. U. Ray, M. F. Jarrold, U. Ray, J. E. Bower, and J. S. Kraus, *J. Chem. Phys.* **91**, 2912 (1989).
80. D. M. Cox, D. J. Trevor, R. L. Whetten, E. A. Rohlfing, and A. Kaldor, *J. Chem. Phys.* **84**, 4651 (1986).
81. M. Acheche, C. Colliex, and P. Trebbia, *Scanning Electron Microsc.* **1**, 25 (1986).
82. R. Hernandez and J. Simons, *J. Chem. Phys.* **96**, 8251 (1992).
83. M. F. Jarrold and J. E. Bower, *J. Chem. Phys.* **87**, 5728 (1987).
84. M. F. Jarrold and J. E. Bower, *J. Am. Chem. Soc.* **110**, 70 (1988).
85. K. Kaya, K. Fuke, S. Nonose, and N. Kikuchi, *Z. Phys. D: At. Mol. Clusters* **12**, 571 (1989).
86. K. Fuke, S. Nonose, N. Kikuchi, and K. Kaya, *Chem. Phys. Lett.* **147**, 479 (1988).
87. P. A. Hintz, S. A. Ruatta, and S. L. Anderson, *J. Chem. Phys.* **92**, 292 (1990).
88. D. M. Cox, K. C. Reichmann, D. J. Trevor, and A. Kaldor, *J. Chem. Phys.* **88**, 111 (1988).
89. D. M. Cox, D. J. Trevor, R. L. Whetten, and A. Kaldor, *J. Phys. Chem.* **92**, 421 (1988).
90. G. Cardenas-Trivino, C. Retamal, and K. J. Klabunde, *J. Appl. Polym. Sci. Appl. Polym. Symp.* **49**, 15 (1991).
91. S. Nonose, Y. Sone, and K. Kaya, *Z. Phys. D: At. Mol., Clusters* **19**, 357 (1991).
92. A. Nakajima, T. Kishi, T. Sugioka, Y. Sone, and K. Kaya, *Chem. Phys. Lett.* **177**, 297 (1991).
93. E. A. Rohlfing, D. M. Cox, R. Petovic-Luton, and A. Kaldor, *J. Phys. Chem.* **88**, 6227 (1984).
94. V. E. Bondybey, G. P. Schwartz, and J. H. English, *J. Chem. Phys.* **78**, 11 (1983).
95. M. Tacke and H. Schnöckel, *Inorg. Chem.* **28**, 2896 (1989).
96. H. Schnöckel, M. Leimkühler, R. Lotz, and R. Mattes, *Angew. Chem. Int. Ed. Engl.* **25**, 921 (1986).
97. C. Dohmeier, R. Mattes, and H. Schnöckel, *J. Chem. Soc. Chem. Commun.*, 358 (1990).
98. C. Dohmeier, C. Robl, M. Tacke, and H. Schnöckel, *Angew. Chem. Int. Ed. Engl.* **30**, 564 (1991).
99. H. Schmidbauer, *Angew. Chem. Int. Ed. Engl.* **24**, 893 (1985); H. Schmidbauer, U. Thewalt, and T. Zafiropoulos, *Angew. Chem. Int. Ed. Engl.* **23**, 76 (1984).
100. M. Tacke, H. Kreinkamp, L. Plaggenborg, and H. Schöckel, *Z. Anorg. Allg. Chem.* **604**, 35 (1991).
101. H. Schnöckel, T. Mehner, H. S. Plitt, and S. Schunck, *J. Am. Chem. Soc.* **111**, 4578 (1989).
102. R. Ahlrichs, L. Zhengyan, and H. Schnöckel, *Z. Anorg. Chem.* **519**, 155 (1984).
103. J. M. Parnis, S. A. Mitchell, T. S. Kanigan, and P. A. Hackett, *J. Phys. Chem.* **93**, 8045 (1989).
104. L. B. Knight, Jr., A. Ligon, S. T. Cobranchi, D. P. Cobranchi, E. Earl, D. Feller, and E. R. Davidson, *J. Chem. Phys.* **85**, 5437 (1986).
105. L. B. Knight, Jr., E. Earl, A. R. Ligon, D. P. Cobranchi, J. R. Woodward, J. M. Bostick, E. R. Davidson, and D. Feller, *J. Am. Chem. Soc.* **108**, 5065 (1986).
106. L. B. Knight, Jr., S. T. Cobranchi, B. W. Gregory, and E. Earl, *J. Chem. Phys.* **86**, 3143 (1987).
107. L. B. Knight, Jr., S. T. Cobranchi, J. T. Petty, and E. Earl, *J. Chem. Phys.* **90**, 690 (1989); and references therein.
108. L. B. Knight, Jr., M. Winiski, P. Miller, and C. A. Arrington, *J. Chem. Phys.* **91**, 4468 (1989).

109. L. B. Knight, Jr., S. T. Cobranchi, J. O. Herlong, and C. A. Arrington, *J. Chem. Phys.* **92**, 5856 (1990).

110. L. B. Knight, Jr. and J. O. Herlong, *J. Chem. Phys.* **91**, 69 (1989).

111. L. B. Knight, Jr. and J. T. Petty, *J. Chem. Phys.* **88**, 481 (1988).

112. L. B. Knight, Jr., D. W. Hill, T. J. Kirk, and C. A. Arrington, *J. Phys. Chem.* **96**, 555 (1992).

113. L. B. Knight, Jr., R. Babb, D. W. Hill, and A. J. McKinley, *J. Chem. Phys.* **97**, 2987 (1992).

114. (a) R. J. Van Zee, S. Li, and W. Weltner, Jr., *J. Chem. Phys.* **98**, 4335 (1993); (b) K. Balasubramanian, *J. Chem. Phys.* **87**, 3518 (1987).

115. L. Lou, L. Wang, L. P. F. Chibante, R. T. Laaksonen, P. Nordlander, and R. E. Smalley, *J. Chem. Phys.* **94**, 8015 (1991).

116. L. Lou, P. Nordlander, and R. E. Smalley, *J. Chem. Phys.* **97**, 1858 (1992).

117. F. L. King, B. I. Dunlap, and D. C. Parent, *J. Chem. Phys.* **94**, 2578 (1991).

118. K. Ito, T. Nakazawa, and K. Osaki, *Thin Solid Films* **151**, 215 (1987).

119. S. Iwana, K. Hayakawa, and T. Arizumi, *J. Cryst. Growth* **56**, 265 (1982).

120. I. Yamada, *Appl. Surf. Sci.* **43**, 23 (1989).

121. I. Yamada, *Mater. Res. Soc. Proc.* **128**, 113 (1989).

122. M. Sosnawski, S. Krommenhoek, J. Sheen, and R. H. Cornely, *J. Vac. Sci. Technol. A* **8**, 1458 (1990).

123. D. L. Barr, *J. Vac. Sci. Technol., B* **5**, 184 (1987).

124. H. Ito, N. Kajita, S. Yamaji, and Y. Minowa, *Jpn. J. Appl. Phys.* **30**, 3228 (1991).

125. R. A. Zuhr, T. E. Haynes, M. D. Galloway, S. Tanaka, A. Yamada, and I. Yamada, *Nucl. Instrum. Methods Phys. Res., B* **59–60**, 308 (1991).

126. I. Yamada, H. Usui, S. Tanaka, and S. Wada, *Nucl. Instrum. Methods Phys. Res., B* **59–60**, 302 (1991).

127. M. Adachi, S. Ikuni, K. Yamada, H. Usui, and I. Yamada, *Nucl. Instrum. Methods. Phys. Res., B* **59–60**, 940 (1991).

128. L. L. Levenson, A. Yahashi, H. Usui, and I. Yamada, *Thin Solid Films* **193–194**, 951 (1990).

129. V. Dietz, P. Ehrhart, D. Guggi, H. C. Haubold, W. Jaeger, M. Prieler, and W. Schilling, *Surf. Coat. Technol.* **43–44**, 963 (1990).

130. M. Ogura, M. Komuro, and K. Shimizu, *Jpn. J. Appl. Phys.* **29**, 2662 (1990).

131. R. E. Hummel, A. Morrone, and E. Lambers, *J. Vac. Sci. Technol., A* **8**, 1437 (1990).

132. I. Yamada, H. Usui, S. Tanaka, U. Dahmen, and K. H. Westmacott, *J. Vac. Sci. Technol., A* **8**, 1443 (1990).

133. A. Yahashi, L. L. Levenson, H. Usui, and I. Yamada, *Appl. Surf. Sci.* **43**, 37 (1989).

134. L. L. Levenson, A. B. Schwartzlander, H. Usui, and I. Yamada, *SIA Surf. Interface Anal.* **15**, 159 (1990).

135. I. Yamada, *Appl. Surf. Sci.* **41–42**, 253 (1989).

136. M. C. Madden, *Appl. Phys. Lett.* **55**, 1077 (1989).

137. L. L. Levenson, H. Usui, I. Yamada, T. Takagi, and A. Swartzlander, *J. Vac. Sci. Technol., A* **7**, 1206 (1989).

138. L. L. Levenson, A. B. Swartzlander, H. Usui, I. Yamada, and J. Takagi, *Mater. Res. Soc. Symp. Proc.* **128**, 131 (1989).

139. I. Yamada, *Nucl. Instrum. Methods Phys. Res., B* **37–38**, 770 (1989).

140. Y. Yamamura, I. Yamada, and T. Takagi, *Nucl. Instrum. Methods Phys. Res., B* **37–38**, 902 (1989).

141. R. E. Hummel and I. Yamada, *Appl. Phys. Lett.* **54**, 18 (1989); **53**, 1765 (1988).

142. I. Yamada, H. Usui, and R. E. Hummel, *Proc. Electrochem. Soc.* **88-19**, 217 (1988).

143. I. Yamada and T. Takagi, *IEEE Trans. Electron Devices* **34,** 1018 (1987).
144. H. S. Jin, A. S. Yapsir, J. M. Lu, W. M. Gibson, I. Yamada, and J. Takagi, *Appl. Phys. Lett.* **50,** 1062 (1987).
145. I. Yamada, H. Takaoka, H. Usui, and T. Takagi, *J. Vac. Sci. Technol., A* **4,** 722 (1986).
146. I. Yamada, H. Usui, H. Inokawa, and T. Takagi, *Surf. Sci.* **168,** 365 (1986).
147. I. Yamada, H. Inokawa, and T. Takagi, *Thin Solid Films* **124,** 179 (1985).
148. I. Yamada, C. J. Palmstrom, E. Kennedy, J. W. Mayer, H. Inokawa, and T. Takagi, *Mater. Res. Soc. Symp.* **37,** 401 (1985).
149. I. Yamada, H. Inokawa, K. Fukushima, and T. Takagi, *Nucl. Instrum. Methods Phys. Res., B* **7–8,** 900 (1985).
150. I. Yamada, H. Inokawa, and T. Takagi, *J. Appl. Phys.* **56,** 2746 (1984).

Carbon Group (Group 14)

I. Carbon Group Atoms and Vapors (C, Si, Ge, Sn, Pb)

A. Occurrence and Techniques

The vapors of these elements are intriguing, particularly carbon, where C_1, C_2, C_3, C_4, and C_5 are all prevalent.[1] Carbon as C_1, C_2, and C_3 is important in astronomy as well as flames, and much of the original literature was reviewed earlier.[1] As a brief summary, carbon has been vaporized (sublimed) in many ways including *e*-beam, laser, resistive heating, and the carbon arc. Very high temperatures are required and the ratios of C_1, C_2, C_3, C_4, and C_5 can vary depending on the method used. However, C_3 is generally the predominant species. Discussion of more recent innovations are given in Section IIA.

Silicon behaves more like a metal during vaporization. It melts first, but unfortunately the melt is very corrosive and so *e*-beam methods are usually preferrable, although Knudsen cell, resistive heating, and arc vaporizations have also been reported. The vapor composition is primarily Si atoms.

Germanium, tin, and lead are relatively easy to vaporize and resistive heating from $W-Al_2O_3$ crucibles works well. Monatomic species make up the vapor.

B. Physical Properties and Theoretical Studies

Collisional quenching of electronically excited carbon atoms has been studied by Husain and Newton.[2] They studied the $^1So(2p^2)$ state, which is 2.684 eV, above the triplet ^3Po state. Quenching by a variety of gases (He, H_2, N_2, Cl_2, CO, H_2O, CH_4, etc.) allowed comparisons of $C(2^1So)$ and $Si(3^1So)$ quenching absolute rate constants. Water and alkanes were among the most effective quenchers.

C. Chemistry

1. Abstraction, Insertion, and Addition Processes of Atoms

In earlier years a wealth of new chemistry resulted from the study of carbon atom abstraction of oxygen, hydrogen, halogen, etc.[1] Recent work has dealt more with detailed mechanistic evaluations. For example, the mechanism of oxygen abstraction from THF by carbon atoms has been evaluated by ab initio theoretical methods.[3] These calculations revealed a low-energy concerted pathway for three-bond cleavage to give the three product molecules directly. Thus, a biradical intermediate may not be involved in this interesting reaction, as was postulated earlier.[4]

Acetylene formation from atomic carbon interacting with CH_4 and other hydrocarbons has also been investigated by theoretical methods for $C(^3P)$ and $C(^1D)$ states, which revealed that an excited state of ethylidene could form from $C(^1D)$, which subsequently eliminates two hydrogen atoms:[5]

$$C(^1D) + CH_4 \rightarrow [HC-CH_3]^{\ddagger} \rightarrow HC \equiv CH + H_2.$$

The barrier for $C(^1D)$ reaction was calculated to be only 11 kcal/mol and exothermic enough to allow $HCCH_3$ decomposition to $HC{\equiv}CH + H_2$. However, $C(^3P)$ would have to cross a substantial barrier (34 kcal/mol) and could only do so if it possesses adequate kinetic energy. The energy required to cross the barriers for $HCCH_3$ decomposition must come from exothermicity of $HCCH_3$ formation:

$$C(^1D) + CH_4 \rightarrow H—C—CH_3 \qquad \Delta H_{rxn} = -82 \text{ kcal/mol}$$

$$C(^3P) + CH_4 \rightarrow H—C—CH_3 \qquad \Delta H_{rxn} = -51 \text{ kcal/mol}.$$

A theoretical investigation of the Si atom -H_2O reaction has revealed that an adduct Si:OH_2 can form without a barrier.[6] Thus, insertions would occur via an early transition state for Si(^1D) and a late transition state for Si(^3P). The crossing point of the triplet and singlet pathways is between transition states. The most favorable reaction appears to be

$$Si + H_2O \rightarrow cis\text{-HSiOH} \rightarrow H_2 + SiO.$$

Under matrix isolation conditions (10 K, Ar) laser-vaporized carbon species C_1, C_2, and C_3 have been allowed to react with CH_4, CH_3Br, and H_2O. A reaction of excited state $C_1(^1D)$ with CH_4 occurred by C–H insertion followed by rearrangement to ethene.[7] Ground-state C(^3P) carbon did not react under these conditions, nor did C_3 and higher clusters. In the case of CH_3Br, C_1, C_2, C_3, and higher clusters reacted upon codeposition. The main reaction mode was C–Br insertion. With water, excited-state $C_1(^1D$ and $^1S)$ reacted to yield CO. Carbon clusters also reacted.

Higher-temperature codeposition (77 K) with isotope labeling and detailed product analysis showed that C_1 reacted by both C–H insertion and hydrogen abstraction, while C_2 and C_3 reacted with CH_4 by hydrogen abstraction.[7]

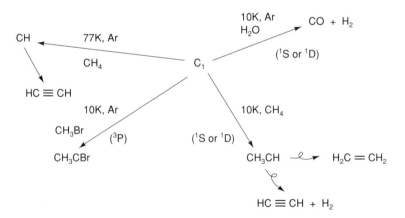

At lower temperature (4 K), carbon and silicon atoms form adducts with CO; ESR spectra were obtained for C_2O and SiCO:[8]

$$C_{atoms} + CO \xrightarrow[\text{Ne}]{4 \text{ K}} C=C=O$$

$$Si_{atoms} + CO \xrightarrow[\text{Ne}]{4 \text{ K}} Si=C=O.$$

These triplet-state molecules exhibited unpaired spin densities as shown below:

$$C=C=O$$
24% 4% 17% $p\pi$ character

$$Si=C=O$$
92% 26% 12% $p\pi$ character.

It seems likely that these ESR active molecules result from reactions of ground-state triplet atoms.

Although carbon atoms are extremely reactive species, selective addition of thermal $^{11}C(^1D)$ atoms to carbon–carbon double bonds has been observed.[9a] Thus, [^{11}C]cyclopentadiene was formed upon reaction of recoil, carbon-11 atoms with 1,3-butadiene, perhaps as shown below:

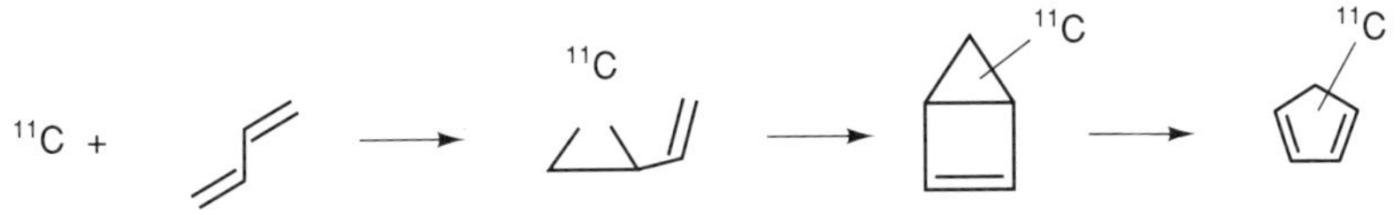

These studies coupled with earlier ones led to conclusions that in the gas phase: (1) $C(^1D)$ will preferentialiy attack olefinic bonds over C–H bonds, (2) translationally hot $C(^1D)$ will undergo direct insertion into an olefinic bond, and (3) thermalized $C(^1D)$ will add across an olefinic bond.[9a] Recently it has also been possible to derive absolute rate data for ground-state $C_1 [2p^2(^3P)]$ reactions with alkenes in the gas phase, using time-resolved atomic resonance absorption spectroscopy in the vacuum UV.[9b] The rate constants for these collisional processes were found to be of the order of collision numbers; for example, ethene $2.0 \times 10^{-10} k_r/cm^3/mole$-cule/sec, propene 4.0×10^{-10}, cyclohexane 4.3×10^{-10}, cycloocta-1,5-diene 4.7×10^{-9}.

Under condensed-phase conditions, acetylenic bonds were found to be very susceptible to attack by Ge and Sn atoms.[10] The codeposition of Ge and Sn atoms with acetylene causeu ɯ.e formation of integral acetylene–metal copolymers of reproducible stoichiometry, i.e., $(C_2H_{2.7}Ge_{0.72})_x$ and $(C_2H_{2.6}Sn0_{0.70})_x$.[10] These addition reactions were free radical-like in nature:

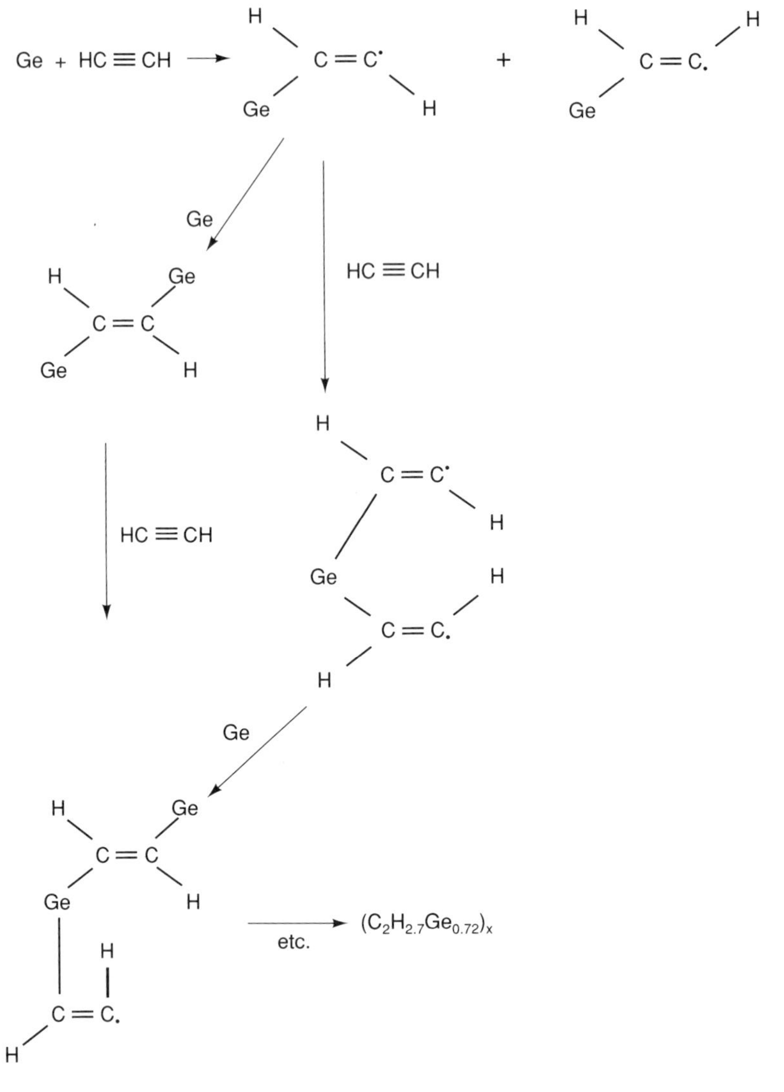

$$Ge + HC \equiv CH \longrightarrow \underset{Ge}{\overset{H}{>}}C = C \cdot \underset{H}{<} \quad + \quad \underset{Ge}{\overset{H}{>}}C = C \cdot \underset{H}{<}$$

Ge

$$\underset{Ge}{\overset{H}{>}}C = C \underset{H}{\overset{Ge}{<}}$$

$HC \equiv CH$

$$\underset{Ge}{\overset{H}{>}}C = C \cdot \underset{H}{<}$$

$$\underset{H}{\overset{Ge}{>}}C = C \cdot \underset{H}{<}$$

$HC \equiv CH$

Ge

$$\underset{Ge}{\overset{H}{>}}C = C \underset{H}{\overset{Ge}{<}}$$

$$\underset{H}{\overset{|}{>}}C = C \cdot$$

$$\xrightarrow[\text{etc.}]{} (C_2H_{2.7}Ge_{0.72})_x$$

These materials were highly cross-linked, insoluble polymers and as pressed pellets exhibited no conductivity. According to ^{119}Sn Mossbauer, the tin species were Sn(II) and Sn(IV) with no Sn(O). Some trapped free radicals were detected by ESR: for Ge/acetylene about 5×10^{17} spins/g and for Sn/acetylene about 1×10^{15} spins/g, or about one unpaired electron for every 20,000 and 3,500,000 carbon atoms, respectively.

A series of metal atoms, including V, Cr, Mn, Fe, Co, Ni, Pb, as well as Ge and Sn, was examined in this work. Only Ge and Sn yielded these polymeric materials. Similar rather complex materials were obtained when Ge and Sn atoms were allowed to react with $CH_3C{\equiv}CCH_3$ and $CF_3C{\equiv}CCF_3$.[11]

2. Free Radical Addition to Metal Atoms

Lagow and co-workers have introduced another reaction type for metal atoms—that of free radical addition.[12,13] In this process alkyl free radicals are generated by dissociation of selected organics in a radiofrequency glow discharge and the resultant radicals cocondensed at 77 K with metal atoms. For example C_2F_6, C_2H_6, and Si_2F_6 have been dissociated to CF_3, CH_3, and SiF_3 radicals and codeposited with Sn, Hg, Te, Bi, and Ge. A series of compounds, some of which are new, including $Hg(CF_3)_2$, $Te_2(CF_3)_2$, $Bi(CF_3)_3$, $Hg(CH_3)_2$, $Cd(CH_3)_2$, $Bi(CH_3)_3$, $Sn(CH_3)_4$, $Ge(CH_3)_4$, and $Hg(SiF_3)_2$, was produced.

Additional work with CF_3 radicals has shown that codeposition with $PbCl_2$ vapor allowed the synthesis of $Pb(CF_3)_4 + Cl_2$. This was the first synthesis of $Pb(CF_3)_4$, which proved to be thermally unstable and reactive with water to yield CHF_3.[13]

$$C_2F_6 \xrightarrow[\text{100 W, 14 MHz}]{\text{rf}} 2CF_3$$

$$PbCl_{2(g)} + 4CF_3 \rightarrow Pb(CF_3)_4 + Cl_2$$

II. Carbon Group Clusters Including Fullerenes C_{60} and Related Cluster Cage Molecules

A. Occurrence and Techniques

Fascinating developments with carbon clusters have occurred in recent years. Clusters of three atoms have been reported as linear or triangular. Larger clusters form linear systems. Still larger clusters of 20 or more atoms (C_{20}–C_{300}) can form cage and tube-like structures, which has led to an exciting new field of chemistry. Herein, these great advances are summarized, beginning with descriptions of the most recent discoveries regarding natural occurrence of carbon group clusters and their preparation in the laboratory.

Indeed it was the initial interest in the chemical properties of small carbon clusters such as C_2, C_3, and C_4[14-16] and the discovery of carbon chains in outer space ($C\equiv C—C\equiv C—C\equiv C—C\equiv\cdots$)[17] that established the field of carbon vapor chemistry as an interesting one and led to the laser vaporization experiments that produced larger clusters. One of the first breakthroughs came from the laboratory of Smalley and co-workers.[18] In 1985 these authors reported that supersonic beams of Si and Ge atoms produced by laser vaporization formed clusters in a He carrier gas. The cluster beams were analyzed by laser ionization and time-of-flight MS (see Chapter 2). Two-photon ionization caused fragmentation, and it was noted that semiconductor (Si and Ge) clusters fragmented by fission processes, rather than by losing one atom at a time, so common of metal clusters. Also, long-lived excited states (about 100 nsec) were detected, again different from metal clusters. This work was a key development in the study of clusters of C, Si, Ge, Sn, and Pb.[18] However, from a historical standpoint, Rohlfing et al.[19] first noted that laser vaporization of carbon led to a series of large clusters (C_1–C_{90}) and proposed long alkyne chains for these species.

In an attempt to simulate carbon-star chemistry, Smalley and co-workers[20-22] began probing the formation of carbon clusters in He environments just after laser pulse vaporization of graphite. It was soon learned that C_{60} and C_{70} were formed preferentially and spontaneously in the process—indeed a series of "magic numbers" was determined as 20, 24, 28, 32, 36, 50, 60, and 70 with 60 being most favored. After much discussion the "geodesic dome" structure for C_{60} with its beautiful architecture, closed-shell structure, and stabilizing aromaticity was proposed (Fig. 8-1).

As more experimental advances were made, the significance of these experiments became clearer. In a paper entitled "Space, Stars, C_{60} and Soot,"[23] Kroto points out the significance of the fact that carbon clusters form spontaneously under laser evaporation, and this has important implications for possible formation under combustion conditions and in certain regions of outer space.

Indeed, Kroto has predicted the possible presence of C_{60} in stars, and a search for C_{60} with an absorption feature near 3860 Å in certain nebula and stars has been made.[25] Mainly C_3 was detected, but a broad residual absorption was observed at the wavelength region expected for C_{60}, although other origins of this absorption cannot be excluded.

Curl and Smalley[26] have also presented a short review of the C_{60} story and point out that the C_{60} molecule may be the most perfectly spherical,

FIGURE 8-1   Structure of buckminster fullerene (C$_{60}$). The two cardboard polyhedra that played key roles in the search for the structure of C$_{60}$. Left, stardome map of the sky, and right, Smalley's model with 60 vertices and 12 pentagonal and 20 hexagonal faces (after Kroto[24]).

edgeless molecule possible. The rapid formation of this species in condensing carbon vapor and its extreme stability would suggest far-reaching implications for fields of combustion and astrophyiscs. And the formation mechanism for C$_{60}$, continuous formation of curved layers, suggests that larger sheets might cover smaller ones eventually yielding imperfect spheroidal layers (Figs. 8-1 and 8-2).[27]

Indeed, the formation mechanism for these large carbon clusters is a fascinating topic. One of the surprises of the century is that highly reactive carbon vapor species can somehow find reaction pathways selective enough to yield perfect symmetry in C$_{60}$. This strange phenomenon has begun to be studied and is discussed later in this chapter.

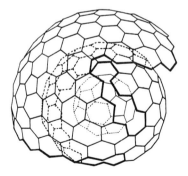

FIGURE 8-2   Spiraling graphitic soot nucleus believed to form from imperfect closure of a growing curved network during carbon condensation (after Curl and Smalley[26]).

The next major breakthrough in the fullerene story was the discovery that *carbon arc* vaporization into He or Ar atmosphere could also produce significant quantities of C_{60} and related spheroidal fullerenes, such as C_{70} and C_{84}. This discovery allowed the synthesis of the fullerenes on a scale where isolation, characterization, and chemical studies were possible. The first report was by Krätschmer *et al.*[28] who collected the soot formed when carbon was arc vaporized into 100 Torr of helium. Dispersion in benzene allowed the C_{60}, C_{70}, and other fullerenes to be dissolved. A flood of reports showing that this method worked well and purification schemes were possible soon followed, and it became feasible to prepare pure C_{60} at a rate of 10 g/day.[29-41] A good yield is about 15% based on carbon evaporated in the pressure range of 100–200 Torr, but yields greater than 3% have been obtained throughout the range of 50 to 760 Torr. A model for the growth of fullerenes that dealt with a competition between annealing of graphitic sheets to curve so that they minimize dangling bonds and further rapid growth to form giant fullerenes was proposed. In support of this competition model, it was found that *laser vaporization* of graphite only yielded fullerenes in reasonable yield if performed in an oven above 1000°C.

Krätschmer[42] remarked that the vapor production of fullerenes, and especially C_{60}-fullerite, is so simple and under such harsh, high-temperature conditions, that fullerene molecules must belong to a very frequently overlooked molecular species in carbon chemistry. And, as predicted, the discovery of C_{60} and related fullerenes from natural sources was not long in coming. For example, both C_{60} and C_{70} have been found in carbon-rich Precambrian rock from Russia.[43] Also fullerenes have been detected in flames, particularly burning benzene.[44] And, in fact, fullerenes can be prepared on synthetic scale from benzene–O_2 flames.[45] Hydrocarbon pyrolysis may also lead to C_{60} according to calculations of McKinnon.[46]

In addition to flames, more and more sources of fullerenes in laboratory experiments are being discovered, for example, in sputtering and electron-beam evaporation processes.[47] And the fullerene series is being rapidly expanded to include isomers of C_{60},[48] larger cages, tunnel-like structures, etc. In fact, Kroto states that we should be more careful in our assumption that graphite layers are innately flat in that aggregates of 32–1000 atoms or more of carbon can exist as closed fullerene cages.[49]

A myriad of characterization methods and spectral probes of C_{60}, C_{70}, and other fullerenes has been applied.[50-53] One of the first electron microscopy observations of C_{60} was reported by Iijima.[54] Notably, this was before the general large-scale synthesis was reported. Scanning tunneling microscopy and atomic force microscopy have also revealed more about the nature of solid-state fullerenes[55] and have confirmed the proposed spherical structures.[56,57] Higher fullerenes and even *chiral* C$_{76}$ have been discussed by Wudl and co-workers.[58]

Obviously, judging from this discussion, the vast majority of reports in recent years have dealt with *carbon clusters*. However, methods for Si, Ge, Sn, and Pb cluster production have also been reported. In particular, resonant neutralization of mass-selected cluster ions provides a universal method for production of neutral cluster beams. Specifically, Pb$_n^+$ could be neutralized by exchange reactions with atomic Na. Very little fragmentation of clusters was observed with this gentle neutralization procedure.[59,60]

B. Physical Properties, Spectra, and Theoretical Work

A flood of papers on the physical properties of clusters of the carbon group has appeared, mostly dealing with the fullerenes. However, first the smaller clusters are discussed, followed by a summary of the fullerenes.

1. Smaller Clusters

Matrix isolation techniques have been employed for trapping and study of C$_2$, C$_2^+$, C$_2^-$, C$_3$, and clusters up to C$_{10}$.[61] Table 8-1 summarizes some of the pertinent information, and an excellent, detailed review has appeared.[62] The reader is referred to that review for coverage of work on all types of carbon molecules, ions, and clusters.

The geometrical structure of C$_3^+$ has been a matter of some conjecture.[81] Coulomb explosion, which involves foil-induced electron stripping and dissociation due to mutual Coulomb repulsion of the nuclei, can be used to determine ion geometries of small molecules. The results indicate that the ground state of C$_3^+$ forms a bent structure. Similar experiments on C$_3$H$_3^+$ indicated a triangular structure, while C$_3$H$_4^+$ a linear geometry. (An alternative explanation of the Coulomb explosion data has been presented as well, however.[82] It was pointed out that the Coulomb explosion data could not unambiguously discriminate between a "cold" cyclic structure and a "hot" floppy, linear structure.) Indeed, theoretical calculations on C$_3^+$ have supported the bent geometry.[83] High level ab initio quantum mechanical methods predict a global minimum energy where C$_3^+$ is strongly bent (CCC bond angle about 70°) and with a substantial barrier (7 kcal/mol) to becoming linear.

Using theoretical methods, small silicon clusters of Si–Si$_5$ and cluster anions have also come under study.[84,85] Structure, binding energies, and

TABLE 8-1
Small Carbon Clusters

Species	Ground state	Comments	Reference
C_2	$^1\Sigma_g^+$		61
C_2^-			61
C_2^+	$^4\Sigma$		63
C_3	$^1\Sigma_g^+$	Quasi-linear vibrational modes observed in gas phase and in a carbon star.	61,64 65
C_4	$^3\Sigma$	Two spins in $p\pi$-orbitals. Cumulene-like rather than acetylene-like. 1586 cm^{-1} computed vibrational frequency.	61,66 67,68 69,70
C_4		Theoretical study of linear and cyclic isomers of C_4 with large basis sets.	71
C_5		Linear, 2164 cm^{-1} stretching frequency. Average bond length of 1.283 Å.	61,72 73,74
C_6–C_{10}		Triplet C_6, C_8, and C_{10} molecules observed by ESR. Probably cumulenic structures, although C_{10} may be bent. However, cyclic structures are energetically feasible.	61,75 76,77 78,79, 80a
C_5		Diode laser absorption spectroscopy of supersonic cluster beams, rovibrational transitions of the cold C_5 cluster; 36 lines were measured with Dopper limited resolution. A linear molecule with C–C bond length of 1.283 Å agrees with data and calculations.	80a
C_6		FTIR of ν_4 and ν_5 stretching modes of linear C_6 in Ar at 10 K	80b

electron affinities were calculated and compared with what limited experimental data were available.[86] Structures for Si_3 (triangle), Si_4 (tetrahedral), and Si_5 (trigonal bipyramid) were used in these calculations. Electron affinities for example, calculated (experiment) were Si, 1.35 eV (1.385), Si_2, 2.25 (2.18), Si_3, 2.24 (2.23), Si_4, 2.06 (2.15), Si_5, 2.36.

The molecules Si_2, SiH_3, and $HSiO$ have been matrix isolated and detected by ESR.[87] The Si_2 species has a $^3\Sigma_g^-$ ground state in an Ar matrix, and its electronic configuration is similar to that of O_2 in that the unpaired electrons occupy πp orbitals in their ground states.

Isotopically labeled Si_2^+ and Ge_2^+ have been produced by pulsed laser vaporization and matrix isolated.[88] The ground states were established as $X^4\Sigma$ for both cations with the three unpaired electrons occupying valence p-type orbitals (according to ESR).

Silicon clusters formed in the gas phase are sensitive to temperature and can change structures, according to annealing/reactivity studies.[89a] These results are of great significance to understanding cluster growth. Usually such growth is by sequential addition of atoms, a process that does not necessarily lead directly to the lowest energy structure. Using chemical reactivity of the clusters with C_2H_4 as a probe, it was found that annealing changed the relative abundance of the structural isomers present. In some cases the more reactive isomer was found to be the most thermodynamically stable (this assumes that annealing drives the clusters to more thermodynamically stable forms). These results indicate that chemical reactivity of Si clusters is not necessarily related to their thermodynamic stability.

For clusters Si_n ($n > 25$) the activation energies for isomer interconversion was 1 to 1.5 eV. It was not possible to anneal most of the clusters to a single form, thus implying that there are only small differences in stability, typically 0.2 eV. Smaller clusters (10–25 atoms) generally required lower E_a values for isomerization (~0.3–0.5 eV). In fact, the processes of electron capture and ionization appear to be quite complex for small Si_n ($n = 3$–10) clusters.[89b]

It is believed that C_2H_4 binds to Si_n in a di-σ-bonded configuration. Obviously, different structural isomers could exhibit variable binding sites. However, these results do not elucidate exactly what structures are involved.

Another significant contribution comes from Smalley's laboratory where photodissociation of semiconductor positive ions Si_n (up to size 80), Ge_n (up to size 40), and GaAs (up to a total of 31 atoms) was investigated.[90] It was found that fragmentations of rather large parts of the clusters took place; for example Ge_n^+ sequentially lost Ge_{10} and to a lesser extent Ge_7. Only $(GaAs)_n^+$ tended to lose an atom at a time.

Since photoionization is widely used to detect (by MS) neutral cluster beams, it is important that the fragmentation processes be well understood. For semiconductor clusters fragmentations can be very serious. However, to minimize fragmentation in favor of simple cluster ionization, data would suggest that the laser-ionizing wavelength should be well above (~2 eV) the ionization threshold and as low a fluence as possible be employed.

Tin and Pb clusters have also been prepared by the laser vaporization/cluster-beam technique. After investigation of the multiphoton-induced fragmentation and photoionization dynamics, relative abundance of different sized clusters could be estimated.[91,92] Especially abundant were Sn_{10} and Pb_{10} which was also true for silicon and germanium. Abundance patterns characteristic of close-packing geometries were observed to be most prominent for lead. Thus, of the C, Si, Ge, Sn, Pb group, carbon behaves uniquely, while Si, Ge, Sn, and Pb behave similarly with regard to cluster abundance. However, the most "metallic" element Pb does show cluster formation tendencies most similar to other metals.

The tetramers of Si_4, Ge_4, Sn_4, and Pb_4 have been studied by theoretical methods,[93] and electron structures and rearrangements of model Si clusters have also been considered.[94]

The structure of the Si_{45} cluster has also come under study by theoretical methods.[95] Two structures in particular seem to be low-energy isomers, are entirely tetracoordinated, and thus are similar to bulk silicon. Clusters smaller than Si_{45} may indeed take on structures less "bulk-like."

Additional information has been gained by considering energy flow dynamics in Si clusters.[96] In this study Si_{39} was considered. It was found that under forced tetrahedral bonding in Si_{39}, energy flow from a locally excited region to the remainder of the cluster should be quite low (nonstatistical behavior in energy flow dynamics is introduced as a result of directional bonding). Thus it might be concluded that Si_{39} and smaller clusters may have a tendency not to take on tetrahedral-like bonding. Indeed, Jarrold and Constant have suggested that smaller silicon clusters may grow by a different mechanism than larger ones, thus leading to different structure types.[97]

2. Large Clusters and the Fullerenes

As pointed out earlier, the fullerenes represent a new form of carbon. These are nonreactive clusters with no dangling bonds, truly closed-shell species. There are no other examples to site of this type of molecular clusters, except stretching the definition to N_2, O_2, F_2 etc.

What types of properties do the fullerenes possess? At this juncture, only the undoped fullerenes are considered—all chemical doping is covered in Section IIIc.

a. Structural Features of Fullerenes   The parent fullerene (C_{60}-fullerite) is made up of six-membered rings and five-membered rings. The five-membered rings are necessary to cause the curvature thus leading to the perfect spherical structure (Fig. 8-1).

Indeed, it is possible to build models of cage structures for C_{36} and higher analogs. However, only the C_{60} species has perfect symmetry. Actually this icosahedral symmetry, the highest finite point group symmetry, is extremely rare. Only the borohydride anion $B_{12}H_{12}^{2-}$, dodecahedrane $C_{12}H_{20}$, and C_{60} share this distinction.[98]

Bond lengths in C_{60} have been determined by studying the ^{13}C NMR of ^{13}C-enriched samples. The ^{13}C–^{13}C dipolar coupling allowed a conclusion that the two types of C–C bond lengths in C_{60} are 1.45 ± 0.015 and 1.40 ± 0.015 Å.[99]

Ring currents in icosahedral C_{60} have also been reported.[100] Using a radius of ~3.5 Å, these authors calculated paramagnetic currents comparable in size to the ones in benzene that flow within the pentagons, whereas weaker currents flow all around the C_{60} molecule. The overall small magnetic susceptibility results from a cancellation of diamagnetic and paramagnetic contributions. Thus, NMR can detect such currents, whereas magnetic measurements (a property of the molecule as a whole) cannot.

Of course there are many members of the fullerene family. Larger species such as C_{70}, C_{84}, and others also form in substantial amounts during the high-temperature carbon vapor synthesis. Note that always even-numbered clusters are favored. This is a consequence of the stable nature of the C_2 building block, and energetic considerations regarding formation of closed-shell structures (where even numbers of carbon atoms are required.)

Simple Hueckel arguments have been used to predict closed electronic shells for one isomer of C_n, where $n = 70 + 30\ m$ (D_{5h} or D_{5d} symmetry). Similarly, a closed shell is predicted for $n = 84 + 36\ m$ for D_{6h} or D_{6d} symmetry (m is a nonnegative integer).[101]

Some fullerenes can, in theory, be chiral species.[102] Qualitative molecular orbital theory predicts a chiral, D_2 symmetry structure for C_{76}. Published data showing 16 ^{13}C NMR lines seem to support the proposed chiral structure.

New, hypothetical carbon allotropes have been proposed.[103] The building principle consists of connecting triptycyl moieties by benzene rings in two dimensions so that two-dimensional hexagonal nets arise. Thus, a mixture of sp^2 and sp^3 carbons arrange in infinite arrays; an example is shown in Fig. 8-3.

Isomerization of C_{60} fullerenes has also been discussed,[104] along with the geometry of small fullerenes C_{20} to C_{70}.[105] The crystal structure of C_{60} has been reported by Burgi *et al.*[106]

Electron counting and the structure of carbon clusters have been considered for C_n, where $n = 60 + 6m$, where m is zero or any positive integer.[107] Within this formula the expected C_n cage-like clusters are expected to have closed shells, and application of simple rules can predict structures. An interesting feature of this "leapfrog" principle is that the structures generated never have adjacent pentagons. In fact C_{60} is the

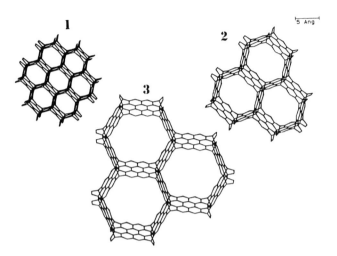

FIGURE 8-3   Hexagonal crystal structures. In structures 1, 2, and 3 exactly zero, one, and two fused benzene rings are located between any two neighboring triptycyl units, respectively. Reprinted with permission from Karfunkel and Dressler, *J. Am. Chem. Soc.* **114**, 2286 (1992). Copyright 1992 American Chemical Society.

smallest "leapfrog" where every pentagonal face is surrounded by hexagons, and the C_n, where $n = 60 + 6m$ predicts new, larger clusters that follow this same rule.

Solid C_{60} and its orientational order[108] can be treated by a microscopic theory of orientational interactions.[108] It was determined that all static and dynamic orientational-dependent physical properties of C_{60} can be described by a same set of rotator functions.

   b. Formation Mechanism   Most workers have proposed that intermediates and models for fullerene synthesis are fullerene fragments and can be visualized as building blocks. However, Jarrold and coworkers[109a] have presented evidence for carbon ring structures as well, which appears to be another dominant structure for large carbon clusters. Thus, during formation, there apparently are many unstable forms that upon mild annealing yield rings and fullerenes. Higher-temperature annealing could cause the rings to form fullerenes—the speculation is that carbon rings may be precursors to fullerenes.

   The extreme stability of the fullerenes has led to theoretical treatments; formation mechanism and disintegration of C_{60} have been modeled using molecular dynamics simulations, which predict that C_{60} should be stable against spontaneous disintegrations up to 5000 K.[109b] The cage formation process was modeled by cooling and compressing 60 carbon atoms from the gas phase. The optimized C$_{60}$ molecule is indeed predicted to be an ideal fullerene. And calculated vibrational properties of this ideal structure agree well with experimental results.

Curling and closure of graphitic ribbons[110] have also been modeled by theoretical methods. At high temperature, graphitic ribbons are predicted to show large deviations from planarity that often could result in the formation of open-ended hollow carbon structures—which would represent good fullerene precursors. Two key parameters were pentagon formation and high temperature.

The response of the scientific community to the fullerene discovery has been phenomenal. The fullerene literature continues to grow at an exponential rate. To briefly summarize some of the physical and spectroscopic investigations of the fullerenes, Table 8-2 was assembled. Hope-

TABLE 8-2
Physical and Spectral Studies of Fullerenes

Species	Technique of study	Comments	Reference
C$_{60}$, C$_{70}$	Selected in flow tube	Bracketed second ionization energies of C$_{60}$ (<11.8 eV) and C$_{70}$ (<11.3 eV).	111
C$_{60}$	Photoexcitation of C$_{60}$	Paramagnetic states in noematic liquid crystals studied by ESR.	112
C$_{60}$, C$_{70}$	Semi-imperical calc.	Polarizability calculated.	113,114
Cn$^+$ ($n \leq 60$)	Size specific fragment loss in MS	C$_3$ loss predominant, but also C$_1$, C$_5$, C$_{10}$, C$_{14}$.	115
C$_{60}$	Photoacoustic calorimetry	Triplet energy of C$_{60}$ 36.0 ± 0.6 kcal/mol.	116
C$_{60}$	ESR	Free radical generated, effect of O$_2$ studied.	117
C$_{vapor}$	Matrix isolation	IR and UV spectra. 1998 cm^{-1} assigned to C$_8$ and UV absorption at 310 nm also assigned to C$_8$.	118
C$_{60}^-$	ESR	Trapped in molecular sieve 13X.	119
C$_{60}$	Simulations of collisions with diamond (111) surface	Up to 150 eV nonreactive. Above that chemisorption of C$_{60}$ or atom exchange should occur.	120

(*continued*)

TABLE 8-2 (*continued*)

Species	Technique of study	Comments	Reference
C_{60}	Ab initio calculations.	Anisotropic NMR shielding in C_{60} calculated.	121
C_{60}	NMR, solid state	Rapid isotropic motion at room temp., but at 77 K sufficiently slow for chemical shift measurements.	122
C_{60}	Various techniques	Photophysics summarized for C_{60} (see Table 8-3)	123
C_{60}	Raman	High pressure (up to 18.0 GPa) Low temp. (4–360 K). Rotational–vibrational coupling. C_{60} rotates freely at 240 K. Rotational glass formed at high pressure.	124
C_{60} films	Luminescence and absorption spectra	Under high-vacuum luminescence in 700–1100 nm range with fine structure at low temperature.	125
C_{60}/C_{70} films	Vibrational Raman	273 cm^{-1} squashing mode. 1469 and 497 cm^{-1} double bond stretching and breathing.	126
C_{60}	IR and UV	Four vibrational bands observed, in agreement with that predicted by theory for soccerball-shaped C_{60}.	127
C_{60}, C_{70}	Solid-state ^{13}C NMR	At 296 K C_{60} molecules rotate rapidly and isotropically, while C_{70} rotates more anisotropically.	128
C_{60}	Solid-state ^{13}C NMR	Rotation slow at 77 K; chemical shift tensors of 220, 186, and 40 ppm determined.	129
C_{60}	Photodissociation	Neutral products detected.	130
C_{60}	Vibrational studies	Isotopic and anharmonic perturbations to dipole active vibrations studied.	131
C_{60}	Induced phosphorescence	External heavy atom-induced phosphorescence emission. Triplet energy determined.	132

(*continued*)

TABLE 8-2 (*continued*)

Species	Technique of study	Comments	Reference
C_{60}	Optical properties	Nonlinear optical properties studied by theory and experiment.	133
C_{60}	Photo detachment	Turnable dye laser used. An electron affinity of 2.650 ± 0.050 eV was determined.	134
C_{60}, C_{70}	Two-photon ionization	Lowest triplet states of C_{60} and C_{70} have lifetimes of 41 and 42 μsec, and energies of 1.7 and 1.6 eV, respectively.	135
C_{60}	Emission and absorption spectra	Interpretation of vibrational structure.	136
C_{60}, C_{70}	Electron spectrum	Cold molecular beam. Resonant two photon ionization spectrum resembles diffuse interstellar band spectrum.	137
C_{100}–C_{60}	FTICR, thermionic emission	Ions trapped in ICR apparatus; laser excitation. Cooling mechanisms are (1) IR emission, evaporative loss of C_2, (3) thermionic emission.	138
C_{60}	Laser-induced fragmentation	Photophysics of C_n^+. Larger clusters ($n > 34$) fragment by loss of C_2, smaller clusters by loss of C_3 or C_1.	139
C_{70}	Absorption, fluorescence	Photophysics of C_{70}. Triplet energy about 33 kcal/mol weak fluorescence.	140
C_{70} anions	Near IR absorption	Absorption for several C_{70} anions are reported.	141
C_{60}	Electron diffraction and electron microscopy at low temperature	Phase transitions in C_{60} and related lattice defects of microstructural films.	142
C_{60}	Absorption	Absolute absorption cross sections at 600°C.	143
C_{60}, C_{70}	Cross-polarization, magic angle spinning NMR study.	In the presence of organic impurities such as toluene ^{1}H–^{13}C cross-polarization facilitated.	144

(*continued*)

TABLE 8-2 (*continued*)

Species	Technique of study	Comments	Reference
C_{60}	Optical absorption	Toluene solutions of C_{60} block the passage of light above a critical intensity.	145
C_n	Electron attachment	Smaller fullerenes	145b
C_{60}	X-ray diffraction	Compressibility at P = O to 1.2 GPa of solid phase cubic C_{60} is three times as much as graphite and 40 times as much as diamond.	146
C_{60}	ESR	Radical reactivity of C_{60}.	147, 148
C_{60}^-	Pople–Pariser–Parr calculations	Low-lying excited states of fullerene anions were studied.	149
C_{76}	Ab initio study	C_{76} fullerene isomers.	150
C_{60}	Multiphoton excitation, dissociation, and ionization of C_{60}	High, internal excitation possible by absorbing 10–20 photons.	151
C_{60}	Electroreduction	In aprotic solvents; supporting electrolyte and temperature effects.	152
C_{60}	Reduction	Detection of C_{60}^{6-} and C_{70}^{6-}	153
C_{60}	Reduction	Production of C_{60}^{5-} in solution.	154
C_{60}	Oxidation	C_{60}^+ produced by photosensitized electron transfer.	155
C_{60}	Auger	Core excited states in C_{60} determined by Auger line shape.	156

fully, the reader can choose topics of interest and then refer to the original literature. Table 8-3 brings together some of the photophysical properties of C_{60}.

Some of the unique features of C_{60} are its comparatively large compressibility, ionization energy, electron affinity, and energy of triplet state. Also note the large number of oxidation states, from 2^+ to 6^-, that are possible with C_{60}. And it has also been determined that C_{60} is capable of absorbing many photons leading to internal vibrational excitation, rather than direct multiphoton ionization.[151] These high internal energies

TABLE 8-3
Photophysical and Spectral Properties of C_{60}[123]

Energy of singlet E_S	46.1 kcal/mol
Energy of triplet E_T	37.5 ± 4.5 kcal/mol
Ext. coefficient $\in_T$ (480 nm)	(2.8 ± 0.2) × 10^3/M/cm
k(quenching by O_2)	(1.9 ± 0.2) × 10^9/M/sec
Triplet lifetime t_T	40 ± 4 μs (in Ar-saturated solution)
k(quenching $^3O_2$)	(5 ± 2) × 10^5/M/sec
Quatum yield $\phi(^1O_2)$ 355 nm	0.76 ± 0.05
Quatum yield $\phi(^1O_2)$ 532 nm	0.96 ± 0.04
Compressibility	$-d(\ln V)/dP = 7.0 ± 1 × 10^{-12}$ cm²/dyne
UV/vis	213, 257, 329 nm ($\in$ = 135000, 175000, 51000) 404, 440–670 with max at 500, 540, 570, 600, 625
IR of solid	1430, 1182, 577, 527 cm^{-1}
^{13}C NMR (benzene)	142.68 ppm
Electron affinity	2.6–2.8 eV
Ionization energy	7.61 ± 0.02 eV
X-ray data	C–C = 1.388(9) Å six–six ring fusion
	C–C = 1.432(5) Å five–six ring fusion
Solubility	1.7 mg/ml (benzene), 7.9(CS_2)
	33(1-methylnapthalene); 0.043 (hexane)
	47 solvents studied[123e]

can reach 50 eV. Thus, multiphoton ionization is in competition with delayed ionization and fragmentation. It is concluded that the majority of ionization for fullerenes results from *thermionic* electron emission, a "heating" process that has a time scale of microseconds.

C. Chemistry

1. Smaller Clusters of C, Si, Ge, Sn, Pb Group

As mentioned earlier, C_2 and C_3, and higher clusters react under matrix isolation conditions (10 K, Ar) with CH_3Br and H_2O by C–Br and O–H insertion.[7] At higher codeposition temperatures (77 K) methane reacted by hydrogen abstraction processes.

$$C_2 \xrightarrow{\quad CH_4 \quad} HC\equiv C \xrightarrow{\quad HC\equiv C \quad} HC\equiv C-C\equiv C-H$$

$$\Big\downarrow CH_4$$

$$HC\equiv CH$$

Alkynes, polyalkynes, and allenes were typical products isolated in small yields.

The C_2 molecule as the smallest "cluster" received a great deal of attention in earlier years.[1,16] More recently it has been trapped as a borane–phosphine complex:[157]

$$(C_6H_5)_3B^-C\equiv C^+PCH_3(C_6H_5)_2.$$

The reaction of C_3 with water under matrix isolation conditions yielded the trapped $C_3(H_2O)$ adduct.[158] Upon photolysis this complex rearranged and formed the hydroxy ethynylcarbene.

$$C_2 + H_2O \xrightarrow[\text{Ar}]{10\text{ K}} C_3(H_2O) \xrightarrow{h\nu} HC\equiv C-C-OH.$$

Matrix reactions of carbon vapor with CO have yielded evidence for the C_4O and C_6O molecules.[159] These triplet-state molecules were detected by ESR and identities substantiated by ^{13}C and ^{17}O substitutions. Simple theory predicts linear structures with triplet ground states for C_nO, where n is even, and singlet ground states when n is odd, $CO(^1\Sigma)$, $C_2O(^3\Sigma)$, $C_3O(^1\Sigma)$.

The production of C_4O and C_6O only occurred when carbon vapor -CO matrices were photolyzed with blue light. Thus, excited-state C_3 reacted with CO or C_3O.

$$C_3 \xrightarrow{h\nu} C_3^{\cdot} \xrightarrow{\quad CO \quad} C_4O$$

$$\Big\downarrow C_3O$$

$$\longrightarrow C_6O$$

It seems likely that singlet C_3 and singlet CO reacted to yield singlet C_4O and this was followed by intersystem crossing:

$$C_3(^1\Pi_u) + CO(^1\Sigma) \rightarrow C_4O(^1\Pi) \rightarrow C_4O(X^3\Sigma).$$

The reactions of gas-phase carbon cluster ions with hydrocarbons have been studied by FTMS.[160] Reactivity patterns suggested that there is a structural change from linear to cyclic for C_n, where $n = 9$ and 10.

Reactions of the small clusters proceeded by carbene insertions into C–H bonds and the unsaturated C–C bonds of C_2H_2 and C_2H_4 reactants. Also, when C_2H_2 and C_2H_4 were used as reactants, even the relatively unreactive larger clusters C_n ($n \geq 10$) were consumed. In similar experiments, C_n^+ reactions with naphthalene have been investigated. Reaction pathways included adduct formation and loss of hydrogen atoms.[161] The reactions were found to be extremely dependent on the kinetic energy of the cluster ion, and such dependency was used to advantage to determine the average kinetic energy of desorbed ions. The results of these experiments provided a plausible mechanism for growth of large polyaromatic hydrocarbons in such environments as interstellar clouds or flames.

In a related vein, high-pressure H_2 reactions with neutral C_n ($n = 4–22$) led to a variety of structures, including both chain and ring structures, simultaneously.[162]

Silicon clusters in the size range of Si_5 to Si_{66} have been prepared by the pulsed laser-beam method and selected clusters trapped and studied by FTICR.[163] The Si_n^+ clusters chemisorbed NH_3 at rates varying over three orders of magnitude. Clusters where $n = 21, 25, 33, 39$, and 45 were particularly unreactive while cluster Si_{43} was the most active. The data suggested that several structural isomers must have been present for certain sized clusters. However, Si_{39} and Si_{45} being particularly unreactive indicates that these were especially stable, crystallized forms. In fact, the Si_{45}^+ cluster may be a "filled fullerene" such that it has all dangling bonds of the surface tied up as much as possible through interconnected five- and six-membered rings and with a Si atom fully coordinated on the inside of the "sphere."

Similar experiments with Me_3N chemisorption on Si_n^+ and Si_n^- ($n = 39$, $43, 48$) provided additional evidence that the annealed clusters adopt well-organized structures. Also, it appears that both positively and negatively charged clusters have similar structures.[164]

Jarrold *et al.* have also probed the chemistry of Si_n^+ clusters.[165] Ammonia chemisorption was studied with annealed cluster ions Si_n^+ (n up to 70). At room temperature, all clusters reacted at close to the collision rate and collisional annealing did not have an influence. At about 400 K equilibria could be established, and binding energies of NH_3 were determined to be above 1 eV, probably indicating molecular adsorption as the dominant process. At higher temperatures dissociative chemisorption occurs, and interestingly, at 700 K the sticking probabilities were two to four orders of magnitude *smaller* than on bulk silicon.

Water reactions with Si_n^+ ($n = 10–65$) yielded a series of $Si_n(D_2O)_m^+$ adducts in the gas phase.[166] Large variations in reactivity were observed for clusters smaller than $n = 40$; $n = 11, 13, 14, 19$, and 23 were particularly unreactive. Larger clusters were considerably *less* reactive than bulk silicon. These results suggest that a two-step process may be involved: adduct formation followed by rearrangement to a more strongly bound species (a slow process).

Analogous results were found for O_2–Si_n^+ reactions; O_2 reacts with Si_n^+ much more slowly than with bulk silicon.[167] Smaller clusters reacted to give Si_{n-2}^+ plus two SiO molecules, while larger clusters yielded $Si_nO_2^+$ adducts. The transition between these reaction pathways occurred at $n = 29$ to 36. Again, a two-step process appears to operate, molecular chemisorption followed by slow dissociative chemisorption. The activation barriers vary for these processes with cluster size.

To account for the lower reactivity of clusters vs bulk silicon, it was proposed that bulk silicon possesses more reactive sites. That is, clusters probably still have close-packed structures with few dangling bonds and resemble closed shells more than bulk silicon surfaces. In this way, silicon clusters do mimic fullerenes, and it is likely that only carbon and silicon would exhibit this rather strange behavior—clusters being less reactive than bulk surface.

Reactions of Si_n^+ ($n = 11$–50) with C_2H_4 indicated that numerous structural isomers must have been present.[168] Virtually all clusters possess different isomers that can react at vastly different rates. C_2H_4 was found to be particularly sensitive to structural changes, and Si_{20}^+ was found to interconvert from one structure to another on the time scale of the experiments. These results make the study of Si clusters particularly challenging when regarding comparisons of reactivity vs cluster size.

Recently the chemistry of deposited, size-selected Si clusters has been reported.[169] Cluster sizes $n = 10$, 13, and 40–50 were studied by XPS during room-temperature treatment with O_2. The size-selected Si_n^+ were deposited on an amorphous carbon substrate at 5 eV kinetic energy and stuck with probabilities of 95%. The striking variations in reactivity of gas-phase clusters vs each other and vs bulk silicon were not found. Instead, all deposited clusters exhibited similar O_2 sticking coefficients. It was proposed that substantial cluster–cluster coalescence and sintering had occurred.

2. Chemistry of Fullerenes and Fullerene Films

a. Addition–Substitution Reactions and Adducts of Fullerenes   Although fullerenes C_{40}–C_{80} are not reactive with small molecules such as NO, SO_2, or O_2 under normal conditions,[27] a host of novel chemical reactions has been achieved. Hydrogenated derivatives were reported early on.[34] The

$C_{60}H_{36}$ adduct was readily prepared by a Birch reduction and could be dehydrogenated back to C_{60} using DDQ:

$$C_{60} \xrightarrow[\text{t-BuOH}]{\text{Li, liq. NH}_3} C_{60}H_{36} \xrightarrow[\text{toluene reflux}]{\text{DDQ}} C_{60}$$

Thus, the hydrogen reduction was fully reversible without change in the C_{60} framework.

The facile loss of H_2 from the C_{60} frame should not be surprising. For example, semiempirical calculations of the dihydrogenated species $C_{60}H_2$ have suggested that only a few isomers should form exothermically by addition of H_2 to C_{60}.[170a] In fact, only recently has the simplest derivative $C_{60}H_2$ been prepared and characterized.[170b] The hydroboration reaction used produced a borane derivative that should be useful in other functionalization:

$$C_{60} \; + \; THF \cdot BH_3 \; \longrightarrow \; [C_{60}\overset{\displaystyle H}{\underset{\displaystyle BH_2}{<}}] \; \xrightarrow{H_2O} \; C_{60}H_2$$

The $C_{60}H_2$ derivative is a static structure at room temperature, on an NMR time scale. The H, H addition appears to be across a 6,6-ring fusion.

A host of derivatives of the fullerenes has been prepared by typical inorganic and organic reactions. For example, isomerically pure monoadducts such as $C_{60}HEt$ were obtained when EtLi was added to solutions of C_{60} followed by protonation.[171] Such strong nucleophiles add to a double bond between two six-membered rings. An interesting morpholine adduct $C_{60}H_6(N(CH_2CH_2)_2O)_6$ has also been prepared. In this compound the amine molecules are covalently added to six pyracylene units, and the six hydrogen atoms are able to migrate on the surface of the C_{60} sphere with a surprisingly low activation barrier.[172]

"Sweetened buckyballs" have been reported—these are fullerene sugars where enantiomerically pure spirolinked C–glycoside adducts of C_{60} were prepared.[173]

Highly fluorinated derivatives of fullerenes have also been reported.[174] By controlled direct F_2 addition a series of $C_{60}F_x$ and $C_{70}F_y$ compounds were prepared, where $x \leq 48$ and $y \leq 56$. According to MS data, defluorination was rather facile. No perfluoroderivative was produced. It appears that $C_{60}F_{48}$ is the most fluorine that can be added.

Olah et al. have reported on chlorination and bromination of fullerenes.[175] A hot tube reaction at 250°C was necessary to get Cl_2 addition to proceed. About 24 chlorine atoms/C_{60} unit were taken up. It was found that polychlorofullerenes at 400°C dechlorinate to the parent fullerene (under argon).

Chlorine substitution was possible, as shown below:

$$C_{60}Cl_n \xrightarrow[\text{reflux}]{\text{MeOH, KOH}} C_{60}(OMe)_n$$

$$C_{60}Cl_n \xrightarrow[\text{AlCl}_3]{\text{C}_6\text{H}_6} C_{60}(Ph)_n .$$

Bromination near room temperature allowed the uptake of 2 to 4 Br atoms:

$$C_{60} \xrightarrow[\text{up to 50°C}]{\text{Br}_2\text{(neat)}} C_6Br_n \qquad n = 2 \text{ or } 4.$$

In a similar vein, it was found that Lewis acid-catalyzed Freidel–Crafts reactions could be carried out such that polyarylated fullerenes (fullerenation of aromatics) could be achieved.[176]

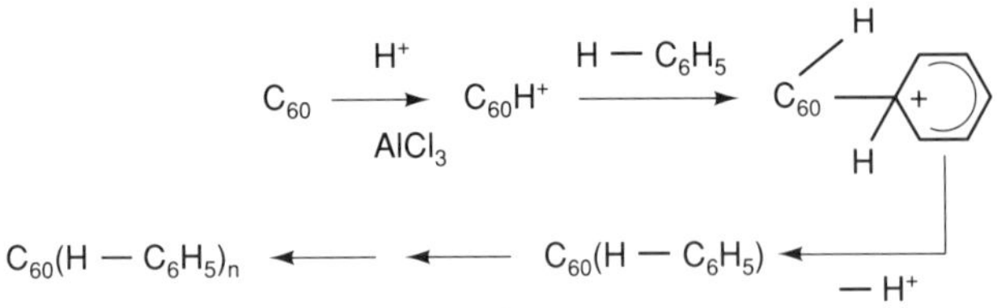

Fullerenes, fullerene radicals, and fullerene cations have all been derivatized.[177] In the case of C_{60}^{2+} and C_{70}^{2+} in the gas phase, unsaturated hydrocarbons and amines were found to add to form adducts, for example:

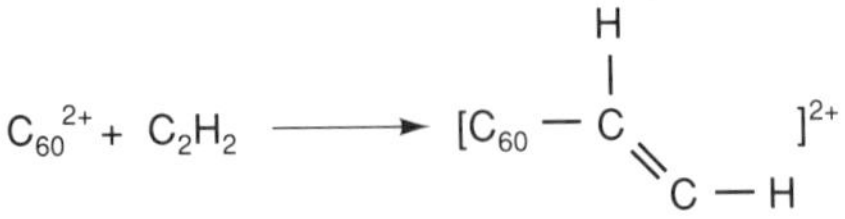

When alkyl radicals add to C_{60}, the resultant RC_{60} radicals dimerize.[178] The intermediate radical $R–C_{60}$ has its unpaired electron mostly confined to two, six-membered rings on the C_{60} surface (extensive delocalization of the unpaired electron was ruled out).

Olah and co-workers have treated C_{60} and C_{70} with Li metal in THF with the aid of ultrasound. Upon reduction the reduced species went into solution as deep red-brown species and most likely contained even numbers of excess electrons. Alkylation of these polyanions with CH_3I yielded light brown solids that, according to FTMS, were methylated derivatives $C_{60}(CH_3)_x$, where $x = 1$–24. Dominant were $x = 6$ and 8.[179]

The monoalkylated t-BuC$_{60}$ anion has also been reported,[180] and an adduct t-BuC$_{60}^-$Li$^+ \cdot 4CH_3CN$ was isolated in pure form. Protonation yielded t-BuC$_{60}$H, which proved to be one of the strongest acids made up of only carbon and hydrogen. The radical t-BuC$_{60} \cdot$ was produced by reaction of the anion with I_2. Dimerization was observed to occur as an equilibrium. The radical also reacted with $(n$-Bu$)_3$SnH. These interesting substitution reactions are shown in Scheme 8-1.

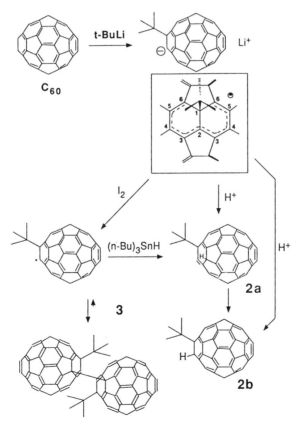

SCHEME 8-1   Some substitution reactions of C_{60} (after Fagan et al.[180]).

Benzyne and fullerenes react in benzene solution to give a series of compounds [C_{60} + $(C_6H_4)_n$], where n = 1, 2, 3, 4.[181] The monoadduct was isolated and characterized, showing that addition had occurred across one double bond of a six-membered ring:

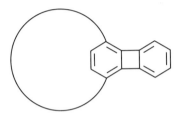

Yellow adduct of C_{2v} symmetry

Even C^+ (gas phase) has been allowed to react with C_{60}.[182] In this case, C_{61}^+ is a metastable product. Carbon atom exchange or decomposition can occur. Single collision conditions over the energy range 2 to 78 eV allowed C^+ to add to C_{60} to form C_{61}^+ with no activation barrier. Decomposition to C_{60} + C^+ with carbon exchange was shown by ^{13}C labeling. At high collision energies fragmentation to C_{60-2n} was observed, again with ^{13}C exchange.

A general overview of substitution patterns for C_{60}[183] was aided by MNDO calculations. The geometries of 1,7- and 1,9-$C_{60}X_2$ (X = H, F, Cl, Br, I) were optimized. The 1,9-structure is energetically preferred for X = H, F while the 1,7-structure is preferred for X = Cl, Br. No preference was detectable for X = I.

Incorporation of C_{60} into polymers as a covalently bonded entity has been accomplished by reaction with the p-xylylene diradical:[184]

The polymer is insoluble and appears to be cross-linked, with a ratio of xylylene to C$_{60}$ of 3.4:1.

Nano-phase-separated, polymer-substituted fullerenes, called "flagellenes" have been prepared by initiating anionic polymerization of styrene in the presence of a C$_{60}$ toluene solution, followed by quenching with CH$_3$I.[185]

$$[R - (CH_2 - \underset{|}{CH}]_x \text{—} C_{60} \text{—} [CH_3]_y$$

The polystyrene: C$_{60}$ ratio could be varied and this method promises to yield tractable C$_{60}$-polymer materials.

Gas-phase polymerization of butadiene has been initiated by fullerene dications.[186] The fullerene cations were produced by electron bombardment of C$_{60}$ and C$_{70}$ vapor entrained in Ar carrier gas. After mass selection these species were thermalized and then exposed to butadiene vapor. Interestingly C$_{60}^{+}\cdot$ and C$_{70}^{+}\cdot$ were found to be unreactive toward the diene, but C$_{60}^{2+}$ and C$_{70}^{2+}$ were observed to react rapidly (nearly unit reaction efficiency). The reaction may be driven by electrostatic forces and the mechanism proposed involved both one- and two-sided additions to the fullerene cages.

Fullerenes have proven to be interesting ligands in organometallic chemistry as well. A series of reports on organometallic adducts have shown that C$_{60}$ weakly coordinates to some metal centers as an "arene"-type ligand.

Perhaps a good starting place for reviewing this literature is the work with metal atoms and C$_{60}$.[187] Silver atoms were deposited with a C$_{60}$-cyclohexane solution on a rotating cryostat at 77 K. ESR analysis at 175 K indicated that a AgC$_{60}$ adduct of silver(0) had formed with most of the unpaired spin density on the C$_{60}$ ligand. Warming above 175 K caused decomposition of the adduct.

Several Ir complexes have been reported.[188,189] The complex (η^5-C$_9$H$_7$)Ir(CO)(η^2-C$_{60}$) undergoes electrochemical reduction such that the C$_{60}$ ligand bears much of the negative charge. Analysis of the spectral data suggests that complexation of C$_{60}$ with the (η^5-C$_9$H$_7$)Ir(CO) moiety causes relatively small perturbations in its electronic structure.

Similarly, when C_{70} was forced to accommodate two Ir moieties, binding took place with each in a dihapto-fashion to 6–6 ring junctions to yield $(C_{70}[Ir(CO)Cl(PPhMe_2)_2]_2) \cdot 3C_6H_6$.[190] NMR studies indicated an isomeric mixture; however, an X-ray crystal structure determination revealed only one isomer in the solid state. These results suggest that the Ir complexation is weak and binding can shift easily from one site to another on the C_{70} unit.

The effect of metal coordination on the photophysical properties of C_{60} has been investigated in the $(\eta^5\text{-}C_9H_7)Ir(CO)(\eta^2\text{-}C_{60})$ adduct.[191] A transient excited-state triplet with a lifetime of 100 nsec was detected. The lifetime of this state in the Ir adduct is 500 times shorter than in C_{60} itself. And in the presence of O_2 a very short 25-nsec lifetime was observed, apparently due to the rapid and efficient quenching to form singlet O_2.

Other organometallic complexes reported involve $(\eta^2\text{-}C_{70})Ir(CO)Cl(PPh_3)_2$[192] and $(\eta^2\text{-}C_{60})Pt(PPh_3)_2$.[193] Weak, dihapto-coordination was evident from X-ray crystal structure determinations.

Gas-phase organometallic complexes have been formed from C_{60}^+ and C_{70}^+ reactions with $Fe(CO)_5$.[194]

$$C_{60}^+ + Fe(CO)_5 \rightarrow Fe(CO)_4C_{60}^+ + CO.$$

Also the reaction of Ni^+ with C_{60} yielded $Ni(C_{60})_2^+$ in the gas phase.[195] Indeed, a series of gas-phase MC_{60}^+ species have been detected (M = V, VO, Fe, Co, Ni, Cu, Rh, La), where M is on the outside of the C_{60} cage.[196,197]

Finally, the solution-phase osmylation reaction should be described, where OsO_4 is allowed to add to C_{60} through oxygen–carbon σ-bonding.[198,199] This represents one of the first examples of adding heteroatom functionality to C_{60} without disrupting the carbon framework

and promises to allow further functionalization by manipulating the osmium grouping.

A detailed NMR (INADEQUATE) analysis of osmylated C_{60} yielded chemical shifts for the 17 types of carbon atoms in $C_{60}(OsO_4)(4\text{-}t\text{-butylpy-ridine})_2$.[200] Using this compound as a model of C_{60} itself, hybridizations for C_{60} were estimated as 31.5% "s" character for six–five ring fusions, 34.0% "s" character for six–six ring fusions, and 3% "s" character for the π-orbitals.

b. Metal Atoms Inside Fullerenes (M @ C_n) and Cage Doped Fullerenes Curl and Smalley have suggested that metal atoms contained inside a fullerene cage should be symbolized as M @ C_n,[201] whereas metal derivatives outside the cage would be simply MC_{60}. Indeed, the first report of a metal atom being trapped inside the fullerene cage was from Smalley's laboratory.[202] This remarkable finding was due to the mixing of La salts with graphite before laser vaporization. During the high-temperature/ plasma process the La ions were reduced to atoms and somehow trapped within the framework during the aggregation, folding, and closing process. Since that time a series of caged species have been reported, which are summarized in Table 8-4. A particularly interesting aspect of this work is that it may be possible to stabilize normally unstable small fullerene cages through binding to a metal atom/ion inside. An example of this is U @ C_{28}.[203] In this case the C_{28} cage may behave as a sort of hollow super atom with an effective valence of *four*. Thus, stable, closed-shell derivatives of C_{28} might be possible by forming $C_{28}H_4$ (outside cage H bonding) or trapping of a Ti atom inside.[204]

Another particularly interesting system is the Sc_3 @ C_{82} species, where ESR evidence suggests that the Sc_3 species is an equilateral triangle similar to that found as a "free" species in rare-gas matrices.[205] Also, the La @ C_{82}, Y @ C_{82}, and Sc @ C_{82} are best described as M^{3+} and C_{82}^{3-} species.[205,206]

It would appear that many new stable, M @ C_n species may be waiting to be discovered. The presence of the metal may serve to stabilize the C_n species and allow a closed-shell structure to exist. In fact, La @ C_{82} was found to be extremely robust in air, much more so than C_{82} itself. Also, the evidence for U @ C_{28} and Ti @ C_{28} discussed above indicates that even "unstable" C_n species can be isolated when complexed to certain metals.

These findings recall the ubiquitous cyclopentadienyl ring, an unstable species as a free anion, cation, or radical, but when complexed to a metal, a very stable entity.

Another integral way of "doping" fullerenes is to substitute a cage carbon for some other atom. If a stable structure is to be obtained, only a very limited number of elements would seem appropriate. Boron is one such element. Indeed, when a graphite pellet containing boron nitride was laser vaporized, several fullerenes that contained one or more substitutions were produced.[207] Predominant were $C_{49}B$, $C_{59}B$, $C_{58}B_2$, $C_{57}B_3$,

TABLE 8-4

Summary of M @ C_n Species (Metal Atoms/Ions Trapped in Fullerene Cages)

Fullerene	Metal atom	Comments	Reference
C_{82}	Y	EXAFS study of YC_{82}.	210
C_{82}	Sc, Y, La	Isomers and ^{13}C hyperfine structures.	211
C_{60}	Ne	Nonlinear vibrational dynamics.	212
C_{60}	Na^+	Endohedral vibrations.	213
C_{80}	La_2	A soluble material.	214
C_{82}	Y, Y_2	MS and ESR.	215
C_n	Rare earth	Endohedral rare-earth complexes.	216
C_{28}	U	Tetravalent C_{28} stabilized by U atom.	203
C_{60}, C_{70}	La	M @ C_n terminology suggested.	202
C_{60}^+, C_{70}^+	He	Host-guest chemistry. Inclusion of He by collision of C_{60}^+ or C_{70}^+ with He.	217
C_{60}^+	He	Formed by collision of kiloelectron volt C_{60}^+ with He.	218
C_{82}	Sc, Sc_2, Sc_3	Likely that Sc_3 in an equilateral within the C_{82} cage (ESR evidence).	205a
C_{60}	Fe	Graphite arc vaporization in the presence of $Fe(CO)_5$ and He. Mossbauer (singlet). EXAFS, MS.	219
C_{76}	La	Extraction of La @ C_{76} and other La @ C_{2n} species.	220
C_{60}	Alkali metal atoms	Interactions between alkali metal atoms and C_{60}.	221
C_{60}^+	He, Ne	Endohedral complexes of fullerene radical cations formed by collision of C_{60}^+ with He or e.	222a
C_{60}, C_{70}, C_{74} C_{80}, C_{84}	La, Y	La @ C_{60}, La @ C_{70}, La @ C_{82}, La_2 @ C_{80}, La @ C_{84}, YLa @ C_{80} produced by covaporization.	222b
C_{82}	Sc_3	Electronic structure.	205b
C_{74}, C_{82}, C_{84}	Sc_2	Isolation and spectroscopic properties.	205c
C_{60}	He, Ne	One He trapped/880,000 fullerene molecules. He does not exchange once trapped. Release of He has an energy barrier of about 80 kcal/mol. A cage-opening mechanism is proposed.	222c

TABLE 8-5
Thin Films of Fullerenes and Adducts (not including superconductors)

System	Comments	Reference
$C_{60} \cdot CH_2I_2 \cdot C_6H_6$	First example of ordered, unmodified C_{60} molecules. Arranged in hexagonally close-packed layers separated by solvent components. NMR studies suggested molecular rotation of C_{60} in a time scale of tens of microseconds.	237
C_{60} and C_{70} at air–water interface	Langmuir trough exp. showed that multilayered domains form in initial dilute state.	238
C_{60}	Electrochemistry.	239 240
C_{60}	Optical properties; C_{60}–toluene solutions transmit low-intensity light, but block light above a critical intensity, apparently a result of higher absorbtivity of excited state C_{60} than ground-state C_{60}.	241
$C_{60}TDAE_{0.86}$	Tetrakis (dimethylamino) ethylene adduct of C_{60} exhibits soft ferromagnetism.	242
C_{60} on GaAs(110)	Locally well ordered, structurally stable arrays of C_{60} observed on GaAs surface.	243
Conducting polymer-C_{60}	Photoinduced electron transfer from polymer to C_{60} observed.	244
CO and NO absorption of C_{60} films	Two absorption sites were detected on C_{60}. Large shifts in V_{CO} and V_{NO} indicated rather strong interactions.	244b
C_{60}/amphiphilic compounds	Morphological study of thin films; electron microscope.	244c
C_{60} polycryst. films	Order–disorder transitions studied by Raman spectroscopy.	244d
C_{60}/C_{70} microcrystals	Lattice vibrations consist predominantly of localized modes. Specific heat and thermal conductivity studied.	244e
C_{60} microcrystals	Irradiation with visible or UV light caused photopolymerization.	244f

TABLE 8-6
An Update on C_{60}/C_{70}/Fullerenes

Species	System of Interest	Comments	Reference
C_{60}	Interaction with graphite	van der Waals binding.	244g
C_{60}	Multiphoton ionization		244h
C_{60}	Diene adduct	Adduct with ortho-Quinodimethane.	244i
C_{60}	Photofragmentation	Mechanism.	244j
C_{60}	$N^+ + C_{60}$ reactive scattering	Substition, charge transfer, fragmentation.	244k
C_{60}, C_{70}	Radical adducts	Laser flash photolysis and pulse radiolysis.	244l
C_{60}	In molecular sieves	Loaded from gas phase into VPI-5 aluminophosphate.	244m
C_{70}	Inelastic neutron scattering	Intramolecular vibrations of C_{70}.	244n
C_{60}	Reduction	Electrochemistry in liquid ammonia.	244o
C_{60}	Organometallic complex	Double addition to form $C_{60}[Ir(CO)Cl(PMe_2Ph)_2]_2$.	244p
C_{60}/C_{70}	Polysilane doping	Fullerene-doped polysilanes are excellent polymeric photoconductors.	244q
C_{60}	Polymer bound	Soluble amino polymers add to the double bonds of C_{60}.	244r
C_{60}	Free-standing C_{60} membranes	van der Waals forces allow fabrication of free-standing C_{60} membranes on silicon wafers.	244s
C_{60}	Addition of azides to fullerenes	Derivatization is possible with RCH_2N_3.	244t
C_{60}	Electrocrystallization and ESR study	Single crystal of $[N(NPh_3)_2]C_{60}$.	244u
C_{60}	Self-assembly	Fullerene self-assembly onto $(MeO)_3Si(CH_2)_3\ NH_2$-modified oxide surfaces.	244v
C_{60}	Electroreduction	Kinetics and thermodynamics in benzonitrile.	244w
C_{60}	Electroreduction	Kinetic parameters studied by scanning electrochemical microscopy and fast scan cyclic voltammetry.	244x
C_{60}	Diels–Alder adduct	Synthesis and X-ray structure of bicyclic derivative.	244y
C_{60}	Debromination	Selective, catalytic debromination by anions.	244z

(*continued*)

TABLE 8-6 (*continued*)

Species	System of Interest	Comments	Reference
C_{60}	ESR study of R–C_{60} derivatives	Hindered internal rotation analyzed for alkyl-C_{60} radicals.	244aa
C_{60}	Mathematic predictions of properties	From first principles, the high symmetry allows many mathematical treatments. Building your own buckball is part of this interesting paper.	244ab
C_{60}/C_{70}	Radiation stability	Pulse radiolysis in hydrocarbon solvents.	244ac
C_{60}, C_n	Laser desorption FT-mass spec.	Fullerene molecular weight distributions in graphite soot.	244ad

$C_{56}B_4$, and $C_{69}B$ (usually detected as NH_3 adducts). These molecules had a tendency to lose C_2 with resultant cage shrinkage. The boron-substituted fullerenes are Lewis acids and readily chemisorbed one NH_3 molecule/surface boron atom.[208] Nitrogen derivatives of C_{60} and C_{70} have also been reported.[209] By carbon vaporization in N_2 gas, species such as $C_{70}N_2$, $C_{59}N_6$, $C_{59}N_4$, and $C_{59}N_2$ were formed. In this case, apparently N substitutes carbon in the framework of the growing cluster similarly to boron. Thus, the two modes of intimately doping fullerene molecules are within the cage such as La @ C_{60}, and by substituting cage atoms, such as $C_{59}B$ or $C_{59}N_2$. Table 8-4 summarizes many of the reports of metal atoms trapped within fullerenes.

   c. **Thin Films of Fullerenes and Fullerene–Metal Adducts**   Certainly one of the strongest driving forces for continued investigation of fullerenes was the early report of superconductivity in M_xC_n adducts.[223] For example, K_xC_{60} films showed an onset of superconductivity $T_c = 18$ K. More detailed analysis[224] by shielding diamagnetism measurements indicted a single superconducting phase ($T_c = 19.3$ K) for composition K_3C_{60}. Similarly, a Rb_xC_{60} sample exhibited evidence for superconductivity with $T_c = 30$ K with a diamagnetic shielding of 7%.
   A series of other alkali metal-doped fullerenes (or fullerites) have appeared in the literature.[225–233] It would appear that the highest T_c reported is for a $Cs_xRb_yC_{60}$ film that exhibited an onset at $T_c = 33$ K with a diamagnetic shielding of over 60%.[234] This is the highest T_c ever observed for a molecular superconductor. It was concluded that the T_c of these fullerites is determined mainly by the density of states at the Fermi level.

Another interpretation of resistivity and superconductivity deals with electron donation into the t_{1g} band of C_{60}, based on experiments with Ca_xC_{60}, Sr_xC_{60}, K_6C_{60}, and Ca_3C_{60}.[235,236] In Ca_3C_{60} and K_6C_{60} the t_{1a} band is filled and these are insulators. Addition of more Ca to yield Ca_5C_{60} allows donation of electrons to the t_{1g} band, which again allows superconductivity. Other interesting properties of C_{60} films are summarized in Table 8-5.

Finally, as an update on some of the most recent work with the fullerenes, Table 8-6 has been added.

III. Binuclear Systems

A. Occurrence and Techniques

Coverage in this section is restricted to those unusual high-temperature species containing atom(s) of the carbon group and another element. Formally, this could include all carbenes and most carbon-based radicals, but such broad coverage is not the intent here. Only those rather unusual species that are generally formed by gas-phase high-temperature processes are covered.

For example, molecular CCN, CS, SiO, SiO_2, $SiCl_2$, metallocarbohedrene clusters, and a few others are discussed. Laser vaporization of bulk solids, hollow cathode discharges, and high voltage discharges are the usual methods of generation.[245-247]

Particularly interesting have been metallocarbohedrene clusters from laser vaporization of Ti and graphite powders yielding a "magic number" of $Ti_8C_{12}^+$ as detected by MS (Fig. 8-4).[248] Indeed, a new class of stable, neutral clusters where M = T, V appears to have been discovered by these cluster-beam experiments[249] and MC_2 fragments may be important building blocks in this high-temperature, gas-phase formation of these new cage-like structures. These metal–carbon combinations promise to rival the fullerenes in diversity and novelty. For example the Ti_8C_{12} structure has been chosen as a model and electronic, magnetic, and geometric structure calculated.[249b] It is predicted to be a distorted dodecahedron with a binding energy of 6.1 eV per atom. The robust bonding is a result of strong covalent C–C and Ti–C bonds with no contributions of Ti–Ti bonds. The Ti sites probably carry a small magnetic moment of 0.35 μ_B/ atom, making the overall cluster weakly magnetic. Macroscopic quantities of both Ti_8C_{12} and V_8C_{12} have now been prepared by an arc dicharge technique (about 1% yield overall).[249c] Interestingly, these products are

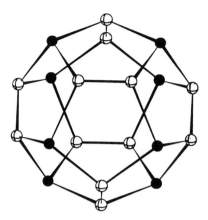

FIGURE 8-4 Geometrical structure of Ti_8C_{12}. The solid circles are Ti sites and the shaded circles represent C sites. (After Castleman and Reddy and co-workers.[249a–c]

stable in air. Indeed, this fascinating work was preceded by a very similar synthetic approach to the synthesis of metal carbide fine particles.[250] In an atmosphere of inert gas a carbon rod was mounted touching the top surface of a refractory metal. By applying a heavy electric current through the carbon–metal junction, fine particles of metal carbides were produced. Cubic SiC as well as carbides of Ti, V, Cr, and Zr were obtained. In the case of Hf, Nb, Mo, Ta, and W crystalline metal carbide particles were not obtained.

B. Physical Properties and Theoretical Studies

The CS molecule and its properties have been thoroughly reviewed earlier and are not discussed in detail here.[245] A few other interesting diatomics have also received attention, such as the CP radial that has been formed upon laser vaporization of carbon/phosphorus elemental mixtures.[251] Isotopes of $^{12}C^{31}P$ and $^{13}C^{31}P$ were trapped in Ne matrices at 4 K and analyzed by ESR. Experimental and theoretical results show that the odd electron in CP resides primarily on carbon in a $2s/2pz$ hybridized orbital.

Molecular silicon–carbon species can also be generated by laser vaporization of the elements. For example, the SiC radical has been generated and spectroscopically analyzed.[252] Likewise, jet-cooled SiC_2 has been analyzed by laser-induced fluorescence spectroscopy.[253] Theoretical and experimental work on Si_2C[254,255] has been possible by vaporizing mixtures of carbon and silicon and quenching the vapors in frozen Ar. The Si_2C molecule exhibits an asymmetric Si–C stretch at 1188 cm^{-1} and a Si–Si stretch at 840 cm^{-1}, but no bending mode was observed. It was concluded that Si_2C has a floppy, bent structure in the ground state.

$$Si - Si \diagdown_C$$

This is in contrast to the earlier work[256] where Si–C–Si was proposed with a C_{2V} symmetry and a bond angle of lower limit 110°.

Similarly, Si_3C has been treated by theoretical and experimental methods.[257,258] A structure of C_{2V} symmetry with two equivalent Si atoms was proposed:

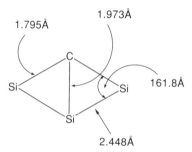

The molecule SiC_4 was produced by vacuum UV photolysis of SiH_4/butadiene mixtures, and matrix isolation of SiC_4 was accomplished.[259] A C=C stretching mode at 2080 cm^{-1} was observed, and the molecule has been proposed to be linear.

Molecular SiO, SiO_2, and S_2 have been studied by Schnöckel and co-workers.[260–262] Dimeric $(SiO_2)_2$ was matrix isolated from the reaction of $(SiO)_2$ with O_2. Its structure was deduced by IR studies of ^{18}O isotopomers. With the help of theoretical calculations, the dimerization energy $2SiO_2 \rightleftarrows (SiO_2)_2$ was estimated as −453 kJ/mol.

$$O = Si \diagup^{O}_{O} \diagdown Si = O$$

In a similar way SiS_2 was generated in a cold matrix by reaction of SiS with S atoms. IR studies and theoretical treatments suggested a nearly symmetrical molecule and with ν_{Si-S} at 918 cm^{-1} and a bond energy of SSi=S of about 533 kJ/mol (compare OSi=O of 622 kJ/mol). Similar experiments and calculations gave a value of 435 kJ/mol of SGe=S.

C. Chemistry

Very little chemistry has been reported for the binuclear high-temperature species of interest herein. The molecule that has received the most attention has been CS, and a recent review has thoroughly documented the chemical properties.[245] Basically CS behaves as a nucleophilic carbene of moderate reactivity. Very efficient insertion reactions with sulfenyl chlorides and strongly basic amines have been observed.[263–265]

$$R-S-H + CS \longrightarrow R-S-\overset{\displaystyle S}{\overset{\displaystyle \|}{C}}-H$$

$$RS-Cl + CS \longrightarrow R-S-\overset{\displaystyle S}{\overset{\displaystyle \|}{C}}-Cl$$

$$RR'N-H + CS \longrightarrow RR'N-\overset{\displaystyle S}{\overset{\displaystyle \|}{C}}-H$$

Addition reactions with electron-rich alkynes have also been observed as well as many other interesting reactions.

$$R-C\equiv C-R' + CS \longrightarrow$$

$$R, R' = Me_2N, Et_2N$$
$$\text{or Me, Et}_2N$$

Close relatives of CS are SiO and SiS molecules. Rowlands and Timms have continued to investigate SiO as a chemical reagent and have found that SiO attacks toluene and iodine to eventually yield an Si–O polymer with aryl, H, and I substitution.[266]

Molecular SiO interacts with Ag atoms in a cold Ar matrix yielding an adduct Ag(SiO) and probably $Ag_2(SiO)$, identified by IR spectra.[267] The ν_{Si-O} band for free SiO (1226 cm^{-1}) shifted to 1163 cm^{-1} upon complexation to Ag. Careful analysis of the isotopic shifts for $^{16}O, ^{18}O, ^{28}Si, ^{29}Si,$

^{30}Si various isotopes of SiO suggested that the Ag(SiO) molecule has an angle of $\leq 90°$, which may imply side-on bonding:

TABLE 8-7

Thin Films of Si, Ge, Sn, and Their Oxides Prepared by Deposition of Atoms/Clusters

Species	Technique	Comments	Reference
SiO_2	Microwave ion source and ICB	SiO vapor and O_2 gaseous ions with O_2 resulted in SiO_2 film formation at low substrate temperatures. Films were transparent and were good insulators.	268
$Si_{1-x}N_x$ SiO_2	Microwave ion source and ICB	Si_n cluster ions reacted with microwave-produced N or O ion beams. Transparent silicon nitride and stoichiometric SiO_2 films could be produced.	269
Ge	Time of fight MS with ICB	Only small cluster ions were detected (a few atoms). Proposed that vapor condensation on nozzle walls plays a key role in formation of larger Ge clusters.	270
Ge	ICB	Epitaxial films were grown at 300–450°C on Si substrates. Computer-controlled phase-modulated ellipsometer for rapid monitoring of nucleation described.	271
Ge	Molecular beam epitaxy (MBE)	A 30- to 80-Å layer of Ge deposited on Si at room temperature, followed by thermal annealing at 300–500°C. The Ge atoms cluster in randomly distributed islands and help release mismatch stress at the interface. Then 10,000-Å Ge film deposited.	272, 273
Ge	ICB, ultrahigh vacuum	Ge doped with Al (a p-type dopant). Epitaxial growth of Ge(100) on Si(100) achieved as low as 300°C.	274
Ge	ICB	Clusters of about 100 atoms were produced by vaporization of Ag or Ge. About 5–25% of the clusters were ionized. The best films were produced by smaller clusters.	275
Si	ICB	Epitaxial Si on Si obtained at 300–730°C.	276
Ge	Vapor deposition	Ge deposited on NaCl gave heavily twinned structures.	277

IV. Films of Si, Ge, Sn, and Their Oxides Prepared by Vaporization and Ionized Cluster Beam Methods

As in other chapters, a brief review of new approaches to film formation is given. These methods generally employ ionized cluster beams (ICB), electron beams, or similar techniques. Table 8-7 gives the pertinent information necessary for Si, Ge, Sn clusters and their oxides.

References

1. K. J. Klabunde, "Chemistry of Free Atoms and Particles," Academic Press, New York, 1980.
2. D. Husain and D. P. Newton, *J. Chem. Soc. Faraday Trans. 2,* **78,** 51 (1982).
3. M. L. McKee, G. C. Paul, and P. B. Shevlin, *J. Am. Chem. Soc.* **112,** 3374 (1990).
4. P. S. Skell, K. J. Klabunde, J. H. Plonka, J. R. Roberts, and D. L. Williams-Smith, *J. Am. Chem. Soc.* **95,** 1547 (1973).
5. M. L. McKee and P. B. Shevlin, *J. Am. Chem. Soc.* **107,** 5191 (1985).
6. S. Sakai, M. S. Gordon, and K. D. Jordon, *J. Phys. Chem.* **92,** 7053 (1988).
7. G. H. Jeong, K. J. Klabunde, O.-G. Pan, G. C. Paul, and P. B. Shevlin, *J. Am. Chem. Soc.* **111,** 8784 (1989).
8. R. J. Van Zee, R. F. Ferrante, and W. Weltner, Jr., *Chem. Phys. Lett.* **139,** 426 (1987).
9. (a) R. A. Ferriere, A. P. Wolf, D. A. Baltuskonis, and Y. N. Tang, *J. Chem. Soc. Chem. Commun.,* 1321 (1982); (b) N. Haider and D. Husain, *J. Chem. Soc., Faraday Trans.* **89,** 7 (1993).
10. R. W. Zoellner and K. J. Klabunde, *Inorg. Chem.* **23,** 3241 (1984).
11. R. W. Zoellner and K. J. Klabunde, *Chem. Rev.* **84,** 545 (1984).
12. T. J. Juhlke, R. W. Braun, T. R. Bierschenck, and R. J. Lagow, *J. Am. Chem. Soc.* **101,** 3229 (1979).
13. T. J. Juhlke, J. I. Glanz, and R. J. Lagow, *Inorg. Chem.* **28,** 980 (1989).
14. P. S. Skell and L. D. Wescott, *J. Am. Chem. Soc.* **85,** 1023 (1963).
15. P. S. Skell and R. F. Harris, *J. Am. Chem. Soc.* **87,** 5807 (1965).
16. P. S. Skell, J. J. Havel, and M. J. McGlinchey, *Acc. Chem. Res.* **6,** 97 (1973).
17. B. F. Coles, P. B. Hitchcock, and D. R. M. Walton, *J. Chem. Soc. Dalton Trans.* **442,** 1975 (1975).
18. J. R. Heath, Y. Lin, S. C. O'Brien, Q. L. Zhang, R. F. Curl, F. K. Tittel, and R. E. Smalley, *J. Chem. Phys.* **83,** 5520 (1985).
19. E. A. Rohlfing, D. M. Cox, and A. Kaldor, *J. Chem. Phys.* **81,** 3322 (1984).
20. H. W. Kroto, J. R. Heath, S. C. O'Brien, R. F. Curl, and R. E. Smalley, *Nature (London)* **318,** 162 (1985).
21. J. R. Heath, Q. Zhang, S. C. O'Brien, R. F. Curl, H. W. Kroto, and R. E. Smalley, *J. Am. Chem. Soc.* **109,** 359 (1987).
22. K. W. Kroto, J. R. Heath, S. C. O'Brien, R. F. Curl, and R. E. Smalley, *Astrophys. J.* **314,** 352 (1987).
23. H. Kroto, *Science* **242,** 1139 (1988).

24. H. W. Kroto, *Angew. Chem, Int. Ed. Engl.* **31**, 111 (1992).

25. W. B. Somerville and J. G. Bellis, *Mon. Not. R. Astron. Soc.* **240**, 41P (1989).

26. R. F. Curl and R. E. Smalley, *Science* **242**, 1017 (1988).

27. Q. L. Zhang, S. C. O'Brien, J. R. Heath, Y. Liu, R. F. Curl, H. W. Kroto, and R. E. Smalley, *J. Phys. Chem.* **90**, 525 (1986).

28. W. Krätschmer, L. D. Lamb, K. Fostiropoulos, and D. R. Huffman, *Nature* **347**, 354 (1990).

29. R. E. Haufler, Y. Chai, L. P. F. Chibante, J. Conceicao, C. Jin, L-S Wang, S. Maryyama, and R. E. Smalley, *Mater. Res. Soc. Symp. Proc.* **206**, 627 (1991).

30. R. Taylor, J. P. Hare, A. K. Abdul-Sada, and H. W. Kroto, *J. Chem. Soc. Chem. Commun.*, 1423 (1990).

31. H. Ajie, M. M. Alvarez, S. J. Anz, R. D. Beck, F. Diederich, K. Fostiropoulos, D. R. Huffman, W. Kratschmer, Y. Rubin, K. E. Schriver, D. Sensharma, and R. L. Whetten, *J. Phys. Chem.* **94**, 8630 (1990).

32. A. S. Koch, K. C. Khemani, and F. Wudl, *J. Org. Chem.* **56**, 4543 (1991).

33. W. A. Scrivens, P. V. Bedworth, and J. M. Tour, *J. Am. Chem. Soc.* **114**, 7917 (1992).

34. R. E. Haufler, J. Conceicao, L. P. F. Chibante, Y. Chai, N. E. Byrne, S. Flanagan, M. M. Haley, S. C. O'Brien, C. Pan, Z. Xiao, W. E. Billups, M. A. Ciufoling, R. H. Hauge, J. L. Margrave, L. J. Wilson, R. F. Curl, and R. E. Smalley, *J. Phys. Chem.* **94**, 8634 (1990).

35. D. H. Parker, P. Wurz, K. Chatterjee, K. R. Lykke, J. E. Hunt, M. J. Pellin, J. C. Hemminger, D. M. Gruen, and L. M. Stock, *J. Am. Chem. Soc.* **113**, 7499 (1991).

36. K. C. Khemani, M. Prato, and F. Wudl, *J. Org. Chem.* **57**, 3254 (1992).

37. K. Chatterjee, D. H. Parker, P. Wurtz, K. R. Lykke, D. M. Gruen, and L. M. Stock, *J. Org. Chem.* **57**, 3253 (1992).

38. (a) Q. Li, F. Wudl, C. Thilgen, R. L. Whetten, and F. Diederich, *J. Am. Chem. Soc.* **114**, 3994 (1992); (b) M. Endo and H. W. Kroto, *J. Phys. Chem.* **96**, 6941 (1992).

39. T. Drewello, K.-D. Asmus, J. Stach, R. Herzschuh, M. Kao, and C. S. Foote, *J. Phys. Chem.* **95**, 10554 (1991); T. Braun, *Angew. Chem. Int. Ed. Engl.* **31**, 588 (1992).

40. (a) Y. K. Bae, D. C. Lorents, R. Malhotra, C. H. Becker, D. Tse, and L. Jusinski, *Mater. Res. Soc. Proc.* **206**, 733 (1991); (b) A. Mittelbach, H. G. von Schnering, J. Carlsen, R. Janiak, and H. Quast, *Angew. Chem. Int. Ed. Engl.* **31**, 1640 (1992).

41. W. A. Scrivens and J. M. Tour, *J. Org. Chem.* **57**, 6932 (1992).

42. W. Krätschmer, *Z. Phys. D: At. Mol. Clusters* **19**, 405 (1991).

43. P. B. Buseck, S. J. Tsipursky, and R. Hettich, *Science* **257**, 167, 215 (1992).

44. J. B. Howard, J. T. McKinnon, Y. Makerousky, A. I. Lafleur, and M. E. Johnson, *Nature (London)* **352**, 139 (1991).

45. J. B. Howard, J. T. McKinnon, M. E. Johnson, Y. Makarousky, and A. L. Lafleur, *J. Phys. Chem.* **96**, 665 (1992).

46. J. T. McKinnon, *J. Phys. Chem.* **95**, 8941 (1991).

47. S. Prakash, H. J. Doerr, L. Isaacs, A. Wehrsig, C. Yenetzian, H. Cynn, and F. Diederich, *J. Phys. Chem.* **96**, 6866 (1992).

48. K. Raghaucchari and C. M. Rohlfing, *J. Phys. Chem.* **96**, 2463 (1992).

49. H. W. Kroto, *J. Chem. Soc. Dalton Trans.*, 2141 (1992).

50. G. Zhennan, Q. Kiuxin, Z. Xihuang, W. Yongqing, Z. Xing, F. Sunqi, and G. Zizhao, *J. Phys. Chem.* **95**, 9615 (1991).

51. F. Diederich and R. L. Whetten, *Angew. Chem. Int. Ed. Engl.* **30**, 678 (1991).

52. H. Ajie, M. M. Alvarez, S. J. Anz, R. D. Beck, F. Diederich, K. Fostiropoulos, D. R. Huffman, W. Kratschmer, Y. Rubin, K. E. Schriver, D. Sengharma, and R. L. Whetten, *J. Phys. Chem.* **94**, 8630 (1990).

53. J. F. Stoddart, *Angew Chem. Int. Ed. Engl.* **30,** 70 (1991).

54. S. Iijima, *J. Phys. Chem.* **91,** 3466 (1987).

55. R. J. Wilson, G. Meijer, D. S. Bethune, R. D. Johnson, D. D. Chambliss, M. S. de Vries, H. E. Hunziker, and H. R. Wendt, *Nature (London)* **348,** 621 (1990).

56. J. L. Wragg, J. E. Chamberlain, H. W. White, W. Krätschemer, and D. R. Huffman, *Nature (London)* **348,** 623 (1990).

57. E. J. Synder, M. E. Anderson, W. M. Tong, R. S. Williams, S. J. Anz, M. M. Alvarez, Y. Rubin, F. N. Diederich, and R. L. Whetten, *Science* **253,** 171 (1991).

58. Q. Li, F. Wudl, C. Thilgen, R. L. Whetten, and F. Diederich, *J. Am. Chem. Soc.* **114,** 3994 (1992).

59. M. Abshagen, F. Traeger, J. Kowalski, M. Meyberg, G. ZuPutlitz, and J. Slaby, *Nater. Res. Soc. Symp. Proc.* **206,** 13 (1991).

60. M. Abshagen, J. Kowalski, M. Meyberg, G. ZuPutlitz, J. Slaby, and F. Traegar, *Z. Phys. D: At. Mol. Clusters* **19,** 199 (1991).

61. W. Weltner, Jr. and R. J. Van Zee, *J. Mol. Struct.* **222,** 201 (1990).

62. W. Weltner, Jr. and R. J. Van Zee, *Chem. Rev.* **89,** 1713 (1989).

63. (a) D. Forney, H. Althaus, and J. P. Maier, *J. Phys. Chem.* **91,** 6458 (1987); (b) L. B. Knight, Jr., S. T. Cobranchi, and E. Earl, *J. Chem. Phys.* **88,** 7348 (1988).

64. K. Matsumana, H. Kanamori, K. Kavaguchi, and E. Hirota, *J. Chem. Phys.* **89,** 3491 (1988).

65. K. H. Hinkle, J. J. Keady, and P. F. Bernath, *Science* **241,** 1319 (1988).

66. W. Krätschmer, *Surf. Sci.* **156,** 814 (1985).

67. W. Krätschmer, *Astrophys. Space Sci.* **128,** 93 (1986).

68. H. M. Cheung and W. R. M. Graham, *J. Chem. Phys.* **91,** 6664 (1989).

69. L. N. Shen and W. R. M. Grahm, *J. Chem. Phys.* **91,** 5115 (1989).

70. J. R. Heath and R. J. SayKalley, *J. Chem. Phys.* **94,** 3271 (1991).

71. J. D. Watts, J. Gauss, J. F. Stanton, and R. J. Bartlett, *J. Chem. Phys.* **97,** 8372 (1992).

72. M. Vala, T. M. Chandrasekhar, J. Szczepanski, R. Van Zee, and W. Weltner, Jr., *J. Chem. Phys.* **90,** 595 (1989).

73. P. F. Bernath, K. H. Hinkle, and J. J. Keady, *Science* **244,** 562 (1989).

74. G. Pacchioni and J. Koutecky, *J. Chem. Phys.* **88,** 1066 (1988).

75. R. J. Van Zee, R. F. Ferrante, K. J. Zeringue, and W. Weltner, Jr., *J. Chem. Phys.* **88,** 3465 (1988).

76. K. Raghavachari and J. S. Binkley, *J. Chem. Phys.* **87,** 2191 (1987).

77. M. Vala, T. M. Chandrasckhar, J. Szczepanski, and R. Pellow, *High Temp. Sci.* **27,** 19 (1990).

78. J. Szczepanski and M. Vala, *J. Phys. Chem.* **95,** 2792 (1991).

79. R. Pellow and M. Vala, *Z. Phys. D: At. Mol. Clusters* **15,** 171 (1990).

80. (a) J. R. Heath, A. L. Cooksey, M. H. W. Gruebele, C. A. Schmuttenmaer, and R. A. SayKally, *Science* **244,** 564 (1989); (b) R. H. Kranze and W. R. M. Graham, *J. Chem. Phys.* **98,** 71 (1993).

81. A. Faibis, E. P. Kanter, L. M. Tack, E. Bakke, and B. J. Zabransky, *J. Phys. Chem.* **91,** 6445 (1987).

82. Z. Vager and E. P. Kanter, *J. Phys. Chem.* **93,** 7745 (1989).

83. R. S. Grev, I. L. Alberto, and H. F. Schaefer, III, *J. Phys. Chem.* **94,** 3379 (1990).

84. L. A. Curtiss, P. W. Deutsch, and K. Raghavachari, *J. Chem. Phys.* **96,** 6868 (1992).

85. C. M. Rohlfing and K. Raghavachari, *J. Chem. Phys.* **96,** 2114 (1992).

86. T. N. Kitsopoulos, C. J. Chick, A. Weaver, and D. Neumark, *J. Chem. Phys.* **93,** 6108 (1990).

87. R. J. Van Zee, R. F. Ferrante, and W. Weltner, Jr., *J. Chem. Phys.* **83,** 6181 (1985).

88. L. B. Knight, Jr., J. O. Herlong, R. Babb, E. Earl, D. W. Hill, and C. A. Arrington, *J. Phys. Chem.* **95,** 2732 (1991).
89. (a) M. F. Jarrold and E. C. Hones, *J. Am. Chem. Soc.* **114,** 459 (1992); (b) W. von Niessen and V. G. Zakrzewski, *J. Chem. Phys.* **98,** 1271 (1993).
90. Q. L. Zhang, Y. Liu, R. F. Curl, F. K. Tittel, and R. E. Smalley, *J. Chem. Phys.* **88,** 1670 (1988).
91. K. Lailting, R. G. Wheeler, W. L. Wilson, and M. A. Duncan, *J. Chem. Phys.* **87,** 3401 (1987).
92. M. C. Heaven, T. A. Miller, and V. E. Bondybey, *J. Phys. Chem.* **87,** 2072 (1983).
93. D. Dai and K. Balasubramanian, *J. Chem. Phys.* **96,** 8345 (1992).
94. D. J. Wales and M. C. Waterworth, *J. Chem. Soc. Faraday Trans.,* 3409 (1992).
95. D. A. Jelski, B. L. Swift, T. T. Rantala, X. Xia, and T. F. George, *J. Chem. Phys.* **95,** 8552 (1991).
96. T. A. Home and W. J. Lee, *J. Phys. Chem.* **96,** 3568 (1992).
97. M. F. Jarrold and V. A. Constant, *Phys. Rev. Lett.* **67,** 2994 (1991).
98. R. P. Johnson, G. Meijer, and D. S. Bethune, *J. Am. Chem. Soc.* **112,** 8983 (1990).
99. C. S. Yannoni, P. P. Bernier, D. S. Betlune, G. Meijer, and J. R. Salem, *J. Am. Chem. Soc.* **113,** 3190 (1991).
100. A. Pasquarello, M. Schlüter, and R. C. Haddon, *Science* **257,** 1660 (1992).
101. P. W. Fowler, *J. Chem. Soc. Faraday Trans.* **86,** 2073 (1990).
102. D. E. Maralopoulos, *J. Chem. Soc. Faraday Trans.,* 2861 (1991).
103. H. R. Karfunkel and T. Dressler, *J. Am. Chem. Soc.* **114,** 2285 (1992).
104. J-Y. Yi and J. Bernhole, *J. Chem. Phys.* **96,** 8634 (1992).
105. B. L. Zhang, C. Z. Wang, K. M. Ho, C. H. Xu, and C. T. Chan, *J. Chem. Phys.* **97,** 5007 (1992).
106. H. B. Burgi, E. Blanc, D. Schwarzenbach, S. Liu, Y. J. Lu, M. M. Kappes, and J. A. Ibers, *Angew. Chemie., Int. Ed. Engl.* **31,** 640 (1992).
107. F. W. Fowler and J. I. Steer, *J. Chem. Soc. Chem. Commun.,* 1403 (1987).
108. K. H. Michel, *J. Chem. Phys.* **97,** 5155 (1992).
109. (a) J. Hunter, J. Fye, and M. F. Jarrold, *J. Phys. Chem.* **97,** 3460 (1993); (b) C. Z. Wang, C. H. Xu, C. T. Chan, and K. M. Ho, *J. Phys. Chem.* **96,** 3563 (1992).
110. D. H. Robertson, D. W. Brenner, and C. T. White, *J. Phys. Chem.* **96,** 6133 (1992).
111. S. Petrei, G. Javabery, J. Wang, and D. K. Bohme, *J. Phys. Chem.* **96,** 6121 (1992).
112. H. Levanon, V. Meiklyar, A. Michaeli, S. Michaeli, and A. Regev, *J. Phys. Chem.* **96,** 6128 (1992).
113. N. Matsuzana and D. A. Dixon, *J. Phys. Chem.* **96,** 6241 (1992).
114. N. Matsuzana and D. A. Dixon, *J. Phys. Chem.* **96,** 6872 (1992).
115. P. P. Radi, T. L. Bunn, P. R. Kemper, M. E. Molchan, and M. T. Bowers, *J. Chem. Phys.* **88,** 2809 (1988).
116. R. R. Hung and J. J. Grabowski, *J. Phys. Chem.* **95,** 6073 (1991).
117. M. D. Pace, T. C. Christidis, J. J. Yin, and J. Milliken, *J. Phys. Chem.* **96,** 6855 (1992).
118. J. Kurtz and D. R. Huffman, *J. Chem. Phys.* **92,** 30 (1990).
119. P. N. Keizer, J. R. Morton, K. F. Preston, and A. K. Sugden, *J. Phys. Chem.* **95,** 7117 (1991).
120. R. C. Mowrey, D. W. Brenner, B. I. Dunlap, J. W. Mintmire, and C. T. White, *J. Phys. Chem.* **95,** 7138 (1991).
121. P. W. Fowler, P. Lazzeretti, M. Malagoli, and R. Zanasi, *J. Phys. Chem.* **95,** 6404 (1991).
122. C. S. Yonnoni, R. D. Johnson, G. Meijer, D. S. Bethune, and J. R. Salem, *J. Phys. Chem.* **95,** 9 (1991).

123. (a) J. W. Arbogast, A. P. Darmanyan, C. S. Foote, Y. Rubin, F. N. Diederich, M. M. Alvarez, S. J. Anz, and R. L. Whetten, *J. Phys. Chem.* **95,** 11 (1991); (b) H. W. Kroto, A. W. Allaf, and S. P. Balm, *Chem. Rev.* **91,** 1213 (1991); (c) J. F. Stoddart, *Angew. Chem. Int. Ed. Engl.* **30,** 70 (1991); (d) R. M. Fleming *et al.,* "Fullerenes" (G. Hammond and V. Kuck, Eds.), ACS Symp. Series 481, p. 24, *Am. Chem. Soc.,* Washington, DC, 1992; (e) R. S. Ruoff, D. S. Tse, R. Malhotra, and D. C. Lorents, *J. Phys. Chem.* **97,** 3379 (1993); (f) R. Honeychuck, T. W. Cruger, and J. Milliken, *J. Am. Chem. Soc.* **115,** 3034 (1993).
124. S. Tolbert, A. P. Alivisatos, H. E. Lorenzana, M. B. Kruger, and R. Jeanloz, *Chem. Phys. Lett.* **188,** 163 (1992).
125. C. Reber, L. Yee, J. McKiernan, J. I. Zink, R. S. Williams, W. M. Tang, D. A. A. Ohlberg, R. L. Whetten, and F. Diederich, *J. Phys. Chem.* **95,** 2127 (1991).
126. D. S. Bethune, G. Meijer, W. C. Tang, and H. J. Rosen, *Chem. Phys. Lett.* **174,** 219 (1990).
127. W. Krätschmer, K. Fostiropoulos, and D. R. Huffman, *Chem. Phys. Lett.* **170,** 167 (1990).
128. R. Tycko, R. C. Haddon, G. Dabbagh, S. H. Glarum, D. C. Douglass, and A. M. Mujsce, *J. Phys. Chem.* **95,** 518 (1991).
129. C. S. Yannoni, R. D. Johnson, G. Meijer, D. S. Bethune, and J. R. Salem, *J. Phys. Chem.* **95,** 9 (1991).
130. K. R. Lykke and P. Wurz, *J. Phys. Chem.* **96,** 3191 (1992).
131. D. E. Weeks, *J. Chem. Phys.* **96,** 7380 (1992).
132. Y. Zeng, L. Biczok, and H. Linschitz, *J. Phys. Chem.* **96,** 5237 (1992).
133. N. Manickam, M. Samoe, M. E. Orczyk, S. P. Karna, and P. N. Prasad, *J. Phys. Chem.* **96,** 5206 (1992).
134. L. S. Wang, J. Conceicao, C. Jin, and R. E. Smalley, *Chem. Phys. Lett.* **182,** 5 (1991).
135. R. E. Haufler, L. S. Wang, L. P. F. Chibante, C. Jin, J. Conceicao, Y. Chai, and R. E. Smalley, *Chem. Phys. Lett.* **179,** 449 (1991).
136. F. Negri, G. Orlandi, and F. Zerbetto, *J. Chem. Phys.* **97,** 6496 (1992).
137. R. E. Haufler, Y. Chai, L. P. F. Chibante, M. R. Fraelich, R. B. Weisman, R. F. Curl, and R. E. Smalley, *J. Chem. Phys.* **95,** 2197 (1991).
138. S. Maruyama, M. Y. Lee, R. E. Haufler, Y. Chai, and R. E. Smalley, *Z. Phys. D: At. Mol. Clusters* **19,** 409 (1991).
139. S. C. O'Brien, J. R. Health, R. F. Curl, and R. E. Smalley, *J. Chem. Phys.* **88,** 220 (1988).
140. J. W. Arbogast and C. S. Foote, *J. Am. Chem. Soc.* **113,** 8886 (1991).
141. D. L. Lawson, D. L. Feldheim, C. A. Foss, P. K. Dorhout, C. M. Elliot, C. R. Martin, and B. Parkinson, *J. Phys. Chem.* **96,** 7175 (1992).
142. G. V. Tendeloo, C. V. Heurck, J. V. Landuyt, S. Amelinckx, M. A. Veheijen, P. M. vanLoosdrecht, and G. Meijer, *J. Phys. Chem.* **96,** 7424 (1992).
143. B. B. Brady and E. J. Beiting, *J. Chem. Phys.* **97,** 3855 (1992).
144. J. V. Hanna and M. A. Wilson, *J. Phys. Chem.* **96,** 6518 (1992).
145. (a) L. W. Tutt, and A. Kost, *Nature (London)* **356,** 225 (1992); (b) H. S. Carman, Jr. and R. N. Compton, *J. Chem. Phys.* **98,** 2473 (1993).
146. J. E. Fischer, P. A. Heiney, A. R. McGhie, W. J. Romanow, A. M. Denenstein, J. P. McCauley, Jr., and A. B. Smith, III, *Science* **252,** 1288 (1991).
147. P. J. Krusic, E. Wasserman, B. A. Parkinson, B. Malone, E. R. Holler, Jr., P. N. Keizer, J. R. Morton, and K. F. Preston, *J. Am. Chem. Soc.* **113,** 6274 (1991).
148. M. Ruebsam, K. P. Dinse, M. Plueschan, J. Fink, W. Kraetschmer, K. Fostiropoulos, and C. Taliani, *J. Am. Chem. Soc.* **114,** 10059 (1992).

149. F. Negri, G. Orlandi, and F. Zerbetta, *J. Am. Chem. Soc.* **114**, 2909 (1992).

150. J. R. Colt and G. E. Scuseria, *J. Phys. Chem.* **96**, 10265 (1992).

151. P. Wurz and K. R. Lykke, *J. Phys. Chem.* **96**, 10129 (1992).

152. D. Dubois, G. Mininot, W. Kutner, M. T. Jones, and K. M. Kadish, *J. Phys. Chem.* **96**, 7137 (1992).

153. Q. Xie, E. Perzez-Cordero, and L. Echegeyen, *J. Am. Chem. Soc.* **114**, 3978 (1992).

154. D. Dubois, K. K. Kadish, S. Flanagan, and L. J. Wilson, *J. Am. Chem. Soc.* **113**, 7773 (1991).

155. S. Norell, J. W. Arbogast, and C. S. Foote, *J. Phys. Chem.* **96**, 4169 (1992).

156. D. E. Ramaker, N. H. Turner, and J. A. Milliken, *J. Phys. Chem.* **96**, 7627 (1992).

157. H. J. Bestmann, H. Behl, and M. Bremer, *Angew. Chem. Int. Ed. Engl.* **28**, 1219 (1989).

158. B. J. Ortman, R. H. Hauge, J. L. Margrave, and Z. H. Kafafi, *J. Phys. Chem.* **94**, 7973 (1990).

159. R. J. Van Zee, G. R. Smith, and W. Weltner, Jr., *J. Am. Chem. Soc.* **110**, 609 (1988).

160. S. McElvary, *J. Chem. Phys.* **89**, 2063 (1988).

161. J. A. Zimmerman and W. R. Creasy, *J. Chem. Phys.* **96**, 1942 (1992).

162. M. Doverstaal, B. Lindgren, U. Sassenberg, and H. Yu, *Z. Phys. D: At. Mol. Clusters* **19**, 447 (1991).

163. (a) J. M. Alford, R. T. Laaksonen, and R. E. Smalley, *J. Chem. Phys.* **94**, 2618 (1991); (b) J. M. Alford and R. E. Smalley, *Mater. Res. Soc. Symp.* **131**, 3 (1989).

164. S. Maruyama, L. R. Anderson, and R. E. Smalley, *Mater. Res. Soc. Symp. Proc.* **206**, 63 (1991).

165. M. F. Jarrold, Y. Ijiri, and U. Ray, *J. Chem. Phys.* **94**, 3607 (1991).

166. U. Ray and M. F. Jarrold, *J. Chem. Phys.* **94**, 2631 (1991).

167. M. F. Jarrold, U. Ray, and K. M. Creegan, *J. Chem. Phys.* **93**, 224 (1990).

168. K. M. Creegan and M. F. Jarrold, *J. Am. Chem. Soc.* **112**, 3768 (1990).

169. J. E. Bower and M. F. Jarrold, *J. Chem. Phys.* **97**, 8312 (1992).

170. (a) N. Matsuzawa, D. A. Dixon, and T. Fukunaga, *J. Phys. Chem.* **96**, 7594 (1992); (b) C. C. Henderson and P. A. Cahill, *Science* **259**, 1885 (1993).

171. A. Hirsch, A. Soi, and H. R. Karfunkel, *Angew. Chem. Int. Ed. Engl.* **31**, 766 (1992).

172. A. Hirsch, Q. Li, and F. Wudl, *Angew. Chem. Int. Ed. Engl.* **30**, 1309 (1991).

173. A. Vasella, P. Uhlmann, C. A. A. Waldraff, F. Diederich, and C. Thilgen, *Angew. Chem. Int. Ed. Engl.* **31**, 1388 (1992).

174. A. A. Tuinman, P. Mukherjee, J. L. Adcock, R. L. Hettich, and R. N. Compton, *J. Phys. Chem.* **96**, 7584 (1992).

175. G. A. Olah, I. Busci, C. Lambert, R. Aniszfeld, N. J. Trivedi, D. K. Sensharma, and G. K. S. Prakash, *J. Am. Chem. Soc.* **113**, 9385 (1991).

176. G. A. Olah, I. Busci, C. Lambert, R. Aniszfeld, N. J. Trivedi, D. K. Sensharma, and G. K. S. Prakash, *J. Am. Chem. Soc.* **113**, 9387 (1991).

177. S. Petrei, G. Javahery, J. Wang, and D. K. Bohme, *J. Am. Chem. Soc.* **114**, 9177 (1992).

178. (a) J. R. Morton, K. F. Preston, P. J. Krusic, S. A. Hill, and E. Wasserman, *J. Am. Chem. Soc.* **114**, 3576, 5454 (1992); (b) J. R. Morton, K. F. Preston, P. J. Krusic, S. A. Hill, and E. Wasserman, *J. Phys. Chem.* **96**, 3576 (1992).

179. J. W. Bausch, G. K. S. Prakash, and G. A. Olah, *J. Am. Chem. Soc.* **113**, 3205 (1991).

180. P. J. Fagan, P. J. Krusic, D. H. Evans, S. A. Lerke, and E. Johnson, *J. Am. Chem. Soc.* **114**, 9697 (1992).

181. S. H. Hoke II, J. Molstad, D. Delettato, M. J. Jay, D. Carlson, B. Kahr, and R. G. Cooks, *J. Org. Chem.* **57**, 5069 (1992).

182. J. F. Christian, Z. Wan, and S. L. Anderson, *J. Phys. Chem.* **96**, 3574 (1992).
183. D. A. Dixon, N. Matsuzawa, T. Fukunaga, and F. N. Tebbe, *J. Phys. Chem.* **96**, 6107 (1992).
184. D. A. Loy and R. A. Assink, *J. Am. Chem. Soc.* **114**, 3977 (1992).
185. E. T. Samulski, J. M. DeSimone, M. O. Hunt, Jr., Y. Z. Menceloglu, R. C. Jarnagin, G. A. York, K. B. Labat, and H. Wang, *Chem. Mater.* **4**, 1153 (1992).
186. J. Wang, G. Javaheny, S. Petrie, and D. K. Bohme, *J. Am. Chem. Soc.* **114**, 9665 (1992).
187. J. A. Howard, M. Tomietto, and D. A. Wilkinson, *J. Am. Chem. Soc.* **113**, 7870 (1991).
188. R. S. Koefod, C. Xu, W. Lu, J. R. Shapley, M. G. Hill, and K. R. Mann, *J. Phys. Chem.* **96**, 2928 (1992).
189. R. S. Koefod, M. F. Hudgens, and J. R. Shapley, *J. Am. Chem. Soc.* **113**, 8957 (1991).
190. A. L. Balch, J. W. Lee, and M. M. Olmstead, *Angew. Chem. Int. Ed. Engl.* **31**, 1356 (1992).
191. Y. Zhu, R. S. Keofod, C. Devadoss, J. R. Shapley, and G. B. Schuster, *Inorg. Chem.* **31**, 3505 (1992).
192. A. L. Balch, V. J. Catalano, J. W. Lee, M. M. Olmstead, and S. R. Parkin, *J. Am. Chem. Soc.* **113**, 8953 (1991).
193. (a) P. J. Fagan, J. C. Calabrese, and B. Malone, *Science* **252**, 1160 (1991); (b) P. J. Fagan, J. C. Calabrese, and B. Malone, *Acc. Chem. Res.* **25**, 134 (1992); (c) for a palladium derivative, see V. V. Bashilov, P. V. Petrovskii, V. I. Sokolov, S. V. Lindeman, I. A. Guzey, and Y. T. Struchkov, *Organometallics* **12**, 991 (1993).
194. Q. Jiao, S. A. Lee, J. R. Gord, and B. S. Freiser, *J. Am. Chem. Soc.* **114**, 2726 (1992).
195. H. Huang and B. S. Freiser, *J. Am. Chem. Soc.* **113**, 8186 (1991).
196. L. M. Roth, Y. Huang, J. T. Schwedler, C. J. Cassady, D. Ben-Amotz, B. Kahr, and B. S. Freiser, *J. Am. Chem. Soc.* **113**, 6298 (1991).
197. Y. Huang and B. S. Freiser, *J. Am. Chem. Soc.* **113**, 9418 (1991).
198. J. M. Hawkins, A. Meyer, T. A. Lewis, U. Bunz, R. Nunlist, G. E. Ball, T. W. Ebbesen, and K. Tanigaki, *J. Am. Chem. Soc.* **114**, 7954 (1992).
199. J. M. Hawkins, T. A. Lewis, S. D. Loren, A. Meyer, J. R. Heath, Y. Shibato, and R. J. SayKally, *J. Org. Chem.* **55**, 6250 (1990).
200. J. M. Hawkins, S. Loren, A. Meyer, and R. Nunlist, *J. Am. Chem. Soc.* **113**, 7770 (1991).
201. R. F. Curl and R. E. Smalley, *Science* **242**, 1017 (1998).
202. Y. Chai, T. Guo, C. Jin, R. E. Haufler, L. P. F. Chibante, J. Fure, L. Wang, J. M. Alford, and R. E. Smalley, *J. Phys. Chem.* **95**, 7564 (1991).
203. T. Guo, M. D. Diener, Y. Chai, M. J. Alford, R. E. Haufler, S. M. McClure, T. Ohno, J. H. Wenver, G. E. Scuseria, and R. E. Smalley, *Science* **257**, 1661 (1992).
204. B. I. Dunlap, O. D. Häberlen, and N. Rösch, *J. Phys. Chem.* **23**, 9095 (1992).
205. (a) C. S. Yannoi, M. Hoinkis, M. S. deVries, D. S. Bethune, J. R. Salem, M. S. Crowder, and R. D. Johnson, *Science* **256**, 1191 (1992); (b) J. R. Ungerer and T. Hughbanks, *J. Am. Chem. Soc.* **115**, 2054 (1993); (c) H. Shinohara, H. Yamaguchi, N. Hayashi, H. Sato, M. Ohkohehi, Y. Ando, and Y. Saito, *J. Phys. Chem.* **97**, 4259 (1993).
206. K. Laasonen, W. Andreoni, and M. Parrinello, *Science* **258**, 1916 (1992).
207. T. Guo, C. Jin, and R. E. Smalley, *J. Phys. Chem.* **95**, 4948 (1991).
208. R. E. Smalley, *Large Carbon Clusters''* (G. Hammond and V. Kuck, Eds.), ACS Symposium Series 481, Am. Chem. Soc., Washington, DC, 1992.
209. T. Pradeep, V. Vijayakrishman, A. K. Santra, and C. N. R. Rao, *J. Phys. Chem.* **95**, 10564 (1991).

210. L. Soderholm, P. Wurz, K. R. Lykke, D. H. Parker, and F. W. Lytle, *J. Phys. Chem.* **96**, 7153 (1992).
211. S. Suzuki, S. Kawata, H. Shiromaru, K. Yamaguchi, K. Kituchi, T. Kato, and Y. Achiba, *J. Phys. Chem.* **96**, 7159 (1992).
212. A. L. R. Bug, A. Wilson, and G. A. Voth, *J. Phys. Chem.* **96**, 7864 (1992).
213. P. P. Schmidt, B. I. Dunlap, and C. T. White, *J. Phys. Chem.* **95**, 10537 (1991).
214. M. M. Alvarez, E. G. Gillan, K. Holczar, R. B. Kaner, K. S. Min, and R. L. Whetten, *J. Phys. Chem.* **95**, 10564 (1991).
215. H. Shinohara, H. Sato, Y. Saito, M. Ohkohchi, and M. Ando, *J. Phys. Chem.* **96**, 3571 (1992).
216. E. G. Gillan, C. Yeretzian, K. S. Min, M. M. Alvarez, R. L. Whetten, and R. B. Kaner, *J. Phys. Chem.* **96**, 6869 (1992).
217. T. Weiske, D. K. Böhme, J. Hrusak, W. Krätschmer, and H. Schwartz, *Angew. Chem., Int. Ed., Engl.* **30**, 884 (1991).
218. M. M. Ross and J. H. Callahan, *J. Phys. Chem.* **95**, 5720 (1991).
219. T. Pradeep, G. U. Kulkarni, K. R. Kannan, T. N. Guru Row, and C. N. R. Rao, *J. Am. Chem. Soc.* **114**, 2272 (1992).
220. S. Bandow, H. Kitagawa, T. Mitani, H. Inokuchi, Y. Saito, H. Yamaguchi, N. Hayashi, H. Sato, and H. Shinohana, *J. Phys. Chem.* **96**, 9609 (1992).
221. B. I. Dunlap, J. L. Ballester, and P. P. Schmidt, *J. Phys. Chem.* **96**, 9781 (1992).
222. (a) K. A. Caldwell, D. E. Giblin, C. S. Hsu, D. Cox, and M. L. Gross, *J. Am. Chem. Soc.* **113**, 8519 (1991); (b) M. M. Ross, H. H. Nelson, J. H. Callahan, and S. W. McElvany, *J. Phys. Chem.* **96**, 5231 (1992); (c) M. Saunders, H. A. Jimenez-Vazquez, R. J. Cross, and R. J. Poreda, *Science* **259**, 1428 (1993).
223. A. F. Hebard, M. J. Rosseinsky, R. C. Haddon, D. W. Murphy, S. H. Glarum, T. T. M. Palstra, A. P. Ramirez, and A. R. Kortan, *Nature (London)* **350**, 600 (1991).
224. K. Holczer, O. Klein, S. M. Huang, R. B. Kaner, K. J. Fu, R. L. Whetten, and F. Diederich, *Science* **252**, 1154 (1991).
225. A. Cheng and M. L. Klein, *J. Phys. Chem.* **95**, 9622 (1991).
226. F. Bensebaa, B. Xiang, and L. Kevan, *J. Phys. Chem.* **96**, 6118 (1992).
227. F. Bensebaa, B. Xiang, and L. Kevan, *J. Phys. Chem.* **96**, 10258 (1992).
228. Y. Chabre, D. Djurado, M. Armand, W. R. Romanow, N. Coustel, J. P. McCanley, Jr., J. E. Fischer, and A. B. Smith, III, *J. Am. Chem. Soc.* **114**, 764 (1992).
229. J. P. McCauley, Jr., Q. Zhn, N. Coustel, O. Zhou, G. Vaughan, S. H. J. Idziak, J. E. Fischer, S. W. Tozer, D. M. Groski, N. Bykovetz, C. L. Lin, A. R. McGhie, and A. B. Smith, III, *J. Am. Chem. Soc.* **113**, 8537 (1991).
230. S. P. Kelty, Z. Lu, and C. M. Lieber, *J. Phys. Chem.* **95**, 6754 (1991).
231. R. Tycko, G. Dabbagh, M. J. Rosseinsky, D. W. Murphy, R. L. Fleming, A. P. Ramirez, and J. C. Tully, *Science* **253**, 884 (1991).
232. C. C. Chen, S. P. Kelty, and C. M. Lieber, *Science* **253**, 886 (1991).
233. (a) X. D. Xiang, J. G. Hou, G. Briceno, W. A. Vareka, R. Mostovoy, A. Zettl, V. H. Crespi, and M. L. Cohen, *Science* **256**, 1190 (1992); (b) H. S. Chen, A. R. Kortan, R. C. Haddon, and N. Kopylov, *J. Phys. Chem.* **97**, 3088 (1993); (c) C. C. Chen and C. M. Lieber, *Science* **259**, 655 (1993).
234. K. Tankigaki, T. W. Ebbesen, S. Saito, J. Mizuki, J. S. Tsai, Y. Kubo, and S. Kuroshima, *Nature (London)* **352**, 222 (1991).
235. R. C. Hadden, G. P. Kochanski, A. F. Hebard, A. T. Fiory, and R. C. Morris, *Science* **258**, 1636 (1992).
236. G. K. Wertheim, D. N. E. Buchanan, and J. E. Rowe, *Science* **258**, 1638 (1992).
237. U. Geiser, S. K. Kumar, B. M. Savall, S. S. Harried, K. D. Carlson, P. R. Mobley,

H. H. Wang, J. M. Williams, R. E. Botto, W. Liang, and M. H. Wangbo, *Chem. Mater.* **4**, 1077 (1992).

238. R. Back and R. B. Lennox, *J. Phys. Chem.* **96**, 8149 (1992).
239. F. Zhou, S. L. Yau, C. Jehoulet, D. A. Laude, Jr., Z. Guan, and A. J. Bard, *J. Phys. Chem.* **96**, 4160 (1992).
240. W. Koh, D. Dubois, W. Kutner, M. T. Jones, and K. M. Kadish, *J. Phys. Chem.* **96**, 4163 (1992).
241. L. W. Tutt and A. Kost, *Nature (London)* **356**, 225 (1992).
242. P. M. Allemand, K. C. Khemani, A. Koch, F. Wudl, K. Holczer, S. Donovan, G. Gruner, and J. D. Thompson, *Science* **253**, 301 (1991).
243. Y. Z. Li, J. C. Patrin, M. Chandler, J. H. Weaver, L. P. F. Chibante, and R. E. Smalley, *Science* **252**, 547 (1991).
244. (a) N. S. Sariciftei, L. Smilowitz, A. J. Heeger, and F. Wudl, *Science* **258**, 1474 (1992); (b) M. Fastow, Y. Kozirovski, M. Folman, and J. Heidberg, *J. Phys. Chem.* **96**, 6126 (1992); (c) M. Iwahashi, K. Kikuchi, Y. Achiba, I. Ikemoto, T. Araki, T. Mochida, S. I. Yokoi, A. Tanaka, and K. Iriyama, *Langmuir* **8**, 2980 (1992); (d) K. Akers, K. Fu, P. Zhang, and M. Moskovits, *Science* **259**, 1152 (1993); (e) J. R. Olson, K. A. Topp, and R. O. Pohl, *Science* **259**, 1145 (1993); (f) A. M. Rao, P. Zhou, K. A. Wang, G. T. Hager, J. M. Holden, Y. Wang, W. T. Lee, X. X. Bi, P. C. Eklund, D. S. Cornett, M. A. Duncan, and I. J. Amster, *Science,* **259**, 955 (1993); (g) R. S. Ruoff and A. P. Hickman, *J. Phys. Chem.* **97**, 2494 (1993); (h) D. Ding, R. N. Compton, R. E. Haufler, and C. E. Klots, *J. Phys. Chem.* **97**, 2500 (1993); (i) P. Belik, A. Gügel, J. Spickermann, and K. Müllen, *Angew. Chem.* **32**, 78 (1993); (j) R. L. De Muro, D. A. Jelski, and T. F. George, *J. Phys. Chem.* **96**, 10603 (1992); (k) J. F. Christian, Z. Wan, and S. L. Anderson, *J. Phys. Chem.* **96**, 10597 (1992); (l) N. M. Dimitrijevic, P. V. Kamat, and R. W. Fessenden, *J. Phys. Chem.* **97**, 615 (1993); (m) A. Gügel, K. Müllen, H. Reichert, W. Schmidt, G. Schön, F. Schüth, J. Spickermann, J. Titman, and K. Unger, *Angew. Chem. Int. Ed. Engl.* **32**, 556 (1993); (n) C. Christides, A. V. Nikolaev, T. J. S. Dennis, K. Prassides, F. Negri, G. Orlandi, and F. Zerbetto, *J. Phys. Chem.* **97**, 3641 (1993); (o) F. Zhou, C. Jehoulet, and A. J. Bard, *J. Am. Chem. Soc.* **114**, 11004 (1992); (p) A. L. Balch, J. W. Lee, B. C. Null, and M. M. Olmstead, *J. Am. Chem. Soc.* **114**, 10986 (1992); (q) Y. Wang, R. West, and C. H. Yuan, *J. Am. Chem. Soc.* **115**, 3844 (1993); (r) K. E. Geckeler and A. Hirsch, *J. Am. Chem. Soc.* **115**, 3850 (1993); (s) C. B. Eam, A. F. Hebard, L. E. Trimble, G. K. Celler, and R. C. Haddon, *Science* **259**, 1887 (1993); (t) M. Prato, Q. C. Li, F. Wudl, and V. Lucchini, *J. Am. Chem. Soc.* **115**, 1148 (1993); (u) H. Moriyama, H. Kobayashi, A. Kobayaski, and T. Watanabe, *J. Am. Chem. Soc.* **115**, 1185 (1993); (v) K. Chen, W. B. Caldwell, and C. A. Mirkin, *J. Am. Chem. Soc.* **115**, 1193 (1993); (w) W. R. Fawcett, M. Opallo, M. Fedurco, and J. W. Lee, *J. Am. Chem. Soc.* **115**, 196 (1993); (x) M. V. Mirkin, L. O. S. Bulhoes, and A. J. Bard, *J. Am. Chem. Soc.* **115**, 201 (1993); (y) Y. Rubin, S. Khan, D. I. Freedberg, and C. Yerctzian, *J. Am. Chem. Soc.* **115**, 344 (1993); (z) Y. Huang and D. D. M. Wayner, *J. Am. Chem. Soc.* **115**, 367 (1993); (aa) P. J. Krusic, D. C. Roe, E. Johnston, J. R. Morton, and K. F. Preston, *J. Phys. Chem.* **97**, 1736 (1993); (ab) F. Chung and S. Sternberg, *Am. Sci.* **81**, 56 (1993); (ac) D. K. Maity, D. K. Palit, H. Mohan, and J. P. Mittal, *J. Chem. Soc. Faraday Trans.,* **95** (1993); (ad) W. R. Creasy, J. A. Zimmerman, and R. S. Ruoff, *J. Phys. Chem.* **97**, 973 (1993).
245. E. K. Moltzen, K. J. Klabunde, and A. Senning, *Chem. Rev.* **88**, 391 (1988).
246. V. E. Bondybey, *J. Phys. Chem.* **86**, 3396 (1982).
247. M. Feher, C. Salud, J. P. Maier, *J. Mole. Spectrosc.* **145**, 246 (1991).
248. Z. Y. Chen, G. J. Walder, and A. W. Castelman, Jr., *J. Phys. Chem.* **96**, 9581 (1992).

249. (a) S. Wei, B. C. Guo, J. Purnell, S. Buzza, and A. W. Castleman, Jr., *J. Phys. Chem.* **96**, 4166 (1992); (b) B. V. Reddy, S. N. Khanna, and P. Jena, *Science* **258**, 1640 (1992); (c) S. F. Cartier, Z. Y. Chen, G. J. Walder, C. R. Sleppy, and A. W. Castleman, Jr., *Science* **260**, 195 (1993).

250. Y. Ando and R. Uyeda, *J. Cryst. Growth* **52**, 178 (1981).

251. L. B. Knight, Jr., J. T. Petty, and S. T. Cobranchi, *J. Chem. Phys.* **88**, 3441 (1988).

252. (a) T. J. Buttenhoff and E. A. Rohlfing, *J. Chem. Phys.* **95**, 3939 (1991); (b) M. Mollaaghababa, C. A. Gottlieb, and P. Thaddeus, *J. Chem. Phys.* **98**, 968 (1993).

253. T. J. Buttenhoff and E. A. Rohlfing, *J. Chem. Phys.* **95**, 1 (1991).

254. C. M. L. Rittby, *J. Chem. Phys.* **95**, 5609 (1991).

255. J. D. Presilla-Marquez and W. R. M. Graham, *J. Chem. Phys.* **95**, 5612 (1991).

256. Z. H. Kafafi, R. H. Hauge, L. Fredin, and J. L. Margrave, *J. Phys. Chem.* **87**, 797 (1983).

257. C. M. L. Rittby, *J. Chem. Phys.* **96**, 6768 (1992).

258. J. D. Presilla-Marquez and W. R. M. Graham, *J. Chem. Phys.* **96**, 6509 (1992).

259. P. A. Withey, and W. R. M. Graham, *J. Chem. Phys.* **96**, 4068 (1992).

260. T. Mehner, H. J. Göcke, S. Schunck, and H. Schnöckel, *Z. Anorg. Allg. Chem.* **580**, 121 (1990).

261. R. Köppe and H. Schnöckel, *J. Mol. Struct.* **238**, 429 (1990).

262. H. Schnöckel and R. Köppe, *J. Am. Chem. Soc.* **111**, 4583 (1989).

263. E. K. Moltzen, M. P. Kramer, A. Senning, and K. J. Klabunde, *J. Org. Chem.* **52**, 1156 (1987).

264. K. Jorgensen, E. K. Moltzen, and A. Senning, *J. Org. Chem.* **52**, 2505 (1987).

265. K. J. Klabunde, E. Moltzen, and K. Voska, *Phosphorus Sulfur Silicon* **43**, 47 (1989).

266. W. N. Rowlands and P. L. Timms, *J. Chem. Soc. Chem. Commun.*, 1432 (1989).

267. T. Mehner, H. Schnöckel, M. J. Almond, and A. J. Downs, *J. Chem. Soc. Chem. Commun.*, 117 (1988).

268. G. H. Takaoka, H. Tsuji, and J. Ishikawa, *Mater. Res. Soc. Symp. Proc.* **157**, 25 (1990).

269. G. H. Takaoka, K. Matsugatani, J. Ishikawa, and T. Takagi, *Nucl. Instrum. Methods Phys. Res., B* **37–38**, 783 (1989).

270. Y. Franghiadakis and P. Tzanetakis, *J. Appl. Phys.* **68**, 2433 (1990).

271. D. J. Heim, T. G. Holesinger, K. M. Lakin, and H. R. Shanks, *Mater. Res. Soc. Symp. Proc. 116*, 523 (1988).

272. G. L. Zhou, K. M. Chen, W. D. Jiang, C. Sheng, X. J. Zhang, and X. Wang, *Appl. Phys. Lett.* **53**, 2179 (1988).

273. M. Asai, H. Ueba, and C. Tatsuyama, *J. Appl. Phys.* **58**, 2577 (1985).

274. J. S. McCalmont, D. Robinson, K. M. Lakin, and H. R. Shanks, *Mater. Res. Soc. Symp. Proc.* **91**, 323 (1987).

275. A. E. T. Kuiper, G. E. Thomas, and W. J. Schouten, *J. Cryst. Growth* **51**, 17 (1981).

276. I. Yameda, F. W. Saris, T. Takagi, K. Matsubara, H. Takaoka, and S. Ishiyama, *Jpn. J. Appl. Phys.* **19**, L181 (1980).

277. H. Hofmeister, A. F. Bardamid, T. Junghanns, and S. A. Nepiiko, *Thin Solid Films* **205**, 20 (1991).

9

Phosphorus and Sulfur Groups (Heavier Elements of Groups 15 and 16)

I. Phosphorus and Sulfur Group Vapors: Atoms and Clusters (P, As, Sb, Bi, Se, Te)

A. Occurrence and Techniques

These elements do not vaporize monatomically and their vapor compositions are complex, although all of them are easy to evaporate.[1] With phosphorus and arsenic, the main vapor components are P_4 and As_4 (with small amounts of As_6 and As_8). Antimony vaporization leads to complex mixtures with Sb_3 as dominant. For bismuth Bi_2 is predominant although Bi atoms are also formed in the 580–680°C range. Bismuth atoms are believed to exist in small amounts in the upper atmosphere.

Sulfur vapor is, of course, mainly cyclic S_8, while selenium vapor consists of Se_2–Se_{10} species with Se_5–Se_{10} clusters most abundant, thought mainly to be ring compounds. Indeed, ring formation in the molten state may be the rate-determining step in the vaporization.[1] Tellurium vapor species are predominantly Te_2 and Te_4.

The generation of P atoms from solid GaP has been studied as a way of producing these atoms, but particular attention was given to the several types of defects on the GaP(110) surface that gave rise to ejection of P atoms.[2] Calculations indicated that the P atoms were more weakly bound at defect sites, and so the authors proposed that a new method of producing perfect surfaces of GaP would be to use laser irradiation to eliminate defects.

In recent years rather elegant extrusion reagents have been devised for the thermal extrusion of S_2.[3,4] Steliou *et al.* have taken advantage of bicyclic aromatic thiones:[3]

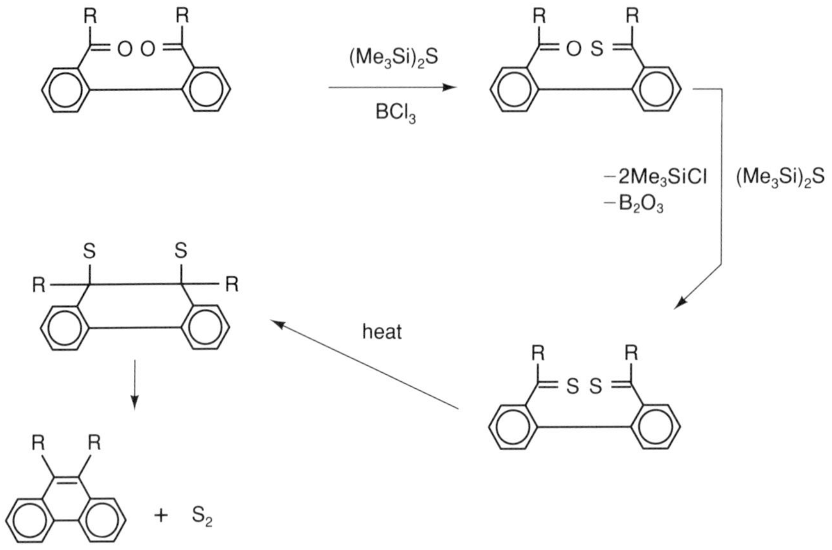

Upon intramolecular dimerization an unstable dithiacyclobutane that smoothly extrudes S_2 is formed. These authors have studied a variety of S_2 addition reactions with dienes to considerable advantage in organic synthesis schemes.

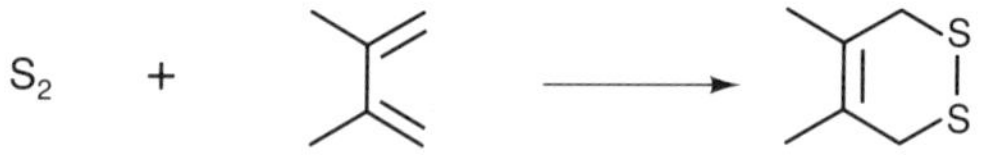

Schmidt and Görl[4] used a different approach which employed a cyclic Se/S compound:

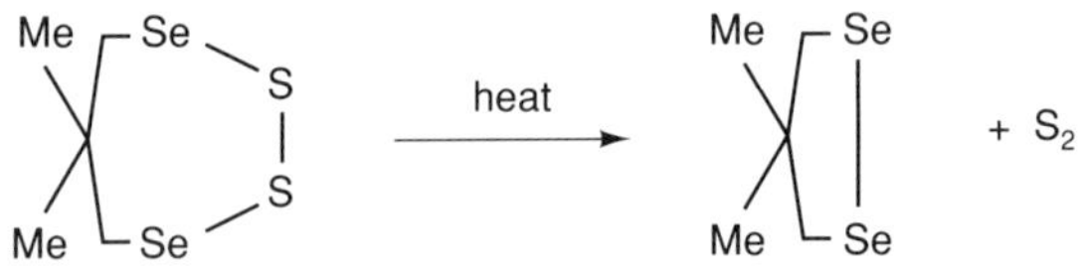

Andrews and co-workers have described a method of generating and matrix isolating S_3 and S_4 species. Sulfur vapor (mainly S_8) was seeded into a microwave-powered argon discharge and the effluent condensed at 12 K.[5] Strong IR bands for S_3 and S_4 were observed at 680, 676, 662, and 642 cm^{-1} and were confirmed by isotopic labeling. The thiozone structure of S_3 was confirmed as being open with C_{2v} symmetry. In a similar way the S_4 species was proposed to be *cis*- or *trans*-planar structures.

Gas-phase cluster ions of Sb and Bi have been produced by laser vaporization in an FTMS, a time-of-flight MS, and by particle bombardment (secondary ion MS).[6] These methods were compared with regard to clusters predominantly formed. For FTMS and TOFMS Sb_{1-5} and Bi_{1-5} were generated, but with SIMS Bi_{1-14} species were also generated. Enhanced abundances were observed at cluster sizes $n = 3, 5, 7, 10$, and 13. It was determined that the dissociations of Bi and Sb clusters are dominated by the losses of stable neutral dimers and tetramers.[6]

Clusters of Pb and Bi, prepared by gas aggregation, have been analyzed by MS.[7] Cluster distributions were recorded for various metals and gas aggregation temperatures. Increasing the temperature of the molten Pb or Bi samples and using cooler gas temperatures resulted in larger clusters, as expected. In He gas smaller clusters were formed, but in Ar, Ne, Kr, N_2, and H_2 the particles formed were very large.

B. Physical Properties and Theoretical Studies

A theoretical investigation of clusters of phosphorus using ab initio methods was applied to 39 species ranging in size up to P_{28}.[8a] Two medium-sized clusters $P_{12}(D_{3d})$ and $P_{16}(C_{2n})$ were established as particularly stable. Indeed, many larger clusters were found to be energetically stable with respect to the remarkably stable tetrahedral P_4. The P_8 system is a particularly interesting one (Figure 9-1), where P_8 (C_{2v}) shown as Fig. 8-1f was found to be more stable than the cubane analog (1d). In general it was found that no P_8 structures were more energetically favorable than two P_4 clusters, but those P_8 systems that were the most stable have symmetries of C_{2v} and D_{2h} (Figs. 8-1f and 8-1g).

The structure of phosphorus clusters has also been examined by theoretical methods using simulated annealing. Jones and Seifert[8b] found that even the small clusters P_9, P_{10}, and P_{11} show tendencies to form tubular structures. Clusters of 10 or more atoms have a large number of local minima in their energy surfaces, many of which have similar energies. Cage molecules such as P_4S_3 and P_7^{3-} have also been considered.[8c,d]

The P_4^+ and P_3 molecules have been matrix isolated in solid Ar.[9] These species were prepared by passing P_4 vapor through a short argon discharge tube and subjected to UV light during matrix deposition. Visible

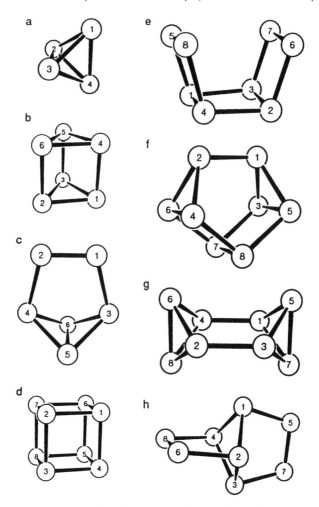

FIGURE 9-1   Phosphorus clusters (after Häser et al.[8a]).

absorption spectra were recorded, which revealed two vibronic band systems. A photosensitive species with an origin at 436 nm and 550 ± 10 cm^{-1} vibrational intervals was assigned as P_4^+. A broader 427 nm origin with 480 ± 20 cm^{-1} intervals was assigned as due to the P_3 radical.

Gas-phase photoelectron spectral studies of cluster anions Sb^-, Sb_2^-, Sb_3^-, and Sb_4^- have been carried out.[10] Anions were formed in an antimony cathode discharge source in He, mass selected, and then photolyzed by 351- or 364-nm laser light. Electron affinity of Sb was determined as 1.046 eV, Sb_2 as 1.282, Sb_3 as 1.85, and Sb_4 as <1.00. The vertical detachment energies of Sb_3^- and Sb_4^- to neutral ground states were found to be 1.90 and 1.57 eV, respectively. Vibrationally resolved electronic structure was only found for the dimer.

Similarly, bismuth cluster anions Bi_2^-, Bi_3^-, and Bi_4^- have been studied, in this case by 351-nm photoelectron spectra.[11] The Bi_2^- species showed at least seven electronic states of Bi_2, four with vibrational resolution. Electron affinity of Bi_2 (1.271 eV), Bi_3 (1.60 eV), and Bi_4 (1.05 eV) were reported. Ab initio calculations were especially helpful in assignments.[12]

The structures of Sb_n ($n = 2 - 8$) have been calculated using the charge self-consistent extended Hueckel method. The results showed that the Sb_4 species was most stable, in agreement with experiment.[13]

Gas-phase Sb, Sb_2, and Sb_4 have been investigated by vacuum–UV photoelectron spectroscopy. These species were produced by superheating Sb_4 vapor. The dissociation energy of Sb_2 was determined as 3.37 ± 0.14 eV.[14]

Clusters of Sb and Bi have also been matrix isolated and analyzed by Raman spectroscopy.[15] Mainly dimers and tetramers were observed, but larger clusters could be prepared by matrix aggregation or inert-gas aggregation prior to codeposition. Trimers Sb_3 and Bi_3 were never observed, however.

Indeed, very large clusters of these species have been prepared by gas aggregation in He.[16] A sequence of Sb tetramers $(Sb_4)_x$ was predominant for antimony, while for bismuth Bi_5, Bi_7 were predominant, with smaller amounts of Bi_6, Bi_9, Bi_{12}, and Bi_{15}. Such clusters have been deposited on amorphous carbon and the 1.5- to 20-nm particles of Bi analyzed by electron microscopy. Single-core crystals surrounded by amorphous Bi were observed.[17,18]

A fascinating "real-time" study of Au, Bi, and Pt clusters up to 100 Å in diameter have been examined at atomic resolution by an electron microscope. As the size of the particles decreased to below 50 Å, the particle shapes changed continually. Thus, internal transformation from a single crystal to a twinned crystal to multiply-twinned crystals and vice versa was observed. These transformations took place in a fraction of a second and were apparently caused by electronic excitation due to electron-beam irradiation during the microscopy experiment.[19]

The S_3 and S_4 clusters have been isolated in frozen Ar and electronic spectra taken.[20] A broad green band centered at 518 nm and a structured red bank in the 560–600 nm range were observed in solid Ar. The species responsible for these absorptions were interconvertable with selected light wavelengths and were assigned as structural isomers of S_4, probably *cis*-planar and branched ring nonplanar isomers. Molecular S_3 showed sharp bands between 350 and 400 nm, which were assigned to three Ar trapping sites.

Isomerization and electron affinities of the open and cyclic forms of Se_3 and Te_3 have been investigated by von Niessen *et al.* by theoretical methods. According to their results Se_3 and Te_3 should exist as C_{2v} or closed D_{3h} forms with nearly the same energy.[21] Both Se_3 and Te_3 should have positive electron affinities for both structural isomers.

Matrix-isolated Se_3 and Te_3 have also been studied, in these cases by resonance Raman.[22,23] The Se_3 species was prepared by cocondensing Se and Se_2 at 15 K followed by annealing to 25 K. The spectral data indicated that Se_3 exists as a triangular species with an angle of about 115°. A similar procedure where Te atoms and Te_2 were cocondensed at 15 K in N_2 followed by annealing yielded matrix-isolated Te_3. The data suggested a triangular C_{2v} structure with an angle of 120 ± 10°, which agrees with thermodynamic considerations.

C. Chemistry of Atoms and Clusters

Due to the mixture of species in the vapors and the relatively low reactivity of these elements, the chemistry reported is limited. Perhaps the most important advance comes from the work of Lagow and coworkers where vapors of Te, Bi, and Hg were cocondensed at 77 K with free radicals. Thus, the first trifluorisilyl σ-bonded metal compounds $Te(SiF_3)_2$, $Bi(SiF_3)_3$, and $Hg(SiF_3)_2$ were prepared by such a codeposition procedure:[24]

$$Si_2F_6 \xrightarrow[\substack{30 \text{ W} \\ \text{discharge}}]{10 \text{ MH}_z} 2 \cdot SiF_3 \xrightarrow[\text{cocondense}]{\text{M vapor}} M(SiF_3)_n$$

In a similar way $\cdot SCF_3$ radicals have been cocondensed with Te vapor.[25]

$$CF_3SSCF_3 \longrightarrow 2 \cdot SCF_3 \xrightarrow[\text{vapor}]{\text{Te}} Te(SCF_3)_2$$

Since in this case the discharge method was not very selective to $\cdot SCF_3$ radicals, and $\cdot CF_3$ radicals were also formed, some $Te(SCF_3)(CF_3)$ was also produced (about 2.7 : 1 $SCF_3 : CF_3$). Nevertheless, this method did yield reasonable amounts of these interesting new materials.

Gas-phase ground-state As atoms have been allowed to react with O_2 in a discharge-flow system.[26] The proposed reaction mechanism involved a cyclic intermediate, and a rate constant of about 8×10^{-14} cm³/molecule/sec was found.

$$As + O_2 \longrightarrow As \underset{O - O}{\overset{/ \; \backslash}{}} \longrightarrow O - As - O^* \longrightarrow OAsO + hv$$

Reactions with NO, NO_2, and O_3 were also studied, with NO_2 and O_3 reacting very fast by oxygen donation:

$$As + NO_2 \longrightarrow AsO + NO$$

$$As + O_3 \longrightarrow AsO^* + O_2.$$

The ozone reaction was sufficiently exothermic to populate the lowest vibrational level of the AsO $A^2\Sigma$ excited state.

The P_2 molecule has been matrix isolated in the presence of O_2 at 12 K.[27] To produce molecular P_2, the method of Bondybey and co-workers was employed,[28] where solid GaP was heated to 700–750°C. The resultant $(P_2)_x(O_2)_y$ adducts were photolyzed, driving the chemical reaction to form oxo-bridged P_2O_3 and two isomers of oxo-bridged P_2O_4. With higher concentrations of O_2 (5% in Ar), P_2O_5, P_4O_4, P_4O_6, P_4O_8, and P_4O_{10} were produced, according to infrared analysis. Wavelength dependence of the photochemistry suggested that forbidden excited electronic states of P_2 were involved (absorption in the 220–380 nm range).

The oxidation of P_2 in the gas phase has also been examined.[29] The P_2 molecule was produced by pyrolysis of P_4 vapor. Reaction with O atoms yielded $PO(X^2\pi) + P(4s)$ with $\Delta H = -99.9$ kcal/mol. In this way the PO molecule was generated and its ionization to PO^+ carried out by vacuum UV irradiation.

In the presence of O_3, P_2 and P_4 yielded PO, PO_2, and PO_3 as well as P_2O and P_2O_n.[30] In this study P_2 was generated by heating solid GaP and by a discharge on P_4 vapor. Using isotopically labeled O_3 allowed the identification of several phosphorus–oxygen products. It was shown that P_2 is considerably more reactive with O_3 under matrix deposition conditions than P_4.

II. Binuclear and Polynuclear Species

Absorption and laser-induced emission spectra of matrix-isolated GeTe (the main component of Knudsen cell evaporation of solid GeTe) showed the presence of broadbands in solid Ar and Kr.[31] The fundamental vibrational frequency of $^{74}Ge^{130}Te$ was found at 318 cm^{-1} with the fundamentals of the other 29 isotopic variations spanning the region from 315 to 327 cm^{-1}. Several previously unreported electronic excited states were characterized.

Knudsen cell methods were also used to produce SeO.[32] The equilibrium $SeO_{(g)} + SeS_{(g)} \rightleftarrows Se_{2(g)} + SO_{(g)}$ in the 1300–1800 K region was studied, and it was possible to determine the dissociation energy of SeO as 426 kJ/mol. The corresponding ΔH_f° for $SeO_{(g)}$ from the elements was determined as 58.9 kJ/mol.

Bimetallic clusters can often be formed during evaporation of alloys. Thus, Cs with Sb and Bi and in some cases with additions of In or Sn were evaporated from Knudsen cells or from a heated source (Langmuir evaporation).[33,34] A series of interesting intermetallic clusters were detected by MS. Highly polar species such as Cs_2Sb_2, Cs_2Sb_4 were detected. Clusters Cs_6Sb_4, Cs_6Sb_3Bi, and $Cs_6Sb_2Bi_2$ were also formed favorably. Other combinations such as CsSnSb, CsSnSb$_3$, CsSnSb$_3$, $Cs_2Sn_2Sb_2$, and CsInSb$_3$ were also favored. The cluster precursors are believed to be Cs atoms, Sb_2, and Sb_4 molecules. The relative abundances of the clusters formed did not depend much on the vaporization method, and thermodynamic equilibrium was probably not established in either case.

Knudsen cell generation of Na_2Te, Na_2Te_2, Na_2Te_3, Cs_2Te, Ce_2Te_2, Cs_2Te_3 clusters (by heating Na or Cs plus Te mixtures) has also been accomplished.[35a] Interestingly, if ionic structures of these gaseous species are assumed to be formed, these particles contain the same number of Te ions as polyions as were observed by E. Zintl in 1931 in liquid NH$_3$.[35b] However, in the gas phase, species with only one Na or Cs atom were also detected. Electron impact studies showed that the prevalent fragmentation pathway was always a loss of a Te atom.

A review of main group mixed clusters regarding their expected stabilities and their possible importance in determining structures of solid-state polar intermetallic phases has appeared.[36] A discussion of electron-precise clusters that require an interstitial atom Z within M_6X_{12}-type systems was presented. Also, an even broader area of cluster chemistry could involve transition-metal-main group element combinations, for example Mn_5Si_3, Zr_5Sb_3, Zr_5Sn_3, Zr_5Pb_3, La_5Ge_3, La_5Pb_3, and M_5B_3 (M = Ca, Sr, or Ba and B = Sb or Bi). Valence precise (Zintl) phases can be achieved with some M_5B_3Z systems. This review effectively connects the field of "naked clusters" (the major topic of this book) with new and exciting solid-state materials.[36]

Returning to gas-phase "naked" clusters, Mandich and co-workers have reported on In_xP_y species.[37] These clusters contained 5–14 atoms, and odd-numbered clusters exhibited stronger and more varied absorptions as compared to even-numbered clusters (resonant two-color and one-color photodissociation spectroscopy). Excitation energies between 0.84 and 1.84 eV were required. An optical gap-like absorption feature was observed at the blue of the spectrum, especially for clusters containing an even number of atoms. This absorption feature shifted with cluster size, but had an onset close to the band gap of bulk cryptalline InP. An empirical relationship of low-energy absorption tails for the clusters and amorphous semiconductors was found. The clusters appear to share a common chromophore with cryptalline InP, and this was proposed to be due to $\sigma-\sigma^*$-excitations of electrons of In–P bonding orbitals.

Even-numbered clusters were found to be in higher abundance and have higher dissociation energies than odd-numbered clusters. It appears that the even-numbered clusters have closed-shell singlet ground states, while odd-numbered have open-shell multiplet ground states. Surprisingly, these clusters showed very similar absorption spectra compared with bulk semiconductors.[38]

Ab initio calculations on P_2O_n ($n > 2$) have been reported.[39] These results indicated that a large number of stable P_2O_n isomers are possible and that oxo-bridged species are particularly energetically favorable. More specifically, the energetically preferred structures were predicted to be P_2O, linear (C_∞); P_2O_2, trans-planar (C_{2h}); P_2O_3, nonplanar oxo-bridged (C_2); P_2O_4, nonplanar oxo-bridged (C_1 but nearly C_s); and P_2O_5, nonplanar oxo-bridged (C_2). It was noted that the oxo-bridged P_2O_3, P_2O_4, and P_2O_5 closely resemble the local structures in the tetraphosphorus oxides P_4O_6 and P_4O_{10}.

Vaporizations of arsenic oxides and sulfides have yielded information about As_4O_n species trapped in matrices.[40,41] Detailed IR studies revealed the presence of As_4O_6, As_4O_7, As_4O_8, As_4O_9, and As_4O_{10}. The spectra were interpreted on the basis of a central As_4O_6 cage with appropriate numbers of terminal As–O bonds (Fig. 9-2).[40]

Solid As_2S_3 vaporized to yield As_2S_3, As_4S_4 (dominant), and S_2 plus As_4S_5 (minor), as determined by MS studies.[41] Interestingly, AsS was not found to be a major species. However, AsS was emitted from a heated mixture of Cu_2S and FeS containing a small amount of arsenic, as well as from As_2S_3 vapor. The dissociation energy of As=S was determined to be 89.8 ± 1.5 kcal/mol.

Other unusual species have also been produced, for example, $Cl_2Si=O$ and $Cl_2Si=S$, which were formed by the reaction of SiO or SiS vapor with Cl_2 in a cold matrix.[42] Infrared spectra showed an absorption of 648 cm^{-1} for $Cl_2Si=O$ and 610 cm^{-1} for $Cl_2Si=S$ for these C_{2v} mole-

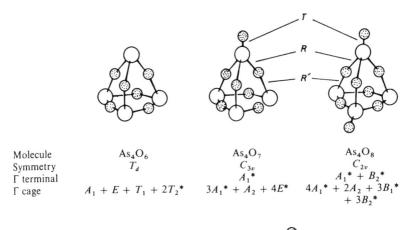

Molecule	As_4O_6	As_4O_7	As_4O_8
Symmetry	T_d	C_{3v}	C_{2v}
Γ terminal		$A_1{}^*$	$A_1{}^* + B_2{}^*$
Γ cage	$A_1 + E + T_1 + 2T_2{}^*$	$3A_1{}^* + A_2 + 4E^*$	$4A_1{}^* + 2A_2 + 3B_1{}^*$ $+ 3B_2{}^*$

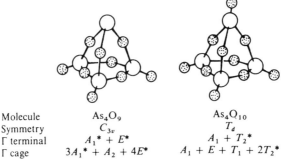

Molecule	As_4O_9	As_4O_{10}
Symmetry	C_{3v}	T_d
Γ terminal	$A_1{}^* + E^*$	$A_1 + T_2{}^*$
Γ cage	$3A_1{}^* + A_2 + 4E^*$	$A_1 + E + T_1 + 2T_2{}^*$

FIGURE 9-2   Structure and IR activities for As_4O_{6-10}. (∗) IR active bands (after Ogden and co-workers[40]).

cules.[43] Schnöckel and co-workers have also produced N≡P—O by the low temperature reaction of PN with O_3.[44]

$$PN + O_3 \longrightarrow PN{-}O_3 \xrightarrow{h\upsilon} PN{-}O + O_2$$

The authors showed that PN—O was favored over the NP—O isomer by about 28 kJ/mol, and the N—O bond is comparable to that in N_2O. The structure of PNO contradicts what might have been predicted; thus an N—O bond that is stronger than the P—O bond in NPO formed.

III. Thin Films

Films of antimony have been prepared by classical molecular-beam deposition and compared with cluster-beam deposition. For the cluster-beam process discontinuous and amorphous films were produced, according to electron microscopy. Particles of Sb were surrounded by an amorphous Sb oxide shell.[45] The high-temperature work on $(SN)_x$ formation is also of interest. Thus, S_4N_4 solid was sublimed at 70–80° and the vapor inlet into a helium radiofrequency plasma.[46] The S_4N_4 molecules were dissociated into SN, which then condensed to form $(SN)_x$ film.

References

1. K. J. Klabunde, "Chemistry of Free Atoms and Particles," p. 211, Academic Press, New York, 1980.
2. N. Itoh, K. Hattori, Y. Nakai, J. Kanasaki, A. Okano, C. K. Ong, and G. S. Khoo, *Appl. Phys. Lett.* **60**, 3271 (1992).
3. K. Steliou, P. Salama, D. Brodeu, and Y. Gareau, *J. Am. Chem. Soc.* **109**, 926 (1987).
4. M. Schmidt and U. Görl, *Angew. Chem. Int. Ed. Engl.* **26**, 887 (1987).
5. G. D. Brabson, Z. Mielke, and L. Andrews, *J. Chem. Phys.* **95**, 79 (1991).
6. M. M. Ross and S. W. McElvany, *J. Chem. Phys.* **89**, 4821 (1988).
7. P. Pfau, K. Sattler, J. Muehlbach, R. Pflaum, and E. Recknagel, *J. Phys. F* **12**, 2131 (1982).
8. (a) M. Häser, U. Schneider, and R. Ahlrichs, *J. Am. Chem. Soc.* **114**, 9551 (1992); (b) R. O. Jones and G. Seifert, *J. Chem. Phys.* **96**, 7564 (1992); (c) R. O. Jones and G. Seifert, *J. Chem. Phys.* **96**, 2942 (1992); (d) G. Seifert and R. O. Jones, *J. Chem. Phys.* **96**, 2951 (1992).
9. L. Andrews and Z. Mielke, *J. Phys. Chem.* **94**, 2348 (1990).
10. M. L. Polak, G. Gerber, J. Ho, and W. C. Lineberger, *J. Chem. Phys.* **97**, 8990 (1992).
11. M. L. Polak, J. Ho, G. Gerber, and W. C. Lineberger, *J. Chem. Phys.* **95**, 3053 (1991).
12. K. Balasubramanian and D. W. Liao, *J. Chem. Phys.* **95**, 3064 (1991).
13. C. Gao and K. Zhang, *J. Mol. Sci., Int. Ed.* **5**, 105 (1987).
14. J. M. Dyke, A. Morris, and J. C. H. Stevens, *Chem. Phys.* **102**, 29 (1986).
15. B. Eberle, H. Sontag, and R. Weber, *Surf. Sci.* **156**, 751 (1985).
16. J. Muehlbach, E. Recknagel, and K. Sattler, *in* "Recent Developments in Condens. Matter Physics" (J. T. Devreese, Ed.), Vol. 2, p. 371, Plenum, New York, 1980; see also *Surf. Sci.* **106**, 188 (1981).
17. G. Fuchs, M. Treilleux, F. Santos Aires, B. Cabaud, A. Hoareau, and P. Melinon, *Philos. Mag. A* **61**, 45 (1990).
18. M. Treilleux, G. Fuchs, F. Santos Aires, P. Melinon, A. Hoareau, and B. Cabaud, *Z. Phys. D: At. Mol. Clusters* **12**, 131, 149 (1989).
19. S. Iijima, *Springer Ser. Mater. Sci.* **4**, 186 (1987).
20. P. Hassanzadeh and L. Andrews, *J. Phys. Chem.* **96**, 6579 (1992).
21. W. von Niessen, L. S. Cederbaum, and F. Tartantelli, *J. Chem. Phys.* **91**, 3582 (1989).
22. H. Schnöckel, H. J. Göcke, and R. Elsper, *Z. Anorg. Allg. Chem.* **494**, 78 (1982).
23. H. Schnöckel, *Z. Anorg. Allg. Chem.* **510**, 72 (1984).
24. T. R. Bierschenk, T. J. Juhlke, and R. J. Lagow, *J. Am. Chem. Soc.* **103**, 7340 (1981).
25. T. R. Bierschenk and R. J. Lagow, *Inorg. Chem.* **22**, 359 (1983); see also *Chem. Eng. News* **April 8**, 23 (1985).
26. M. E. Frasër, *J. Phys. Chem.* **88**, 3383 (1984).

27. M. McCluskey and L. Andrews, *J. Phys. Chem.* **95**, 2679 (1991).
28. L. A. Heimbrook, M. Rasanen, and V. E. Bondybey, *Chem. Phys. Lett.* **120**, 233 (1985).
29. J. M. Dyke, A. Morris, and A. Ridha, *J. Chem. Soc., Faraday Trans. 2* **78**, 2077 (1982).
30. Z. Mielke, M. McCluskey, and L. Andrews, *Chem. Phys. Lett.* **165**, 146 (1990).
31. M. A. Epting, J. R. Sweigart, and E. R. Nixon, *J. Mol. Spectrosc.* **78**, 277 (1979).
32. S. Smoes and J. Drowart, *J. Chem. Soc., Faraday Trans. 2*, 1171 (1984).
33. A. Hartmann and K. G. Weil, *High Temp. Sci.* **27**, 31 (1990).
34. A. Hartmann and K. G. Weil, *Z. Phys. D: At. Mol. Clusters* **12**, 11 (1989).
35a. A. Hartmann, L. Poth, and K. G. Weil, *High Temp. Sci.* **31**, 121 (1991); see also *Pure Appl. Chem.* **62**, 103 (1990).
35b. E. Zintl, J. Goubeau, W. Dullenkop, *Z. Phys. Chem. Abt.* **154A**, 1 (1931).
36. J. D. Corbett, E. Garcia, Y. U. Kwon, and A. Guloy, *High Temp. Sci.* **27**, 337 (1990).
37. K. P. Rinnen, K. D. Kolenbrander, A. M. De Santolo, and M. L. Mandich, *J. Chem. Phys.* **96**, 4088 (1992).
38. K. D. Kolenbrander and M. L. Mandich, *Phys. Rev. Lett.* **65**, 2169 (1990).
39. L. L. Lohr, Jr., *J. Phys. Chem.* **94**, 1807 (1990).
40. A. K. Brisdon, R. A. Gomme, and J. S. Ogden, *J. Chem. Soc., Dalton Trans.*, 2725 (1986).
41. K. L. Lau, R. D. Brittain, and D. L. Hildenbrand, *J. Phys. Chem.* **86**, 4429 (1982).
42. H. Schnöckel, H. J. Göcke, and R. Köppe, *Z. Anorg. Allg. Chem.* **578**, 159 (1989).
43. H. Schnöckel, *Z. Anorg. Allg. Chem.* **460**, 37 (1980).
44. R. Ahlrichs, S. Schunek, and H. Schnöckel, *Angew. Chem. Int. Ed. Engl.* **27**, 421 (1988).
45. G. Fuchs, M. Treilleux, F. Santos Aires, B. Cabaund, P. Milionon, and A. Hoareau, *Philos. Mag. B* **63**, 715 (1991).
46. M. W. R. Witt, W. I. Bailey, Jr., and R. J. Lagow, *J. Am. Chem. Soc.* **105**, 1668 (1983).

Lanthanides and Actinides

I. Lanthanide and Actinide Atoms

A. Occurrence and Techniques

Earlier the vaporization properties of the lanthanides and actinides were summarized.[1] The lanthanides actually have quite convenient vaporization properties, while the actinides Th, Pa, U, Nb, and Pu are much more difficult. However, since uranium isotope separation has become such a huge technological problem, U vaporization on a very large scale has been developed. Both lanthanide and actinide metals vaporize as monomers (atoms).

B. Physical Properties and Theoretical Studies

The low ionization energies of the lanthanide metal atoms makes them similar in properties and chemistry to the alkali or alkaline earth metals. Ionic bonding is almost always favored.

C. Chemistry

Evans *et al.* have developed a great deal of chemistry of low-valent lanthanide species, and metal vapor chemistry has been quite useful in this regard. For example, Yb, Sm, and Er atoms (vapors) cocondensed with alkenes led to a complex series of reactions,[2] for example, insertion of the metal atom into C–H bonds, cleavage of carbon–carbon bonds, homologation, oligomerization, and dehydrogenation of ethene, propene, 1, 2-propadiene, and cyclopropane. Although pure organometallic compounds could not be isolated, hydrolysis and other derivitization schemes gave strong evidence that lanthanide–carbon bonding was prevalent:

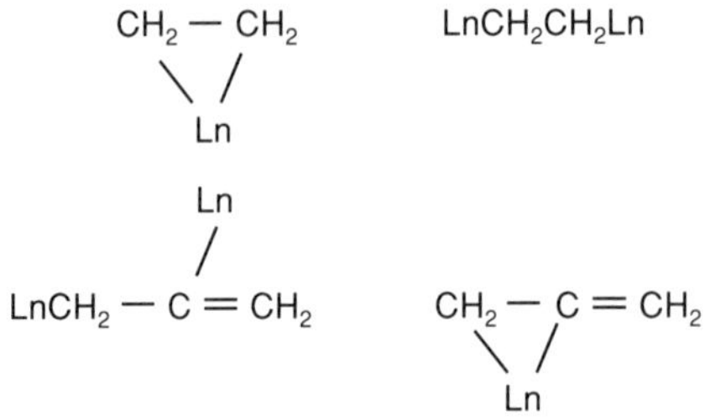

$$CH_2 - CH_2 \qquad LnCH_2CH_2Ln$$
$$\diagdown \diagup$$
$$Ln$$

$$Ln$$
$$\diagup$$
$$LnCH_2 - C = CH_2 \qquad CH_2 - C = CH_2$$
$$\diagdown \diagup$$
$$Ln$$

The most common overall reactions were two-electron reduction of unsaturated carbon–carbon bonds and oxidative addition of C–H linkages.

Alkynes also yielded interesting results in cocodensation reactions with Er vapor.[3] The complex organometallic mixture produced from 3-hexyne served as a catalyst for alkene hydrogenation and analyzed as Er_2 $(C_6H_{10})_3$ (also supported by molecular weight measurements and IR). Similarly, 1-hexyne with Yb atoms caused oxidative addition of the terminal C–H to yield divalent alkynide complexes.[4a] Samarium and Er reacted similarly to produce trivalent species,

$$Yb + HC\equiv CR \longrightarrow H-Yb-C\equiv C-R$$

$$3H-Yb-C\equiv CR \longrightarrow H\ Yb_2(C\equiv CR)_3 + [YbH_2]$$

and oligomerization of unsaturated species apparently took place:

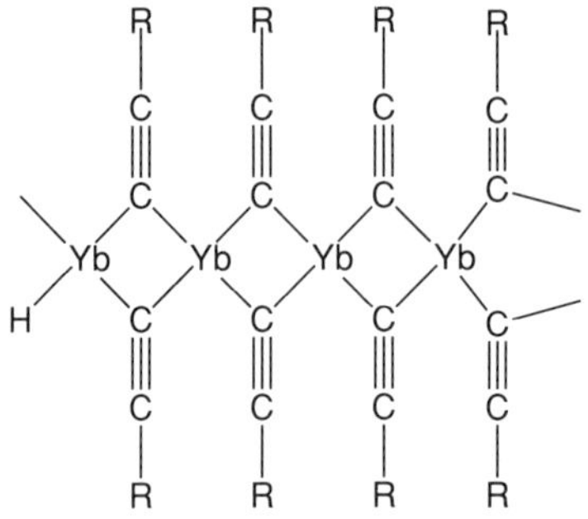

Actually, a series of interesting catalytic processes have been found to proceed utilizing lanthanide vapors.[4b,c,d] Erbium, Lu, Nd, and Yb de-

rivatives have been found to catalyze the hydrogenation of internal alkynes to *cis*-alkenes with selectivities of over 90%. Also, Yb and Sm
slurries in THF catalyze the selective dimerization of ethene to 2-butenes.
And finally, Yb vapor cocondensed with 2-methylbutadiene or (E), (Z)-1,
3-pentadiene promote the polymerization of these dienes in good yield.[4d]
Some selectivity to *cis*-alkenyl polymers was found.

Matrix isolation of Eu–C_2H_4 complexes has allowed further definition
of initial bonding.[5a] Either neat or in Ar or Xe, a species was formed with
thermal stability to about 50 K. At low C_2H_4 concentrations, a monoadduct π-complexed Eu–C_2H_4 was observed by UV–VIS and IR spectroscopies, and the spectral changes resembled those of Fe, Co, Ni–C_2H_4 or
Cu, Ag, Au–C_2H_4 complexes. Odd electron lanthanide–C_2H_4 complexes
closely resembled coinage metal systems, and intense visible metal-to-
ligand charge-transfer transitions were observed. The green absorption of
the Eu(C_2H_4)$_n$ complexes may be due to excitations of the Eu f electrons
into the alkene. These results coupled with molecular orbital calculations
suggested a model C_{2v} Eu(C_2H_4) species that is only weakly bound by Eu f
orbital donation into the π*-system of ethene and with little donation of
alkene σ- or π-electron density onto the metal atom. Higher concentrations of C_2H_4 led to Eu(C_2H_4)$_n$ species.

One report on U atom reactions with N_2 under matrix isolation conditions has appeared.[5b] Laser-vaporized U atoms were codeposited with N_2/
Ar mixtures at 12 K. The only product observed was the insertion product
NUN, according to FTIR analysis. The yield of this product increased
threefold under UV irradiation, which shows that electronically excited
U* reacts more efficiently. However, it is proposed that pulsed laser-
evaporated U atoms have very high kinetic energy, and this provides the
activation energy necessary for the insertion reaction to occur on codeposition. Thus, we see one more example where laser vaporization leads to
extremely high kinetic energies that can drive chemical reactions at low
temperatures (see Chapter 6 for more discussion).

The metal vapor synthesis of an isolable, pure pentamethylcyclopen-
tadienyl–Sm complex was a significant advance by Evans *et al.*[6]

$$Sm + C_5H(Me)_5 \xrightarrow[\substack{-120°C \\ \text{rotary reactor}}]{\substack{\text{hexane} \\ \text{THF}}} [C_5(Me)_5]_2Sm(THF)_2$$

This proved to be the first preparative-scale route to a soluble divalent
organosamarium complex, which has exhibited a rich chemistry. Further
work allowed synthesis of $(C_5Me_4Et)_2Sm(THF)_2$ and $(C_5Me_5)Sm(C_6H_5)$
THF,[7] and a new field or organolanthanide chemistry was derived from
this initial work.[8–12]

A further significant advance due to metal vapor reaction chemistry was the recent synthesis of the first Ln(O) organometallic species, bis(η-1,3,5-tri-tert-butylbenzene) sandwich complexes of Y and Gd:[13]

Cocondensation of Y or Gd vapor with excess arene ligand yielded these complexes in 50% yields based on metal vaporized. These compounds are deep purple in color and can be sublimed at 100°C with partial decomposition.

The Y complex is paramagnetic but is ESR silent at room temperature. The authors believe that this result is in accord with a sandwich structure with an E_2 ground state in which the large g-value anisotropy would be expected to cause the ESR signal to be difficult to detect. However, an ESR spectrum was obtained at 77 K in frozen methycyclohexane.

The X-ray structure of the Gd complex was reported and shows two planar-eclipsed arene ligands with the t-butyl groups bent out of the plane by 6–10°.

Ytterbium atoms did not yield such a complex and only colloidal metal was formed. This observation led to a study of all the lanthanide atoms and a comparative study of stability and bonding.[14] 1,3,5-tri-t-butylbenzene yielded stable sandwich complexes with Nd, Tb, Dy, Ho, Er, Lu, and Gd, while unstable complexes for La, Pr, Sm, and unisolable materials for Ce, Eu, Tm, and Yb.

It was proposed that the metal atom must have an easily accessible d^2s^1 state. If the promotion energy to get to this state is large, then there would be an insufficient gain in metal–arene bond energy to offset the promotion energy. However, the instability of the La, Ce, and Pr complexes could not be explained in this way, and this instability may instead be due to the greater covalent radii of these metals.

The magnetic properties of these complexes gave some support to these ideas, and a bonding scheme where three of the valence electrons of the lanthanide are involved in bonding to the arene rings while the rest remain in the f-shell seems reasonable.

II. Lanthanide and Actinide Clusters

A. Occurrence and Techniques

No recent reports were found.

B. Physical Properties and Theoretical Studies

Theoretical studies of free actinide dimers by X_α–SW calculations have suggested the existence of ϕ-bonds between metal atoms utilizing $5f$–$5f$ orbital interactions.[15] For ϕ-bonds to exist in U_2 and Np_2 free dimers, two conditions must be satisfied: (1) the $5f$ orbitals must be low enough in energy that the ϕ-bonds are lower in energy than the $6d$–$6d$ bonding interaction, and (2) the diatomic interactions must be strong enough to give an energetic separation between the ϕ-bonds and the antibonds. It was concluded that for U(O) and Np(O) this can be the case and that $5f$–$5f$ interactions are likely.

Samarium clusters exhibit unusual behavior regarding oxidation state. Mixed valence behavior was observed by X-ray photoemission and UPS, for example, small clusters exhibited primarily a divalent state, and the trivalent state becomes more dominant for increasing cluster size.[16] Metal particles of diameter 10–90 Å have been produced for Au, Yb, and Eu. The plasmon frequency shift was calculated as a function of particle size, with consideration of electron density change arising from particle lattice contraction due to strong surface-tension effects. These theoretical calculations seemed to agree with experimental data available in the literature.[17]

C. Chemistry

By codeposition of Sm atoms and ketones at 77 K, reduction of the ketones took place under cryochemical conditions.[18] By varying the metal vapor deposition rate, the relative concentration of larger colloidal Sm_x was controlled. It was found that atoms and small clusters tend to reduce ketones to hydrocarbons, while bigger particles cause the formation of pinaconates.

$$\text{Sm} + \text{R} - \overset{\overset{\displaystyle O}{\|}}{\text{C}} - \text{R} \xrightarrow{\text{condense, warm}} \longrightarrow \longrightarrow \text{hydrocarbons}$$

$$\text{Sm}_x \text{ colloidal} + \text{RCR}\overset{\overset{\displaystyle O}{\|}}{} \longrightarrow \longrightarrow \text{R} - \overset{\overset{\displaystyle O^-}{|}}{\underset{\underset{\displaystyle R}{|}}{\text{C}}} - \overset{\overset{\displaystyle O^-}{|}}{\underset{\underset{\displaystyle R}{|}}{\text{C}}} - \text{R}$$

The solvated metal atom dispersion (SMAD) technique described in Chapter 2 has also been employed for preparing La–Ni bimetallic supported catalysts. La and Ni were vaporized simultaneously and deposited with organic solvent. Upon warming from 77 K the La and Ni clustered together and then deposited on powdered SiO_2. The average particle size was about 4.0 nm, and XPS analysis indicated that the Ni was metallic while the La existed as La metal and La_2O_3. Catalytic activities were improved by the presence of La.[19]

Also using the SMAD method, a series of rare-earth metallic catalysts were prepared and exhibited enhanced activities for hydrogenation catalysis.[20] Thus, Sm and Yb particles were formed by clustering of the atoms in cold organic solvents and extensive catalytic studies carried out. Unusual selectivities were realized. For example, alkenes, dienes, and aromatics were readily hydrogenated, but alkynes were not hydrogenated at all. Kinetic data suggested that these hydrogenation processes were controlled by the H_2 chemisorption rate. Another interesting feature was that H–D scrambling did not take place readily over these catalysts. The selectivity patterns and H–D exchange results suggest that these rare-earth systems function differently from transition-metal catalysts mechanistically.

Synthetic production of the heavy, unstable actinides an atom at a time via nuclear synthesis has been carried out so that the metal halides could be formed and their physical properties studied. Element 104 with a half-life of 65 sec was generated and allowed to react with HCl or HBr to form the chloride or bromide. Element 104 formed more volatile bromides than its homolog Hf. Element 105 (half-life = about 30 sec) formed a bromide less volatile than Nb or Ta. Both element 104 and Hf chlorides were more volatile than their respective bromides.[21]

References

1. K. J. Klabunde, "Chemistry of Free Atoms and Particles," Academic Press, New York, 1980.
2. W. J. Evans, K. M. Coleson, and S. C. Engerer, *Inorg. Chem.* **20,** 4320 (1981).
3. W. J. Evans, S. C. Engerer, P. A. Piliero, and A. L. Wayda, *J. Chem. Soc. Chem. Commun.* **22,** 1007 (1979).
4. (a) W. J. Evans, S. C. Engerer, and K. M. Coleson, *J. Am. Chem. Soc.* **103,** 6672 (1981); (b) W. J. Evans, *J. Catal.* **84,** 468 (1983); (c) H. Himamura, K. Kitajima, and S. Tsuchiya, *J. Chem. Soc., Faraday Trans.* **85,** 1647 (1989); (d) G. Vitulli, M. Giampietri, and P. Salvadori, *J. Molec. Catal.* **65,** L21 (1991).
5. (a) M. P. Andrews and A. L. Wayda, *Organometallics* **7,** 743 (1988); (b) R. D. Hunt, J. T. Yustein, and L. Andrews, *J. Chem. Phys.* **98,** 6070 (1993).
6. W. J. Evans, I. Bloom, W. E. Hunter, and J. L. Atwood, *J. Am. Chem. Soc.* **103,** 6507 (1981).
7. W. J. Evans, I. Bloom, W. E. Hunter, and J. L. Atwood, *Organometallics* **4,** 112 (1985).
8. W. J. Evans, T. A. Ulibarri, and J. W. Ziller, *J. Am. Chem. Soc.* **110,** 6877 (1988).
9. W. J. Evans, L. A. Hughes, and T. P. Hanusa, *J. Am. Chem. Soc.* **106,** 4270 (1984).
10. W. J. Evans, S. C. Engerer, P. A. Piliero, and A. L. Wayda, *in* "Fundamental Research in Homogeneous Catalysis" (M. Tsutsui, Ed.), Vol. 3, p. 941, Plenum, New York, 1979.
11. W. J. Evans, I. Bloom, and S. C. Engerer, *J. Catal.* **84,** 468 (1983).
12. W. J. Evans, *Polyhedron,* 803 (1987).
13. J. G. Brennan, F. G. N. Cloke, A. A. Sameh, and A. Zaikin, *J. Chem. Soc. Chem. Commun.,* 1668 (1987).
14. D. M. Anderson, F. G. N. Cloke, P. A. Cox, N. Edelstein, J. C. Green, T. Pang, A. A. Sameh, and G. Shalimoff, *J. Chem. Soc. Chem. Commun.,* 53 (1989).
15. B. E. Bursten and G. A. Ozin, *Inorg. Chem.* **23,** 2910 (1984).
16. M. G. Mason, S. T. Lee, G. Apai, R. F. Davis, D. A. Shirley, A. Franciosi, and J. H. Weaver, *Phys. Rev. Lett.* **47,** 730 (1981).
17. F. Parmigiani, *Phys. Lett. A* **92,** 419 (1982).
18. V. V. Zagorskii, S. E. Kondakov, A. M. Kosolapov, G. B. Sergeev, and V. N. Solovev, *Metallaorg. Khim.* **5,** 533 (1992). [In Russian]
19. S. Wu, W. Huang, W. Zhao, X. Wang, S. Zhang, and X. Wang, *Zhongguo Xitu Xuebao* **9,** 224 (1991). [In Chinese]
20. H. Imamura, A. Ohmura, E. Haku, and S. Tsuchiya, *J. Catal.* **96,** 139 (1985).
21. A. Turler, H. W. Gaeggeler, K. E. Gregorich, H. Barth, W. Bruechle, K. R. Czerwinski, M. K. Gober, N. J. Hannick, R. Henderson *et al., J. Radioanal. Nucl. Chem.* **160,** 327 (1992).

Index